T0201823

Distributed Source Coding

Distributed Source Coding

Theory and Practice

Shuang Wang, Yong Fang, and Samuel Cheng

This edition first published 2017
© 2017 John Wiley & Sons Ltd

Registered office John Wiley & Sons Ltd, The Atrium, Southern Gate, Chichester, West Sussex, PO19 8SQ, United Kingdom

For details of our global editorial offices, for customer services and for information about how to apply for permission to reuse the copyright material in this book please see our website at www.wiley.com.

The right of Shuang Wang, Yong Fang, and Samuel Cheng to be identified as the authors of this work has been asserted in accordance with the Copyright, Designs and Patents Act 1988.

Library of Congress Cataloging-in-Publication data applied for

ISBN: 9780470688991

A catalogue record for this book is available from the British Library.

Cover design by Wiley
Cover image: Top numbers image: a_Taiga/Gettyimages
         Bottom molecular image: KTSDESIGN/SCIENCE PHOTO LIBRARY/Gettyimages

Set in 10/12pt WarnockPro by SPi Global, Chennai, India
Printed and bound in Malaysia by Vivar Printing Sdn Bhd

10 9 8 7 6 5 4 3 2 1

Contents

Preface

This book intends to provide fundamental knowledge for engineers and computer scientists to get into the topic of distributed source coding, and is targeted at senior undergraduate or first-year graduate students. It should also be accessible for engineers with a basic knowledge of calculus and applied probability. We have included short chapters on information theory, channel coding, and approximate inference in the appendix. The goal is to make the content as self-contained as possible.

Excluding the appendix, the book is divided into three parts. First-time readers who would like to get into applications quickly should work through Chapters 1, 2, 3, 5, and 7, and then go straight to Part III, the application chapters. Chapters 4 and 6 are more advanced and may require more math maturity from readers.

Distributed source coding is an active research area and the authors have not been able to cover every aspect of it. We have included materials that are geared more toward our own research interests. However, we have striven to include the most interesting applications in this book.

Acknowledgment

Book writing is a long endeavor. It is especially so in this case. This book project grew out of several distributed source coding tutorials presented in the US, Taiwan and Switzerland around 2009 by Dr. Vladimir Stankovic, Dr. Lina Stankovic, and the third author. Many materials in this book reflect the immense influence of these tutorials. Drs. Stankovic have been very kind in offering extensive unconditional help and allowed us to include many of their ideas and even their own words in this book. For this we wish to express our most sincere and deepest gratitude.

The third author would also like to express special thanks to his Master and PhD advisor, Dr. Zixiang Xiong, who taught him not just distributed source coding, but also everything relating to research and academia.

Over the years, the authors have received help from many colleagues, friends, and students, including Angelos Liveris, Costas Georghiades, Yang Yang, Zixin Liu, Zongmin Liu, Qian Xu, Shengjie Zhao, Nikoloas Deligiannis, Xiaoqian Jiang, Adrian Munteanu, Peter Schelkens, Andrei Sechelea, Feng Chen, and Lijuan Cui. In particular, we would like to offer special thanks to Liyan Chen and Xiaomin Wang for generating some of the figures. We could not possibly include everyone in the list above, and we would like to ask for forgiveness for any omission. Finally, the authors would like to acknowledge support from the National Science Foundation, the National Institutes of Health (NHGRI K99HG008175), and the National Nature Science Foundation of China.

About the Companion Website

Don't forget to visit the companion website for this book:

www.wiley.com/go/cheng/dsctheoryandpractice

There you will find valuable material designed to enhance your learning, including:

- Example codes

Scan this QR code to visit the companion website

1

Introduction

Imagine a dense sensor network consisting of many tiny sensors deployed for information gathering. Readings from neighboring sensors will often be highly correlated. This correlation can be exploited to significantly reduce the amount of information that each sensor needs to send to the base station, thus reducing power consumption and prolonging the life of the nodes and the network. The obvious way of exploiting correlation is to enable neighboring sensors to exchange data with one another. However, communication among sensors is usually undesirable as it increases the complexity of the sensors, which in turn leads to additional cost and power consumption. How then is it possible to avoid information exchange among sensors but still be able to exploit the statistical dependency of the readings in different sensor nodes? The solution lies in *distributed source coding*.

Distributed source coding (DSC) enables correlation to be exploited efficiently without the need for communications among sensors. Moreover, in some specific cases it does not incur any loss compared to the case when the sensors communicate.

DSC has been receiving significant attention since the beginning of the 21st century from academics and industrial researchers in different fields of electrical and computer engineering, and mathematical and computer science. Indeed, special sessions and tutorials dedicated to DSC are given at major communications, signal processing, multimedia, and computer engineering conferences. This is no wonder as DSC has potential applications ranging from wireless sensor networks, ad-hoc networks, surveillance networks, to robust low-complexity video coding, stereo/multiview video coding, high-definition television, and hyper-spectral and multi-spectral imaging. This book is intended to act as a guidebook for engineers and researchers to grasp the basic concepts quickly in order to understand and contribute to this exciting field and apply the emerging applications.

Distributed Source Coding: Theory and Practice, First Edition. Shuang Wang, Yong Fang, and Samuel Cheng.
© 2017 John Wiley & Sons Ltd. Published 2017 by John Wiley & Sons Ltd.
Companion Website: www.wiley.com/go/cheng/dsctheoryandpractice

1.1   What is Distributed Source Coding?

DSC deals with the source coding or compression of correlated sources. The adjective *distributed* stresses that the compression occurs in a distributed or noncentralized fashion. We can, for example, assume that the sources to be compressed are distributed across different nodes in a network. The task is to compress these sources and communicate compressed streams to a decoder for joint decompression. The basis of DSC is that the compressions take place *independently*, that is, the nodes do not exchange their information, whereas decompression is *joint*.

This is illustrated in Figure 1.1 for the simple case of two nodes. Each node has access only to its source, and does not have information about sources present at other nodes. Therefore, "distributed" in this context refers to separate compression at each node. Note that if decompression were also separate for each of the sources, then the problem would boil down to multiple conventional compressions. Throughout the book, distributed compression will always refer to separate encoding and joint decoding, whereas joint compression will refer to joint encoding and joint decoding, if not stated otherwise.

The search of achievable rates of DSC is a major information-theoretical problem and lies in the framework of network information theory, a branch of information theory that tries to find the compression and communication limits of a network of nodes. In the case of discrete sources and perfect reconstruction at the decoder, DSC extends Shannon's Source Coding Theorem, in information theory, from the point-to-point to multipoint scenario. This is referred to as lossless DCS. When we allow for some distortion in reconstruction, the DSC problem becomes a rate-distortion problem and is referred to as lossy DSC.

1.2   Historical Overview and Background

DSC started as an information-theoretical problem in the seminal 1973 paper of Slepian and Wolf [1]. Slepian and Wolf considered lossless separate compression of two discrete sources, and showed that roughly speaking there is no

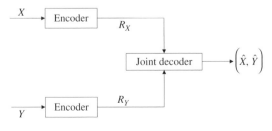

Figure 1.1 DSC concept with two separate encoders who do not talk to each other and one joint decoder. X and Y are discrete, correlated sources; R_X and R_Y are compression rates.

performance loss compared to joint compression as long as joint decompression is performed.

This remarkable result triggered significant information-theoretical research resulting in solutions – in the form of achievable rate regions – for more complicated lossless setups. In 1976, Wyner and Ziv [2] considered a lossy version, with a distortion constraint, of a special case of the Slepian–Wolf (SW) problem, where one source is available at the decoder as side information. Wyner and Ziv showed that for a particular correlation where source and side information are jointly Gaussian, there is no performance loss due to the absence of side information at the encoder. The lossy case of the generalized SW setup, known as multiterminal (MT) source coding, was introduced by Berger and Tung in 1977 [3, 4].

A possible realization of DSC via the use of conventional linear channel codes to approach the SW bound was known as early as 1973, but due to the lack of any potential application of DSC, work on code designs, that is, how to code the sources to approach given bounds, started only at the end of the last century. The first practical design was reported in 1999 [5], followed by many improved solutions. One key insight of these designs is that conventional channel coding can be used for compression. Indeed, correlation between the sources is seen as a virtual communication channel, and as long as this virtual channel can be modeled by some standard communication channel, for example Gaussian, channel codes can be effectively employed. Capacity-approaching designs [6] based on quantization followed by advanced channel coding, for example with turbo codes [7] and low-density parity-check (LDPC) codes [8], come very close to the bounds for two jointly Gaussian sources.

1.3 Potential and Applications

The launch of wireless sensor networks (WSNs) ignited practical DSC considerations in the early years of this century since WSNs naturally call for distributed processing. Closely located sensors are expected to have correlated measurements; thus in theory the DSC setup fulfills the requirement of power-efficient compression for distributed sensor networks. However, many practical problems remain to be solved before DSC is used in mainstream commercial networks. The challenges include the complex correlation structure of real signals, nonGaussian sources, mandatory long codeword lengths, and the complexity of current designs.

Though WSN triggered renewed interest in DSC, another application has emerged: low-complexity video, where the DSC paradigm is used to avoid computationally expensive temporal prediction loop in video encoding. Indeed, loosely speaking, a conventional video encoder needs to find the best matching block to the current one by examining all possible candidates, for example a previous frame. Then, the difference between the two blocks is encoded and motion vectors sent to the decoder. This architecture imposes heavy encoding and light decoding, exactly what is needed for television broadcasting! However, the fact that compression performance is not degraded by the lack of side information at the encoder [2] brings an incredible possibility of avoiding the most resource-consuming motion search at the encoder and shifting it to the decoder side. This idea had remained for more than 20 years in a US Patent [9]. It was only recently with the developed DSC designs that Wyner–Ziv (WZ) or distributed video coding was revisited.

Distributed video coding (DVC) was rediscovered in 2002 [10, 11, 23, 150, 151] and taken forward by many researchers. DVC was proposed as a solution for unconventional setups, such as video surveillance with tiny cameras and cell-to-cell communications, where low encoding complexity is a must. However, other criteria (e.g., the inbuilt robustness of DVC coders to error impairments as there is no error accumulation due to the mismatch of reference frame[1] and a unique opportunity of using a single code for both compression and error protection) have influenced the design philosophy of DVC for broader applications such as robust scalable video transmission over wireless networks and video streaming from multiple sites. Yet another application is multiview and stereo video, where DSC is used to exploit the correlation between different camera views when encodings must be independent.

DVC has also been used to exploit correlation between different bands in multi-spectral and hyper-spectral image coding, then in biometrics, and for compression in microphone array networks. And other applications are emerging, such as spectrum sensing in cognitive radio, compress-and-forward in cooperative communications, and more. These will be discussed in more detail later in the book.

1.4 Outline

While some might prefer to think that DSC is driven by applications, others are fascinated by the information-theoretical challenges it poses. Thus, this book attempts to reconcile both schools of thought by discussing practical code designs and algorithms, together with the theoretical framework for DSC as well as the potential applications various DSC setups have and will impact. The authors attempt to strike a balance between mathematical rigor and intuition. This book is intended to be a guide for both senior/graduate students and engineers who are interested in the field. We have tried our best to ensure the manuscript is self-contained. While we assume no background knowledge in

---

1 Because the encoder does not use a reference frame.

information and coding theory, we expect the reader to be familiar with calculus, probability theory, and linear algebra as taught in a first- or second-year undergraduate course. The book comprises working examples where possible and includes references to further reading material for more advanced or specialized topics where appropriate.

The book is organized into three parts: Theory, Algorithms, and Applications of DSC.

Part I, Theory, starts with a brief and essential background of information theory and data compression. The background is followed by a comprehensive review of the roots of DSC, namely Slepian and Wolf's theorem and its extension to multiple sources and lossless MT networks. Next, rate-distortion problems will be visited, that is, WZ and lossy MT source coding problems.

Part II, Algorithms, introduces and discusses designs of algorithmic solutions for DSC problems. Most of this part is dedicated to the evolution of code designs for the three most important DSC problems, namely SW, WZ, and MT source coding, introduced in Part I. We start with a background chapter on coding theory, followed by designs for the SW coding problem and elaborate developments from the early work of Wyner in 1974 to the most recent achievements with advanced channel codes, Belief Propagation, and arithmetic codes. Similarly, for the WZ problem we start with the 1999 work of Pradhan and Ramchandran [5] and describe subsequent developments.

Part III is dedicated to potential DSC applications, grouped as multimedia applications and wireless communications applications. We start by providing the necessary background on image/video processing. DVC is discussed first, followed by biometric applications and multi-spectral/hyper-spectral image compression. The two final chapters are dedicated to wireless communications: data gathering in wireless sensor networks, spectrum sensing in cognitive radio, compress-and-forward in cooperative communications scenario, and DSC's relationship to compressive sampling.

Part III will be of great interest to practitioners. Each topic has an educational description and approaches latest developments critically, discussing their strengths and shortcomings, and suggesting possible extensions and future work.

Part I

Theory of Distributed Source Coding

2

Lossless Compression of Correlated Sources

In this chapter we consider the setup when the reconstructed signal after lossless compression is identical to the original input. Because the limit of compressing a source is governed by its entropy, lossless compression is often called entropy coding. The most popular entropy coding algorithms are Huffman coding, which can operate very close to the theoretical limit provided that we allow the codeword length to be arbitrarily long. Of course, one may apply entropy coding on each source independently but possible dependency or correlation among the sources is not captured in this case. In contrast, lossless DSC can offer much efficient compression. We will develop the idea of lossless DSC by means of the following application example that we will build upon throughout the chapter.

Example 2.1 *(A network of temperature sensors).* Temperature sensors scattered over a geographically small area need to report their readings regularly. All readings must be compressed and sent to a central unit for processing. Since closely located sensors will have similar readings, the problem boils down to compression of correlated sources. Each sensor can, of course, compress its measurements using any entropy coding algorithm up to the source entropy limit, but this way it will not exploit redundancy across the neighbouring sensors, and thus it cannot achieve the highest possible compression.

Lossless DSC refers to any method that independently and losslessly compresses the sources and jointly decompresses them by exploiting mutual correlation among them. This chapter will explain how this is done and outline the main information-theoretical results, including key proofs following the notation and concepts introduced in the previous chapter. The chapter answers how much we can independently compress correlated sources so that a joint decoder can still recover them with very low probability of error. We assume that all sources are discrete-time random variables taking values from finite alphabets for lossless compression. The simplest lossless DSC case is SW coding

Distributed Source Coding: Theory and Practice, First Edition. Shuang Wang, Yong Fang, and Samuel Cheng.
© 2017 John Wiley & Sons Ltd. Published 2017 by John Wiley & Sons Ltd.
Companion Website: www.wiley.com/go/cheng/dsctheoryandpractice

[1], where only two sources are to be compressed separately and decompressed jointly. We start with the SW coding problem, which is not only the simplest but was also the first considered distributed compression problem. Towards the end of the chapter, an extension to multiple sources is given.

2.1 Slepian–Wolf Coding

In 1973 Slepian and Wolf considered the separate compression and joint lossless decompression of two discrete sources [1]. This is the simplest DSC problem with only two encoders and one decoder, as shown in Figure 2.1. As this chapter focuses on lossless compression, the sources need to be reconstructed at the decoder exactly, without any distortion, with probability approaching one.

Let X and Y be two random variables denoting two correlated discrete-time memoryless sources drawn i.i.d. from joint probability distribution $p(x, y)$. Let $p(x)$ and $p(y)$ be the marginal probability distribution of X and Y, respectively. The encoders of X and Y will independently convert length-n sequences X^n and Y^n into symbols U and V, respectively, whereas a joint decoder will receive both U and V and convert them back to the estimate of the sequences $\hat{X}^n$ and $\hat{Y}^n$. Denote $\mathcal{U}$ and $\mathcal{V}$ as the alphabet of U and V, respectively. Then, the code rates for the sources X and Y, R_X and R_Y, are given by $\frac{1}{n}\log_2|\mathcal{U}|$ and $\frac{1}{n}\log_2|\mathcal{V}|$, respectively. For example, if we compress $n = 10{,}000$ samples at code rate $R_X = 0.5$ bits per sample, then we would have $nR_X = 5000$ bits to send.

From Shannon's Source Coding Theorem (see Theorem A.2 in Appendix A), it follows that each X and Y can be compressed at a rate higher than their entropy. Thus, if we have separate decoding, the total compression rate will be $R = R_X + R_Y \geq H(X) + H(Y)$, regardless of whether joint or separate encoding takes place. If the two sources are compressed and decompressed together, the total rate will be $R \geq H(X, Y)$, which also directly follows from Shannon's Source Coding Theorem by taking (X, Y) as a single source to be compressed.

The question Slepian and Wolf wanted to answer was what is the minimum achievable rate if X and Y are to be compressed separately and reconstructed jointly. Obviously, due to potential gain of joint decoding, the minimum needed rate has to be less than or equal to $H(X) + H(Y)$. On the other hand, having both joint encoding and decoding will certainly be at least as good as joint decoding but separate encoding, meaning that the required rate has to be larger than $H(X, Y)$. If the sources are uncorrelated, from Appendix A, the two bounds overlap and thus the minimum rate will simply be $H(X, Y) = H(X) + H(Y)$, which is expected since joint coding cannot bring any gain in this case. However, in the case of correlated sources, we have $H(X, Y) < H(X) + H(Y)$, and thus it is not clear what the minimum needed rate will be.

Figure 2.1 Classical SW setup with two separately encoded sources and a joint decoder.

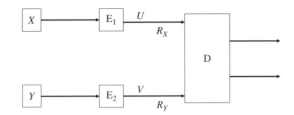

Slepian and Wolf showed a surprising result that under mild conditions the minimum required rate is still $H(X, Y)$, that is, separate encoding is as efficient as joint encoding provided that joint decoding takes place. This suggests that joint encoding brings no benefit over separate encoding, and thus opens up the possibility of distributed processing without any loss.

Before formally stating the SW Theorem and its proof, we go back to our temperature sensor example to illustrate the idea of Slepian and Wolf.

Example 2.2 *(A network of temperature sensors continued).* Let us consider for simplicity only two sensors with readings X and Y, and suppose that each reading can only take values *uniformly* between 0 and 31. Hence, the alphabets of X and Y, $\mathcal{X}$ and $\mathcal{Y}$, are given by $\mathcal{X} = \mathcal{Y} = \{0, 1, \cdots, 31\}$. Thus, each reading can be represented using at least 5 bits. Moreover, let X and Y be correlated in such a way that given Y, X can only take values *uniformly* from $\{Y - 1, Y, Y + 1, Y + 2\}$. In other words, X cannot be lower than 1 degree below Y and cannot be higher than 2 degrees above Y.

Let one realization of X and Y be $y = 11$ (01011 in binary) and $x = 12$ (01100 in binary). Suppose that all five bits of y (01011) are sent to the decoder. Thus, the decoder can decode Y perfectly to $\hat{y} = 11$.

Scenario 1: Suppose that the transmitters of X and Y can communicate, that is, X knows the value of Y. What is the minimum number of bits it needs to use, so that the decoder can recover X without errors? The answer to this question is simply 2, since the encoder can compute the difference between X and Y, which can only take one of four possible values: $(X - Y) \in \{-1, 0, 1, 2\}$, and hence it can be encoded with 2 bits. More specifically, suppose we use a code $\{-1, 0, 1, 2\} \rightarrow \{10, 00, 01, 11\}$, then the X sensor will send 01. The decoder knows that 01 corresponds to $x - y = 1$, and recovers correctly x as $\hat{x} = y + (x - y) = 11 + 1 = 12$. The total number of bits sent by X and Y in this case is $2 + 5 = 7$ bits, which is $H(X, Y)$. Indeed, there are a total of $32 \cdot 4 = 2^7$ possible (x, y) pairs, which are all equally likely. Note that in this scenario, the X encoder exploited the knowledge of y!

Scenario 2: Suppose now, that the two encoders cannot communicate. We pose again the same question: What is the minimum number of bits the X encoder needs to use, so that the decoder can recover X without errors?

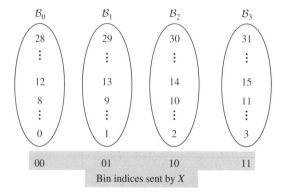

Figure 2.2 Four bins of a network of temperature sensors.

The SW Theorem tells us that still only 2 bits are needed. A trick to do this is to partition all the possible outcomes of $X \in \mathcal{X} = \{0, 1, \cdots, 31\}$ into subsets or *bins*. Instead of passing the value of X directly, only the index of the bin that X lies in is passed to the decoder. If for all y, each bin contains only one unique x satisfying the correlation constraints, then this unique x can be identified given y and the bin index at the decoder. This means that perfect decoding is possible. From our example, if $y = 7$, x can only be 6, 7, 8, or 9. Therefore, from the above argument, these four consecutive values cannot fall into the same bin to avoid ambiguous element at the decoder. Extending this argument to other values of x and y will force the difference between any two x inside the same bin to be no smaller than 4. This gives us a hint that the elements inside a bin should be as far away as possible. Moreover, if SW coding really has no loss, we expect that only 2 bits are needed to compress X and thus $2^2 = 4$ bins should be sufficient. Therefore, a good guess for the four bins will be $\mathcal{B}_0 = \{x | x \bmod 4 \equiv 0\} \cap \mathcal{X}$, $\mathcal{B}_1 = \{x | x \bmod 4 \equiv 1\} \cap \mathcal{X}$, $\mathcal{B}_2 = \{x | x \bmod 4 \equiv 2\} \cap \mathcal{X}$, and $\mathcal{B}_3 = \{x | x \bmod 4 \equiv 3\} \cap \mathcal{X}$. This is basically the optimum way in constructing four bins with the elements inside each bin as far apart as possible. The four bins are shown in Figure 2.2.

Following from our previous example, since $12 \bmod 4 \equiv 0$, bin index 00 will be transmitted from the X encoder. Knowing the bin index 00, the decoder knows that x is divisible by 4. Since $y = 11$ (thus potential $x \in \{10, 11, 12, 13\}$), hence the decoder can conclude that $x = 12$.

The procedure of distributing all possible outcomes of X into bins is called *binning*. Note that the total code rate is again $5 + 2 = H(X, Y) = H(Y) + H(X|Y)$. Thus, X was compressed at the rate $H(X|Y)$ (indeed, given Y, there are in total four possible values that X is equally likely to take). It is worth noting that each bin contains eight values, which is in fact $I(X; Y)$, information that Y gives us about X. That is, knowing Y, we can distinguish among these eight values of x and find the right one.

Before formally stating the SW Theorem, an SW code is defined. A code is usually defined by encoder and decoder mapping functions. The encoder function maps the set of all possible values of a source (source alphabet) into a set of codewords. The decoder function does the opposite. Suppose that R_X and R_Y are code rates for compressing X and Y, respectively, and let n be the length of the source samples that encoders will handle at one time. Then, the encoders will map n samples of X and Y into indices U and V, which are then transmitted losslessly using nR_X and nR_Y bits (provided that $\log_2 |U| \leq nR_X$ and $\log_2 |V| \leq nR_Y$) instead of the samples of X and Y, respectively. Finally, the decoder maps the indices back to the samples of X and Y. We designate this code as an SW ($2^{nR_X}, 2^{nR_Y}$) code and have the following formal definition.

Definition 2.1 (SW code). Let X and Y be two correlated random sources drawn i.i.d. from joint distribution $p(x, y)$ with alphabets $\mathcal{X}$ and $\mathcal{Y}$, respectively. A SW code for compressing X and Y denoted by ($2^{nR_X}, 2^{nR_Y}$) compresses independently source sequences X^n and Y^n at rates R_X and R_Y, respectively, so that a joint decoder can recover them perfectly with probability approaching one. The code is defined with two encoder functions:

$$e_X: \mathcal{X}^n \to \{1, \ldots, 2^{nR_X}\} = U$$
$$e_Y: \mathcal{Y}^n \to \{1, \ldots, 2^{nR_Y}\} = V$$

and one decoder function:

$$d: U \times V \to \mathcal{X}^n \times \mathcal{Y}^n.$$

Thus, each n-bit realization, x^n and y^n, is independently encoded into nR_X and nR_Y bit codeword $u = e_X(x^n)$ and $v = e_Y(y^n)$, respectively. The decoder recovers the sources as $(\hat{x}^n, \hat{y}^n) = d(u, v)$.

Example 2.3 (Compression rate of sensor outputs). Continuing with our sensor example, the scheme we described codes one pair of samples each time and thus $n = 1$. $R_Y = 5$ bits per sample as we used $nR_Y = 5$ bits to represent Y. On the other hand, $R_X = 2$ bits per sample as a total of $nR_X = 2$ bits are sent for X.

Definition 2.2 (Probability of error). The probability of error for the above SW code is defined as $P_e^n = Pr\left(d(e_X(x^n), e_Y(y^n)) \neq (x^n, y^n)\right)$.

Finally, we define the achievable rate and the achievable rate region for compressing X and Y.

Definition 2.3 (Achievable rate). A rate pair (R_X, R_Y) is called achievable if there exists a sequence of SW codes ($2^{nR_X}, 2^{nR_Y}$) with probability of error $P_e^n \to 0$.

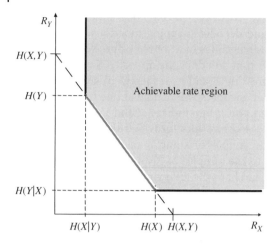

Figure 2.3 SW rate region.

Definition 2.4 *(Achievable rate region).* The set of pairs of rates that contains all achievable rates is called the achievable rate region.

In Definition 2.3, $P_e^n \rightarrow 0$ should be understood as approaching zero probability of error as n approaches infinity. Thus a sequence of codes with increasing n is used for the analysis. Now we state formally the SW Theorem and show a rigorous proof. Our notation closely follows that of [12].

Theorem 2.1 *(Slepian–Wolf).* Let X and Y be two correlated random sources drawn i.i.d. from joint distribution $p(x, y)$. Then, the achievable rate region of compressing X and Y is given by:

$$R_X \geq H(X|Y),$$
$$R_Y \geq H(Y|X),$$
$$R = R_X + R_Y \geq H(X, Y),$$

where R_X and R_Y are rates used for compressing X and Y, respectively.

The achievable SW rate region is illustrated in Figure 2.3. The SW bound comprises of three segments: one horizontal and one vertical line which come from the first and second inequality, and one diagonal line which bounds the rate sum and is also the limit for joint compression.

Note that if the first source is compressed conventionally (up to its entropy), the second source can be compressed as low as its conditional entropy given the first source, and vice versa. Then if we linearly increase the rate used for the second source, we can decrease the rate used for the first one, that is, we can compress the first source more. At one point, the coding rates will meet, and both sources will be compressed at $R_X = R_Y = H(X, Y)/2$.

If the sources are not correlated, then $R_X \geq H(X|Y) = H(X)$ and $R_Y \geq H(Y|X) = H(Y)$, thus we obtain classic Shannon's Source Coding Theorem. If the sources are, on the other hand, identical, we get $R_X \geq H(X|Y) = 0$, $R_Y \geq H(Y|X) = 0$, and $R = R_X + R_Y \geq H(X,Y) = H(X)$. That is, it is enough to send only one source conventionally compressed.

The above rate region gives us bounds on the amount that the two sources can be compressed. These bounds can be reached only by using SW codes of infinite length, and are thus not useful in practice. However, the importance of this result, similar to the capacity definition in communications theory, is that it gives us bounds to aim for.

Two computational examples are given below.

Example 2.4   *(Binary symmetric channel (BSC) correlation).*
Suppose that X is a binary random source drawn i.i.d. from uniform distribution $p(x)$. Let Y be given by $Y = X \oplus E$, where $\oplus$ is modulo-2 sum and E is binary i.i.d. random variable with $p = p(e = 0) = 0.9$. Thus, one can see Y as the output of a binary symmetric correlation channel whose input is X. What is the minimum total rate R needed for SW coding of X and Y?

Solution:
From the SW Theorem, $R = R_X + R_Y \geq H(X,Y) = H(X) + H(Y|X) = 1 + H(X \oplus E|X) = 1 + H(E) = 1 + H(p) = 1.469$ bits.

Example 2.5   *(Gauss–Markov correlation).*
Suppose that X is a binary random source drawn i.i.d. from uniform distribution $p(x)$. Let Y be given by $Y = X \oplus E_1$ with probability $p_{good} = 0.9$ and $Y = X \oplus E_2$ with probability $p_{bad} = 0.1$, where $p(e_1 = 0) = 0.999$ and $p(e_2 = 0) = 0.9$. Thus, we can see this correlation as a Gauss–Markov channel with one good and one bad state. Suppose that we want to transfer $n = 10,000$ samples of X and Y. What is the total minimum number of bits we need to send?

Solution:
From the SW Theorem, we have $R = R_X + R_Y \geq H(X,Y) = H(X) + H(Y|X) = 1 + p_{good}H(E_1) + p_{bad}H(E_2) = 1 + 0.9H(0.001) + 0.1H(0.1) = 1 + 0.0572 = 1.0572$. Thus we need $nR = 10{,}572$ bits instead of 20,000 bits.

2.1.1 Proof of the SW Theorem

In this section, we give a formal proof of the SW Theorem. We will present the proof given in [12].

The proof of Theorem 2.1 [1] consists of two parts: (1) the achievability or the forward part and (2) the converse part. The forward part shows that there exist

SW codes if the rate pair lies in the achievable rate region. The converse part shows that there exists no code if the rate pair lies outside the achievable rate region.

Achievability of the SW Theorem

To prove achievability, we need to provide encoding and decoding methods, and show that they can achieve an arbitrary small probability of error $P_e^n \to 0$ when R_X and R_Y satisfy the theorem's constraints.

Following the idea outlined in Example 2.2, we want to distribute all possible sources' realizations into *bins*. Each encoder simply sends the index of the bin to which its realization belongs. Since only nR_X and nR_Y bits are transmitted, we need to form 2^{nR_X} and 2^{nR_Y} bins for X and Y, respectively.

The code generation is based on *random binning* arguments, where typical sequences of X and Y are randomly assigned to one of 2^{nR_X} and 2^{nR_Y} bins, respectively. Then the decoder reconstructs the sequences as the jointly typical sequences that match with the received bin indices. The overall coding procedure consists of three steps:

Code preparation Randomly and uniformly partition the typical sets of X and Y, $T_\epsilon^n(X)$ and $T_\epsilon^n(Y)$, into 2^{nR_X} and 2^{nR_Y} bins, respectively. Denote the bin indices of sequences x^n and y^n as $B_X(x^n)$ and $B_Y(y^n)$, respectively.

Encoding Receiving the input sequences x^n and y^n, the two encoders separately output the bin indices $b_1 = B_X(x^n)$ and $b_2 = B_Y(y^n)$.

Decoding Receiving bin indices b_1 and b_2 from the two encoders, the joint decoder searches for sequences $\hat{x}^n$ and $\hat{y}^n$ such that $(\hat{x}^n, \hat{y}^n) \in T_\epsilon^n(X, Y)$, $B_X(\hat{x}^n) = b_1$ and $B_X(\hat{y}^n) = b_2$. If there is more than one jointly typical sequence in the bin (b_1, b_2), declare a decoding error.

Next, we show that for the above code construction, the probability of decoding error goes to zero as n tends to infinity, that is, $P_e^n \to 0$ as $n \to \infty$.

Intuitively, if the binning is truly random, each bin will have approximately the same number of sequences. The total number of possible bin pairs is $2^{nR_X} \times 2^{nR_Y} = 2^{n(R_X + R_Y)}$. If this number is larger than the total number of typical sequences $2^{nH(X,Y)}$ (see Lemma A.23 in Appendix A), then each bin pair (b_1, b_2) will have approximately only one typical sequence which can therefore be uniquely identified. Similarly, consider the case when y^n is known using index b_2 alone, then there are only approximately $2^{nH(X|Y)}$ sequences of X jointly typical with y^n (see Lemma 2.1). Therefore, there will be on average $2^{n(H(X|Y)-R_X)}$ such sequences in each bin. And thus when $R_X \geq H(X|Y)$, each bin will have less than one such sequence on average and the sequence can be uniquely identified. A similar argument forces that $R_Y \geq H(Y|X)$.

More rigorously, decoding errors may happen if the following error events occur:

1) E_1: x^n and y^n are not jointly typical. (The decoder will then fail to find x^n and y^n since it restricts its search to the jointly typical sequences.)
2) E_2: there exist $\tilde{x}^n \neq x^n$ and $\tilde{y}^n \neq y^n$, such that $B_X(\tilde{x}^n) = b_1$, $B_Y(\tilde{y}^n) = b_2$, and $(\tilde{x}^n, \tilde{y}^n) \in T_\epsilon^n(X, Y)$. (Decoding error due to more than one jointly typical sequence in the bin, $(\tilde{x}^n, \tilde{y}^n)$ in addition to (x^n, y^n).)
3) E_3: there exists $\tilde{x}^n \neq x^n$ such that $B_X(\tilde{x}^n) = b_1$ and $(\tilde{x}^n, y^n) \in T_\epsilon^n(X, Y)$. (Decoding error due to more than one jointly typical sequence in the bin, $(\tilde{x}^n, y^n)$ in addition to (x^n, y^n).)
4) E_4: there exists $\tilde{y}^n \neq y^n$ such that $B_Y(\tilde{y}^n) = b_2$ and $(x^n, \tilde{y}^n) \in T_\epsilon^n(X, Y)$. (Decoding error due to more than one jointly typical sequence in the bin, $(x^n, \tilde{y}^n)$ in addition to (x^n, y^n).)

From the union bound, the total decoding error probability is bounded by:

$$P_e^n = Pr(E_1 \cup E_2 \cup E_3 \cup E_4) \leq Pr(E_1) + Pr(E_2) + Pr(E_3) + Pr(E_4). \quad (2.1)$$

That is, the overall probability of error is smaller than the summation of the probabilities of the four erroneous events. Thus we can consider each event independently and bound the probability of its occurrence.

The probability of E_1 will go to zero as n goes to infinity, which follows from the fact that almost all pairs of sequences drawn from the sources will be jointly typical. Thus, for any $\epsilon > 0$, there exists sufficiently large n such that $Pr(E_1) < \epsilon$.

For the second error event E_2, and for $(\tilde{x}^n, \tilde{y}^n) \neq (x^n, y^n)$, we have

$$Pr(E_2) = Pr(\exists (\tilde{x}^n, \tilde{y}^n) \text{ s.t. } B_X(\tilde{x}^n) = b_1, B_Y(\tilde{y}^n) = b_2, (\tilde{x}^n, \tilde{y}^n) \in T_\epsilon^n(X, Y))$$

$$\overset{(a)}{\leq} \sum_{(\tilde{x}^n, \tilde{y}^n) \in T_\epsilon^n(X,Y)} Pr(B_X(\tilde{x}^n) = b_1, B_Y(\tilde{y}^n) = b_2)$$

$$\overset{(b)}{=} \sum_{(\tilde{x}^n, \tilde{y}^n) \in T_\epsilon^n(X,Y)} 2^{-n(R_X + R_Y)}$$

$$\overset{(c)}{\leq} 2^{-n(R_X + R_Y - H(X,Y) - \epsilon)},$$

where (a) follows from the fact that we sum over all jointly typical sequences (including (x^n, y^n)), (b) comes from the fact that $B_X(x^n)$ and $B_Y(y^n)$ are uniformly distributed random variables taking 2^{nR_X} and 2^{nR_Y} values, and (c) comes from the fact that the total number of jointly typical sequences is $|T_\epsilon^n(X, Y)| \leq 2^{n(H(X,Y)+\epsilon)}$ (see Lemma A.23 in Appendix A). Note that ϵ can be chosen to be arbitrarily small as long as n is sufficiently large. Therefore, if $R_X + R_Y > H(X, Y)$, $Pr(E_2)$ goes to zero as n goes to infinity. In particular, for sufficiently large n, we can make $Pr(E_2) < \epsilon$.

For the third and fourth error event, we will need the following lemma.

Lemma 2.1 *(Size of conditional typical set).* Let X and Y be two correlated random sources drawn i.i.d. from the joint distribution $p(x^n, y^n)$. Let $y^n \in$

$T_\epsilon^n(Y)$ and denote the set of X that is jointly typical with y^n as $T_\epsilon^n(X|y^n)$. Then, $|T_\epsilon^n(X|y^n)| \doteq 2^{nH(X|Y)}$. More precisely,

$$|T_\epsilon^n(X|y^n)| \leq 2^{n(H(X|Y)+2\epsilon)} \tag{2.2}$$

and

$$(1-\epsilon)2^{n(H(X|Y)-2\epsilon)} \leq \sum_{y^n \in T_\epsilon^n(Y)} p(y^n)|T_\epsilon^n(X|y^n)|. \tag{2.3}$$

Proof.

$$1 \geq \sum_{x^n \in T_\epsilon^n(X|y^n)} p(x^n|y^n)$$

$$= \sum_{x^n \in T_\epsilon^n(X|y^n)} \frac{p(x^n,y^n)}{p(y^n)}$$

$$\overset{(a)}{\geq} \sum_{x^n \in T_\epsilon^n(X|y^n)} \frac{2^{n(H(X,Y)-\epsilon)}}{2^{n(H(Y)+\epsilon)}}$$

$$= |T_\epsilon^n(X|y^n)|2^{-n(H(X|Y)+2\epsilon)}|,$$

where (a) comes from the definition of typicality and joint typicality, and this shows (2.2). Similarly, for a sufficiently large n and by Lemma A.22, we have

$$1 - \epsilon \leq \sum_{(x^n,y^n) \in T_\epsilon^n(X,Y)} p(x^n,y^n)$$

$$= \sum_{y^n \in T_\epsilon^n(Y)} p(y^n) \sum_{x^n \in T_\epsilon^n(X|y^n)} p(x^n|y^n)$$

$$= \sum_{y^n \in T_\epsilon^n(Y)} p(y^n) \sum_{x^n \in T_\epsilon^n(X|y^n)} \frac{p(x^n,y^n)}{p(y^n)}$$

$$\overset{(a)}{\leq} \sum_{y^n \in T_\epsilon^n(Y)} p(y^n) \sum_{x^n \in T_\epsilon^n(X|y^n)} 2^{-n(H(X|Y)-2\epsilon)}$$

$$= 2^{-n(H(X|Y)-2\epsilon)} \sum_{y^n \in T_\epsilon^n(Y)} p(y^n)|T_\epsilon^n(X|y^n)|,$$

where (a) again follows from the definition of typicality, and this shows (2.3). □

Now, back to the proof of achievability,

$$Pr(E_3) = Pr(\exists \tilde{x}^n \neq x^n \text{ s.t. } \tilde{x}^n \in T_\epsilon^n(X|y^n), B_X(\tilde{x}^n) = b_1)$$

$$\overset{(a)}{\leq} \sum_{\tilde{x}^n \in T_\epsilon^n(X|y^n)} Pr(B_X(\tilde{x}^n) = b_1)$$

$$\overset{(b)}{=} \sum_{\tilde{x}^n \in T_\epsilon^n(X|y^n)} 2^{-nR_X}$$

$$= |T_\epsilon^n(X|y^n)|2^{-nR_X}$$
$$\leq 2^{-n(R_X - H(X|Y) - 2\epsilon)},$$

where (a) comes from that we sum over all typical sequences (including x^n) and (b) is true if binning is random and uniform. Thus, $Pr(E_3)$ goes to zero as n goes to infinity if $R_X > H(X|Y)$. In particular, for sufficiently large n, $Pr(E_3) < \epsilon$. By swapping X and Y above, we get for sufficiently large n, $Pr(E_4) < \epsilon$.

By replacing the obtained inequalities for $Pr(E_1)$, $Pr(E_2)$, $Pr(E_3)$, and $Pr(E_4)$ in (2.1) we get for the average probability of decoding error: $P_e^n < 4\epsilon$. Thus, there exists an SW code that gives arbitrarily small probability of decoding error. Hence, we can construct a sequence of codes with $P_e^n \to 0$, which concludes the achievability proof.

Like most proofs in information theory, the proof here is nonconstructive in nature. It is not clear how the decoding step can be performed in an effective manner. Practical implementations of SW coding will be discussed in Chapter 5.

Converse of the SW Theorem

The converse proof relies on Fano's inequality and its variation. The proof of one of its most basic forms that considers the lossless compression of i.i.d. source is given in Lemma A.33 in Appendix A. To prove the converse of the SW Theorem, we need two variations of Fano's inequality. We give these variations in the following lemma and leave the proof as an exercise.

Lemma 2.2 *(Fano's inequalities for the SW proof).* For the SW setup as shown in Figure 2.1 and using already established notation, we have

$$\frac{1}{n}H(X^n|U, Y^n) \leq \frac{H(P_e)}{n} + P_e \log_2 |\mathcal{X}|, \tag{2.4}$$

$$\frac{1}{n}H(Y^n|V, X^n) \leq \frac{H(P_e)}{n} + P_e \log_2 |\mathcal{Y}|, \tag{2.5}$$

and

$$\frac{1}{n}H(X^n, Y^n|U, V) \leq \frac{H(P_e)}{n} + P_e \log_2 |\mathcal{X} \times \mathcal{Y}|. \tag{2.6}$$

The exact expressions on the right-hand sides of the equations in Lemma 2.2 are not very important. The main point is that when P_e goes to zero, so do the right-hand sides. These results are actually very intuitive. For example, the last equation essentially means that the average information content of X^n and Y^n per symbol knowing U and V becomes insignificant when the error probability goes to zero. This makes sense because if the probability of error goes to zero, then $\hat{X}^n$ and $\hat{Y}^n$, reconstructed deterministically from U and V, are essentially the same as X^n and Y^n. In other words, X^n and Y^n will be perfectly known given U and V.

Now, we are ready for the converse proof. We need to show that if the probability of error goes to zero, the rates need to satisfy the constraints specified by the SW Theorem. Since the total sum rate R has to be larger than or equal to the joint entropy of U and V in order to pass these indices to the decoder correctly, we have

$$R = R_X + R_Y \geq \frac{1}{n} H(U, V)$$

$$= \frac{1}{n}(H(U, V|X^n, Y^n) + I(U, V; X^n, Y^n))$$

$$\overset{(a)}{=} \frac{1}{n} I(U, V; X^n, Y^n)$$

$$= \frac{1}{n}(H(X^n, Y^n) - H(X^n, Y^n|U, V))$$

$$\overset{(b)}{\geq} \frac{1}{n} H(X^n, Y^n) - \epsilon$$

$$\overset{(c)}{=} H(X, Y) - \epsilon,$$

where (a) is due to the fact that U and V are deterministic given X^n and Y^n, that is, given X^n (Y^n) we know exactly U (V), (b) is due to Fano's Inequality (2.6) that an arbitrary small ϵ exists for sufficiently large n, and (c) is due to the chain rule for entropies and the fact that X and Y are i.i.d. sources.

For the code rate of X, we have

$$R_X \geq \frac{1}{n} H(U)$$

$$\overset{(a)}{\geq} \frac{1}{n} H(U|Y^n)$$

$$= \frac{1}{n}(H(U|X^n, Y^n) + I(U; X^n|Y^n))$$

$$\overset{(b)}{=} \frac{1}{n} I(U; X^n|Y^n)$$

$$= \frac{1}{n}(H(X^n|Y^n) - H(X^n|U, Y^n))$$

$$\overset{(c)}{\geq} \frac{1}{n} H(X^n|Y^n) - \epsilon$$

$$\overset{(d)}{=} \frac{1}{n} \sum_i H(X_i|Y^n, X_1, X_2, \cdots, X_{i-1}) - \epsilon$$

$$\overset{(e)}{=} \frac{1}{n} \sum_i H(X_i|Y_i) - \epsilon$$

$$= H(X|Y) - \epsilon,$$

where (a) is due to that conditioning reduces entropy, (b) is because U is deterministically given X^n, (c) is from Fano's inequality (2.4) that an arbitrary small ϵ exists for sufficiently large n, (d) follows from the chain rule for entropies, and (e) is because X_i is conditionally independent with X_j and $Y_j, j \neq i$ given Y_i, since both X and Y are an i.i.d. source.

One can similarly show that $R_Y \geq H(Y|X)$ as n goes to infinity using Fano's inequality (2.5).

2.2 Asymmetric and Symmetric SW Coding

The SW rate region contains two corner points, namely $(H(X), H(Y|X))$ and $(H(X|Y), H(Y))$, as shown in Figure 2.3. A scheme that attempts to approach one of these two corner points is called the *asymmetric SW coding scheme*. All other schemes are non-asymmetric schemes. The scheme that tends to approach the middle point is called the *symmetric SW coding scheme*.

Note that in asymmetric coding, one source is compressed conventionally, at its entropy rate, whereas the other one is compressed at the minimum possible rate. The former source is first conventionally decompressed at the decoder, and then it is used for decompression of the latter.

From the point of view of the latter source, this case is equivalent to the case when the former source is already present at the decoder, which is called *source coding with decoder side information*. The source present at the decoder is called *side information* and it is used for decoding the unknown source. For completeness, we give the following corollary, which directly follows from the SW Theorem.

Corollary 2.1 *(Source coding with decoder side information).* Let X and Y be two correlated random sources drawn i.i.d. from joint distribution $p(x^n, y^n)$. Suppose that Y is known to the decoder. Then, the achievable rate of compressing X given Y at the decoder is $R_X \geq H(X|Y)$.

This is illustrated by the example below.

Example 2.6 *(Asymmetric SW coding with BSC correlation).*
Suppose that X is a binary random source drawn i.i.d. from distribution $p(x^n)$. Let Y be given by $Y = X \oplus E$, where $\oplus$ is modulo-2 sum and E is binary i.i.d. random variable with $p = p(e = 0) = 0.9$. Thus, one can see Y as the output of a binary symmetric correlation channel whose input is X. What is the minimum rate needed for compressing Y if X is already known at the decoder?

Solution:
From Corollary 2.1, $R_Y \geq H(Y|X) = H(X \oplus E|X) = H(E) = H(p) = 0.469$ bits.

Actually, if we have two schemes at rate (R_X, R_Y) and (R'_X, R'_Y), we will be able to generate a scheme at rate $(\lambda R_X + (1 - \lambda)R'_X, \lambda R_Y + (1 - \lambda)R'_Y)$, $\lambda \in [0, 1]$ by simply operating the first scheme λ fraction of time and the second scheme for the remaining $(1 - \lambda)$ fraction of time. If both schemes are achievable (error probability equal to zero), then the combined scheme should be achievable as well. Such a trick is known as time sharing. We will state this formally as a theorem below but will leave the rigorous proof to the reader as an exercise.

Theorem 2.2 *(Time-sharing).* Suppose that rate-pairs (R_X, R_Y) and (R'_X, R'_Y) are achievable. Then a rate-pair

$$R''_X = \lambda R_X + (1 - \lambda)R'_X, R''_Y = \lambda R_Y + (1 - \lambda)R'_Y,$$

for $0 \le \lambda \le 1$ is also achievable.

In particular, we may pick (R_X, R_Y) and (R'_X, R'_Y) as $(H(X), H(Y|X))$ and $(H(X|Y), H(Y))$, that is, the corner points of the SW bound and these correspond to the asymmetric coding scenarios. By varying λ from 0 to 1, we can achieve all points on the SW bound by time-sharing. Therefore, at least in principle, any rate near the SW bound can be achieved if we can implement a perfect asymmetric SW coding schemes. In practice, there are other limitations that lead one to consider non-asymmetric approaches, as we will see in the application chapters.

2.3 SW Coding of Multiple Sources

The SW Theorem gives the achievable rate region for compressing two i.i.d. correlated sources. The result can easily be extended to multiple i.i.d. sources, as given in [13]. The network setup is shown in Figure 2.4. We state below the theorem following the exposition in [12] and provide the sketch of a proof.

Theorem 2.3 *(SW coding of multiple sources).* Let $X_1, \ldots, X_L$ be L correlated random sources drawn i.i.d. from joint distribution $p(x_1^n, \ldots, x_L^n)$. Then, the achievable rate region of independently compressing and jointly decompressing these sources is given by:

$$R_S \ge H(X_S | X_{S^C}),$$

for all sets $S \subseteq \{1, \ldots, L\}$, where S^C is the compliment of the set S. For a set $A = \{a_1, \ldots, a_k\}$, we denote $X_A = (X_{a_1}, \ldots, X_{a_k})$, $k \le L$. Moreover,

$$R_S = \sum_{i \in S} R_i$$

with R_i being the rate used for compressing X_i.

First we give a simple example with three sources.

Figure 2.4 SW network with multiple sources.

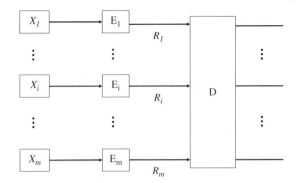

Example 2.7   *(Rate region for three sources).*

Suppose that we have three random binary correlated sources (X, Y, Z) drawn i.i.d. from the joint distribution $p(x^n, y^n, z^n)$ shown in Table 2.1. What is the achievable rate region (R_X, R_Y, R_Z) for lossless compression of these three sources?

Solution:

From Theorem 2.3, the following rates are achievable:

$$R_X \geq H(X|Y, Z),$$
$$R_Y \geq H(Y|X, Z),$$
$$R_Z \geq H(Z|Y, X),$$
$$R_X + R_Y \geq H(X, Y|Z),$$
$$R_Y + R_Z \geq H(Y, Z|X),$$
$$R_X + R_Z \geq H(X, Z|Y),$$
$$R_X + R_Y + R_Z \geq H(X, Y, Z).$$

Table 2.1 Joint probability distribution.

x	y	z	p(x,y,z)
0	0	0	0.45
0	0	1	0
0	1	0	0
0	1	1	0.03
1	0	0	0.01
1	0	1	0.03
1	1	0	0.03
1	1	1	0.45

From the above equations we get $R_X \geq 0.38$, $R_Y \geq 0.35$, $R_Z \geq 0.35$, $R_X + R_Y \geq 0.71$, $R_X + R_Z \geq 0.71$, $R_Y + R_Z \geq 0.71$, $R_X + R_Y + R_Z \geq 1.71$.

Proof of Theorem 2.3. The proof is the same as that of the SW Theorem. It consists of achievability part and converse. For the achievability part, we need to form $2^{nR_1}, \ldots, 2^{nR_L}$ bins for encoding $X_1, \ldots, X_L$, respectively. Each encoder simply sends the index of the bin to which its realization belongs.

Code preparation For each $i \in \{1, \ldots, L\}$, form 2^{nR_i} bins, and index them with $\{1, \ldots, 2^{nR_i}\}$. Assign each possible value of X_i^n, $x_i \in \mathcal{X}_i^n$, into one of the bins randomly, that is, according to uniform distribution.

Encoding For each $i \in \{1, \ldots, L\}$, for a source realization x_i send the index of the bin $b_i = e_{X_i}(x_i)$ to which it belongs.

Decoding Given a received L-tuple $b_1, \ldots, b_L$, declare $(\hat{x}_1, \ldots, \hat{x}_L)$ as decoded L-tuple if it is the only jointly typical $(X_1^n, \ldots, X_L^n)$ in the bins indexed by $(b_1, \ldots, b_L)$, that is, $b_1 = e_{X_1}(\hat{x}_m), \cdots, b_m = e_{X_m}(\hat{x}_m)$. Otherwise, declare a decoding error.

The remainder of the proof follows from the SW Theorem proof. The converse of the theorem uses Fano's inequality in the same way as the SW Theorem converse, so it is omitted here and left as an exercise. $\square$

In the next chapter, we extend DSC setups to more than one decoder with two or more correlated sources.

3

Wyner–Ziv Coding Theory

WZ coding, or source coding with side information at the decoder, can be considered as both a generalized and a degenerated case of SW coding. It generalizes SW coding from the lossless to the lossy case, which is similar to the generalization of the classic source coding problem to the rate-distortion problem. However, it degenerates from SW coding in that it only considers an asymmetric case in which only one of two sources is compressed. The discussion of the generalization of WZ coding to the non-asymmetric case will be deferred to the following chapter.

Let X and Y be two joint i.i.d. sources. The WZ problem attempts to compress a block of X, X^n, lossily and recovers an estimate $\hat{X}^n$ at the decoder with the help of Y^n. More precisely, the encoder is a mapping

$$f_E : \mathcal{X}^n \to \{1, \cdots, M\}$$

and the decoder is a mapping

$$f_D : \{1, \cdots, M\} \times \mathcal{Y}^n \to \hat{\mathcal{X}}^n.$$

The code rate is given by

$$R = \frac{\log_2 M}{n},$$

which can be interpreted as the average number of bits needed to compress a symbol X. In this chapter we will denote the output of the WZ encoder as W, that is, $W = f_E(X^n)$. The setup is summarized in Figure 3.1.

As mentioned earlier, WZ considers a lossy problem, where in general $Pr(\hat{X} \neq X) > 0$. One can characterize the fidelity of $\hat{X}$ from X using a distortion measure $d(\cdot, \cdot)$. In essence, we are interested in finding the smallest rate required to ensure the average distortion $E[d(X, \hat{X})]$ is smaller than some predetermined value D.

Distributed Source Coding: Theory and Practice, First Edition. Shuang Wang, Yong Fang, and Samuel Cheng.
© 2017 John Wiley & Sons Ltd. Published 2017 by John Wiley & Sons Ltd.
Companion Website: www.wiley.com/go/cheng/dsctheoryandpractice

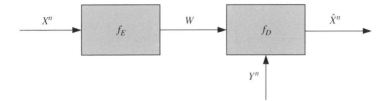

Figure 3.1 Setup of the WZ problem.

The main theorem we will present in this chapter is that the rate-distortion function of the WZ setup, $R_{WZ}(D)$, can be found by solving the following optimization problem:

$$R^*(D) \triangleq \min_{q(z|x), \hat{x}=g(y,z)} I(X;Z|Y)$$

s.t. $\quad E[d(X, \hat{X})] \leq D.$ $\hspace{4cm}$ (3.1)

In other words, we have $R_{WZ}(D) = R^*(D)$.

Before deriving the rate-distortion function, let us first show that $R^*(D)$ is a convex function of D.

Proof. Note that we can write $I(X;Z|Y)$ as

$$
\begin{aligned}
I(X;Z|Y) &= \sum_{x,y,z} p(x,y,z) \log \frac{p(x,z|y)}{p(x|y)p(z|y)} \\
&= \sum_{x,y} p(x,y) \sum_z q(z|x) \log \frac{q(z|x)}{p(z|y)} \\
&= \sum_{x,y} p(x,y) D(q(z|x) \| p(z|y)).
\end{aligned}
$$ $\hspace{2cm}$ (3.2)

Now let $R_1 = R^*(D_1)$ and $R_2 = R^*(D_2)$. From the definition of $R^*(\cdot)$ and (3.2), there exists two coder pairs (one with $q_1(z|x)$ and $g_1(y,z)$, and the other with $q_2(z|x)$ and $g_2(y,z)$) such that

$$
\begin{aligned}
R_1 &= \sum_{x,y} p(x,y) D(q_1(z|x) \| p(z|y)), \\
R_2 &= \sum_{x,y} p(x,y) D(q_2(z|x) \| p(z|y)), \\
D_1 &= E_{p(x,y)q_1(z|x)}[d(X, g_1(Y,Z))], \\
D_2 &= E_{p(x,y)q_2(z|x)}[d(X, g_2(Y,Z))],
\end{aligned}
$$

where $E_{p(x,y)q(z|x)}[d(X, g(Y,Z))] \triangleq \sum_{x,y,z} p(x,y) q(z|x) d(x, g(y,z))$.

Let us apply "time sharing" to the two coder pairs. More precisely, we will use the first coder pair (with $q_1(z|x)$ and $g_1(y,z)$) for λ fraction of the time and the other pair (with $q_2(z|x)$ and $g_2(y,z)$) for $(1-\lambda)$ fraction of the time.

Figure 3.2 Time sharing of two coder pairs results in a (R, D) couple below the linear interpolation of the two end couples (shown as * in the figure). $R^*(D)$ will be smaller than the time-sharing rate since $R^*(D)$ is minimized over all possible distributions.

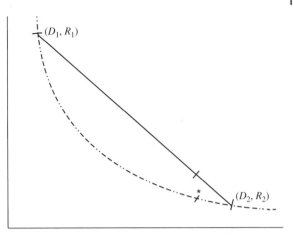

Of course, both transmitter and receiver have to be "synchronized" to use the same coder pair.

After time sharing, it is easy to verify that the distortion is just the weighted average of D_1 and D_2, that is, $D = \lambda D_1 + (1 - \lambda)D_2$. On the other hand, the corresponding rate $\tilde{R}$ will be given by

$$\tilde{R} = \sum_{x,y} p(x,y)D(\lambda q_1 + (1 - \lambda)q_2 \| p)$$

$$= \sum_{x,y} p(x,y)D(\lambda q_1 + (1 - \lambda)q_2 \| \lambda p + (1 - \lambda)p)$$

$$\overset{(a)}{\leq} \sum_{x,y} p(x,y)[\lambda D(q_1 \| p) + (1 - \lambda)D(q_2 \| p)]$$

$$= \lambda R_1 + (1 - \lambda)R_2,$$

where (a) is from Lemma A.12.

Now, by the definition that $R^*(D)$ is the minimum $I(X; Z|Y)$ among all allowed distributions, we must have $R^*(\lambda D_1 + (1 - \lambda)D_2) \leq \tilde{R} \leq \lambda R_1 + (1 - \lambda) R_2 = \lambda R^*(D_1) + (1 - \lambda)R^*(D_2)$ (also see Figure 3.2). Therefore, $R^*(D)$ is convex. □

3.1 Forward Proof of WZ Coding

The forward proof shows that we can always find a code at rate $R \geq R^*(D)$ such that the average distortion is less than D. We will try to offer insight through a sketch proof here. Interested readers are directed to the original paper for a rigorous proof [14].

Consider a block of joint sources x^n, y^n. The objective of WZ is to encode x^n into an index variable w and then reconstruct the source $\hat{x}^n$ with the help of both

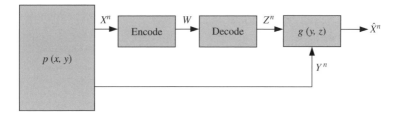

Figure 3.3 Rather than reconstructing $\hat{X}^n$ immediately, we could first construct an intermediate auxiliary output Z^n at the decoder. If enough information is passed through W, we can ensure that X^n, Y^n, and Z^n always have the same statistics as sequences sampled from $p(x,y)q^*(z|x)$ and consequently we will have $E[d(X,\hat{X})] = D$ as desired.

w and y^n. Assume that $p^*(x,y,z) = p(x,y)q^*(z|y)$ and $g^*(z,x)$ are the distribution and function that optimize (3.1). Instead of reconstructing $\hat{x}^n$ through y^n and w directly, we can first construct z^n at the decoder and recover $\hat{x}^n$ as $g(z^n, y^n) \triangleq [g(z_1,y_1), g(z_2,y_2), \cdots, g(z_n,y_n)]$ (see Figure 3.3). If the rate of transmission is sufficiently high, we can ensure that (x^n, y^n, z^n) is a jointly typical sequence from the joint source $p(x,y,z)$. Therefore, the average distortion $E[d(X,\hat{X})] = D$.

The only question left is how many bits are needed to recover Z^n losslessly. Note that we can potentially decrease the rate by further compressing W. Let us imagine that rather than transmitting W directly, we will further encode W and transmit a coded W, say $\tilde{W}$ instead. Moreover, with the help of the side information Y^n, we can compress W into $H(W|Y^n)$ instead of $H(W)$ by SW coding. Therefore, the required rate is

$$R = \frac{H(W|Y^n)}{n}$$
$$= \frac{1}{n}[H(W,Y^n) - H(Y^n)]$$
$$= \frac{1}{n}[H(W) + H(Y^n|W) - H(Y^n)]$$
$$= \frac{1}{n}[H(W) + H(Y^n|W)] - H(Y)]. \tag{3.3}$$

Note that we divide $H(W|Y^n)$ by n since we are processing n samples of X for each transmitted W. Now, since $Y^n \leftrightarrow W \leftrightarrow Z^n$, we have

$$I(Y^n; W) \geq I(Y^n; Z^n)$$
$$\Rightarrow H(Y^n) - H(Y^n|W) \geq H(Y^n) - H(Y^n|Z^n)$$
$$\Rightarrow H(Y^n|W) \leq H(Y^n|Z^n) = \sum_k H(Y_k|Z^n, Y^{k-1})$$
$$\Rightarrow H(Y^n|W) \leq \sum_k H(Y_k|Z_k) = nH(Y|Z). \tag{3.4}$$

On the other hand, if we interpret W as the message required to construct Z^n from X^n, by conventional rate-distortion theorem, the minimum rate required is $I(X^n; Z^n)$ and thus

$$\frac{H(W)}{n} \leq \frac{I(X^n; Z^n)}{n} = I(X; Z). \tag{3.5}$$

Substituting (3.4) and (3.5) into (3.3), we have

$$R \leq I(X; Z) + H(Y|Z) - H(Y) = I(X; Z) - I(Y; Z) = I(X; Z|Y), \tag{3.6}$$

where we will leave the last equality as an exercise. So this shows that the required rate can be smaller than $I(X; Z|Y)$ and this concludes the forward proof. Combining this with the converse proof, shown in the next section, we can prove that the required rate is exactly $I(X; Z|Y)$.

3.2 Converse Proof of WZ Coding

Unlike some other converse proofs of theorems in information theory, the converse proof of WZ coding does give some additional insight. In particular, it suggests the precise candidate of the mysterious auxiliary random variable Z. Now, let us start the proof.

Converse Proof of WZ Coding.

$$R = \frac{\log M}{n} \geq \frac{H(W)}{n} \geq \frac{I(X^n, W)}{n}$$
$$= \frac{H(X^n) - H(X^n|W)}{n}.$$

Note that

$$H(X^n|W) = I(X^n; Y^n|W) + H(X^n|Y^n, W)$$
$$= H(Y^n|W) - H(Y^n|X^n, W) + H(X^n|Y^n, W)$$
$$\overset{(a)}{\leq} H(Y^n) - H(Y^n|X^n) + H(X^n|Y^n, W)$$
$$= I(X^n; Y^n) + H(X^n|Y^n, W),$$

because (a) is due to conditioning reducing entropy and $H(Y^n|X^n) = H(Y^n|X^n, W)$ as $Y^n \leftrightarrow X^n \leftrightarrow W$. Therefore, we have

$$nR \geq H(X^n) - I(X^n; Y^n) - H(X^n|Y^n, W)$$
$$= \sum_{k=1}^{n} H(X_k) - I(X_k; Y_k) - H(X_k|Y^n, W, X^{k-1})$$

Now, if we define the auxiliary random variable by $Z_k \triangleq (X^{k-1}, Y^{k-1}, Y_{k+1}^n, W)$ (note that $Z_k \leftrightarrow X_k \leftrightarrow Y_k$ is satisfied as desired), then

$$nR \geq \sum_{k=1}^{n} H(X_k) - I(X_k; Y_k) - H(X_k|Z_k, Y_k). \tag{3.7}$$

Note that

$$H(X|Y,Z) = H(X,Y|Z) - H(Y|Z)$$
$$= H(X|Z) + H(Y|X,Z) - H(Y|Z)$$
$$\geq H(X|Z) + H(Y|Z) - H(Y|Z).$$

Substituting the above into (3.7), we have

$$nR \geq \sum_k H(X_k) - I(X_k;Y_k) - H(X_k|Z_k) - H(Y_k|X_k) + H(Y_k|Z_k)$$
$$= \sum_k I(X_k;Z_k) - H(Y_k) + H(Y_k|Z_k)$$
$$= \sum_k I(X_k;Z_k) - I(Y_k;Z_k)$$
$$\geq \sum_k R^*(D_k) = n\frac{1}{n}\sum_k R^*(D_k)$$
$$\overset{(a)}{\geq} nR^*\left(\sum_k \frac{1}{n}D_k\right)$$
$$= nR^*(D),$$

where (a) is due to the convexity of $R^*(D)$. $\qquad\qquad\square$

3.3 Examples

We will consider two very common examples in this section. The first one is the *doubly symmetric binary source* and the second one is the *quadratic Gaussian source*.

3.3.1 Doubly Symmetric Binary Source

Problem Setup

In this example, both X and Y are binary. Moreover, the distribution of Y is symmetric, that is, $p_Y(0) = p_Y(1) = 0.5$, and the source input X is the output of Y passing through a binary symmetric channel (BSC) with crossover probability p. As a result, the distribution of X is also symmetric. Since the input behaves like a binary symmetric source passing through a BSC, it is sometimes known as the doubly symmetric binary source[1].

The distortion measure we used here is the simple Hamming distance, namely, one unit of distortion will result if the recovered source differs from

---

1 Because of the symmetry, we can treat Y as the output of a symmetric X passing through a BSC with crossover probability p as well.

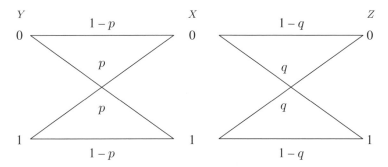

Figure 3.4 An auxiliary random variable Z is generated by passing input X through a binary symmetric channel with crossover probability q.

the original source and zero distortion otherwise. More precisely, $d(1,0) = d(0,1) = 1$ and $d(0,0) = d(1,1) = 0$.

Without loss of generality, we can assume $p \leq 0.5$ since if $p > 0.5$, we can treat the complement of Y, $1 - Y$, as side information instead. From discussion in previous sections, we need to solve the optimization problem in (3.1) to find the rate distortion for this source. Luckily, it turns out that it is not very difficult to guess the solution. We just need to decide the conditional distribution $q(z|x)$ and the deterministic decoding function $g(z, y)$.

A Proposed Scheme

We will start with a very simple guess for $q(z|x)$, as shown in Figure 3.4. Basically, Z is the output of X passing though a BSC with some crossover probability q. Moreover, we will assume that $q \leq p$. The reason for this will become clear as we try to gauge the performance of this scheme below. Given $q(z|x)$, we need to decide the optimum decoding function $g^*(z, y)$. Since X is binary, the best estimate of X given Y and Z is simply Z since we assume $q \leq p$, that is, $g^*(z, y) = z$. This may look a bit surprising at first since this optimum scheme seems to ignore the side information Y. However, we have already used the side information Y when we try to transmit Z to the decoder. It turns out this alone is sufficient to make the scheme optimum.

To summarize, our tentative scheme is

$$q(z|x) = \begin{cases} 1 - q & x = z, \\ q & x \neq z, \end{cases}$$

and

$$g^*(z, y) = z.$$

Now, let us evaluate the performance of this scheme. Basically, we need to find the rate distortion function $R(D)$. Let us begin with the distortion D.

$$D = E[d(X, \hat{X})]$$
$$= E[d(X, g^*(Z, Y))]$$

$$= E[d(X, Z)]$$
$$= E[d(X, Z)|X = Z]Pr(X = Z) + E[d(X, Z)|X \neq Z]Pr(X \neq Z)$$
$$= 0 \cdot (1 - q) + 1 \cdot q = q.$$

We see now why q should be less than p. Even if we do not send anything, the decoder can just give a wild guess of X as the side information Y. In that case, the distortion will be simply $E[d(X, Y)] = p$. So if we allow q to be larger than p, we will still have distortion bounded by p and thus it is better not to send Z at all.

As for the rate R, we need to transmit Z to the decoder with rate at least

$$I(X; Z|Y) = H(Z|Y) - H(Z|X, Y)$$
$$\overset{(a)}{=} H(Z|Y) - H(Z|X)$$
$$= H(p * q) - H(q)$$
$$= H(p * D) - H(D) \triangleq \Gamma(D),$$

where (a) comes from the Markov chain $Z \leftrightarrow X \leftrightarrow Y$.

From above, we have obtained a rate-distortion function $\Gamma(D) = H(p * D) - H(D)$, but we are not done yet. Remember that the decoder can recover an estimated source with distortion p even if the encoder is not sending anything. We can potentially increase the performance of our scheme by time shared with this trivial scheme (with rate 0 and distortion p). Therefore, the final rate-distortion function for this scheme should be

$$\hat{R}(D) = \min_{q, \theta} \theta(H(p * q) - H(q)) + (1 - \theta)0$$
$$s.t. \quad D = \theta q + (1 - \theta)p,$$
$$= \min_{q, \theta} \theta(H(p * q) - H(q)) \triangleq \Gamma^*(D)$$
$$s.t. \quad D = \theta q + (1 - \theta)p,$$

where we put a hat on top of R to emphasize that it has not been shown to be the actual WZ rate-distortion function yet. The remainder of this subsection will verify that this scheme is indeed optimum, that is, $\Gamma^*(D) = R_{WZ}(D)$.

Verify the Optimality of the Proposed Scheme

Since the proposed scheme is just one of all possible schemes, we know that $\Gamma^*(D) \geq R_{WZ}(D)$, as specified in (3.1). To show that the scheme is optimal, we need to show $\Gamma^*(D) \leq R_{WZ}(D)$.

Since the decoding function $g(z, y)$ can be readily determined once the choice of $q(z|x)$ is fixed, a naive thought would be to simply try to show that the proposed symmetric choice of Z is optimal among all binary Z. Unfortunately, even though both X and Y are binary, the WZ Theorem described in the previous sections didn't guarantee that a binary Z is sufficient.[2] Instead, we will just assume that the alphabet of Z is countable and divide the alphabet into two sets: the first one contains z where the decoder will ignore the side information, namely, $g(z, 0) = g(z, 1)$, and the second one contains the remaining z. We call the first set A such that $A = \{z | g(z, 0) = g(z, 1)\}$. Now, for some $z \notin A$ and $g(z, 0) = 0$ (and thus $g(z, 1) = 1$ since $z \notin A$ and so $g(z, 0) \neq g(z, 1)$) we have

$$E[d(X, \hat{X}) | Z = z]$$
$$= Pr(X = 1, Y = 0 | Z = z) + Pr(X = 0, Y = 1 | Z = z)$$
$$= Pr(X = 1 | Z = z)Pr(Y = 0 | X = 1, Z = z)$$
$$\quad + Pr(X = 0 | Z = z)Pr(Y = 1 | X = 0, Z = z)$$
$$\overset{(a)}{=} Pr(X = 1 | Z = z)Pr(Y = 0 | X = 1) + Pr(X = 0 | Z = z)Pr(Y = 1 | X = 0)$$
$$= Pr(X = 1 | Z = z)p + Pr(X = 0 | Z = z)p = p,$$

where (a) is due to Markov chain $Y \leftrightarrow X \leftrightarrow Z$. Similarly, if $z \notin A$ but $g(z, 0) = 1$ (and $g(z, 1) = 0$),

$$E[d(X, \hat{X}) | Z = z]$$
$$= Pr(X = 0, Y = 0 | Z = z) + Pr(X = 1, Y = 1 | Z = z)$$
$$= Pr(X = 0 | Z = z)Pr(Y = 0 | X = 0) + Pr(X = 1 | Z = z)Pr(Y = 1 | X = 1)$$
$$= 1 - p \geq p,$$

where the last inequality is due to the assumption that $p \leq 0.5$. Therefore,

$$E[d(X, \hat{X}) | Z \notin A] \tag{3.8}$$
$$= Pr(g(Z, 0) = 0 | Z \notin A)E[d(X, \hat{X}) | Z \notin A, g(Z, 0) = 0)$$
$$\quad + Pr(g(Z, 0) = 1 | Z \notin A)E[d(X, \hat{X}) | Z \notin A, g(Z, 0) = 1) \tag{3.9}$$
$$\geq p. \tag{3.10}$$

From this, given a distortion constraint D, we have

$$D \geq E[d(X, \hat{X})]$$
$$= Pr(Z \in A)E[d(X, \hat{X}) | Z \in A] + Pr(Z \notin A)E[d(X, \hat{X}) | Z \notin A]$$
$$\geq Pr(Z \in A) \sum_{z \in A} \frac{Pr(Z = z)}{Pr(Z \in A)} E[d(X, \hat{X}) | Z = z] + Pr(Z \notin A)p$$
$$\triangleq D'.$$

---

2 It turns out that it is and generally one can show that the alphabet size of Z is bounded by $|\mathcal{X}| + 1$ using the Caratheodory Theorem [15].

D' is essential for the following discussion. To simplify the notation a little, we introduce the following symbols: $\theta \triangleq Pr(Z \in A)$, $\lambda_z \triangleq \frac{Pr(Z=z)}{Pr(Z \in A)}$, and $d_z \triangleq E[d(X, \hat{X})|Z = z]$. Then, we can rewrite

$$D' = \theta \sum_{z \in A} \lambda_z d_z + (1 - \theta)p. \tag{3.11}$$

Now,

$$I(X;Z) - I(Y;Z) = H(Y|Z) - H(X|Z)$$
$$\geq \sum_{z \in A}(H(Y|Z = z) - H(X|Z = z))Pr(Z = z)$$
$$= \theta \sum_{z \in A} \lambda_z(H(Y|Z = z) - H(X|Z = z)).$$

Recall that for $z \in A$, the decoded output is the same regardless of y. Let us denote the output as γ_z $(= g(z, 0) = g(z, 1))$. Since X is binary,

$$H(X|Z = z) = H(Pr(X = 1 - \gamma_z|Z = z))$$
$$= H(Pr(X \neq \gamma_z|Z = z)) = H(d_z).$$

Similarly, as Y is also binary,

$$H(Y|Z = z) = H(Pr(Y = 1 - \gamma_z|Z = z))$$
$$\overset{(a)}{=} H(Pr(Y = 1 - \gamma_z|X = \gamma_z)Pr(X = \gamma_z|Z = z)$$
$$+ Pr(Y = 1 - \gamma_z|X = 1 - \gamma_z)Pr(X = 1 - \gamma_z|Z = z))$$
$$= H(p(1 - d_z) + (1 - p)d_z) = H(p * d_z), \tag{3.12}$$

where (a) is due to the Markov chain $Y \leftrightarrow X \leftrightarrow Z$.

Therefore,

$$I(X;Z) - I(Y;Z)$$
$$\geq \theta \sum_{z \in A} \lambda_z[H(p * d_z) - H(d_z)]$$
$$= \theta \sum_{z \in A} \lambda_z g(d_z)$$
$$\overset{(a)}{\geq} \theta g\left(\sum_{z \in A} \lambda_z d_z\right)$$
$$= \theta[H(p * q') - H(q')] \geq \Gamma^*(D'),$$

where (a) is from Jensen's inequality and $g(\cdot)$ is convex[3], $q' \triangleq \sum_{z \in A} \lambda_z d_z$ and recall that $D' = \theta q' + (1 - \theta)p$ as in (3.11). Finally, since $D' \leq D$ and $g^*(\cdot)$ is

---

3 This can be easily proved by showing the second derivative to be positive.

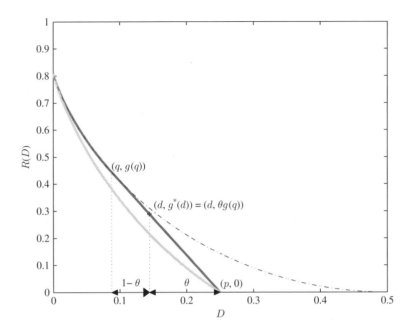

Figure 3.5 WZ limit of doubly binary symmetric source. $\Gamma(D) \triangleq H(p * D) - H(D)$ is shown as the dashed-dotted curve, and the dark solid curve, $\Gamma^*(D)$, is the lowest convex hull of $\Gamma(D)$ and $\{(p, 0)\}$. In particular, $(d, g^*(d))$ on curve $\Gamma^*(D)$ is obtained through time sharing (linear combination) between $(q, g(q))$ on curve $\Gamma(D)$ and $(p, 0)$ (i.e., when no rate is transmitted). It can be shown that $\Gamma^*(D)$ is the actual WZ limit. In contrast, the light solid curve is the theoretical limit when side information is available at both encoder and decoder. It is clear that there is *rate loss* in this case.

monotonically decreasing[4], $I(X; Z) - I(Y; Z) \geq \Gamma^*(D') \geq \Gamma^*(D)$. Since this holds for any distribution of Z, $R_{WZ}(D) \geq \Gamma^*(D)$, $\Gamma^*(D)$ is indeed the WZ limit.

3.3.2 Quadratic Gaussian Source

Problem Setup

In this example, we will consider a jointly Gaussian source. Moreover, we will use a quadratic distortion measure in this problem, namely, $d(X, Y) = (X - Y)^2$. Without loss of generality, we can assume X and Y to be zero mean as otherwise the mean and the mean-subtracted X would be transmitted separately. Since the mean does not change, transmitting it does not consume any rate on average over a long period of time. For a similar reason, since the mean of Y does not change, it does not provide meaningful information to X over a long period

---

4 Again, this can be easily shown by taking the derivative. We will leave this as an exercise for readers.

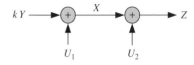

Figure 3.6 For joint Gaussian zero-mean sources X and Y, we can express X as $kY + U_1$, where k is some constant and U_1 is a zero-mean Gaussian random variable independent of Y. To construct the auxiliary random variable Z used in the forward proof of WZ coding, we write Z as $U_2 + X$, where U_2 is another zero-mean Gaussian random variable independent of X. It turns out that such construction is optimum, as shown in this subsection.

of time. Therefore, we can assume $\begin{pmatrix} X \\ Y \end{pmatrix} \sim \mathcal{N}(0, \Sigma)$ for a covariance matrix $\Sigma = \begin{pmatrix} \sigma_X^2 & \sigma_{XY} \\ \sigma_{XY} & \sigma_Y^2 \end{pmatrix}$. Readers who are not familiar with multivariate Gaussian variables or need to be refreshed on the topic can refer to Appendix D.

Let us denote the correlation coefficient $\frac{\sigma_{XY}}{\sigma_X \sigma_Y}$ as ρ. Note that given Y we may consider X as a scaled Y added with an independent Gaussian noise. More precisely, let us write $X = kY + U_1$, where it is easy to show that $k = \frac{\sigma_{XY}}{\sigma_Y^2} = \rho \frac{\sigma_X}{\sigma_Y}$ and $U_1 \sim \mathcal{N}(0, (1 - \rho^2)\sigma_X^2)$ is independent of Y. This model is illustrated in part of Figure 3.6.

Proposed Scheme

Similar to the doubly symmetric binary source , we need first to determine the auxiliary variable Z. We will construct Z simply as X added with an independent Gaussian noise U_2. This is as shown in Figure 3.6.

Now, we need to find a way to construct the estimate $\hat{X}$ from Y and Z. Let us construct $\hat{X}$ linearly as $aY + bZ$ for some constants a and b. In order to have an optimum estimator, the reconstruction error has to be uncorrelated with the observations, that is, Y and Z. Therefore,

$$E[(\hat{X} - X)Y] = E[(\hat{X} - X)Z] = 0. \tag{3.13}$$

and

$$\begin{aligned} E[(\hat{X} - X)Y] &= E[(aY + bZ - X)Y] \\ &= aE[Y^2] + bE[kY^2] - E[kY^2] \\ &= E[(a + bk - k)Y^2] = 0. \end{aligned}$$

This gives us $a + bk - k = 0$. Similarly, we have

$$\begin{aligned} E[(\hat{X} - X)Z] &= E[(aY + bZ - X)Z] \\ &= aE[kY^2] + bE[k^2Y^2 + U_1^2 + U_2^2] - E[k^2Y^2 + U_1^2] \\ &= kE[(a + bk - k)Y^2] + E[b(U_1^2 + U_2^2) - U_1^2] \\ &= E[b(U_1^2 + U_2^2) - U_1^2] = 0. \end{aligned}$$

This gives us

$$a = k\frac{\sigma_2^2}{\sigma_1^2 + \sigma_2^2} \qquad \text{and} \qquad b = \frac{\sigma_1^2}{\sigma_1^2 + \sigma_2^2}, \tag{3.14}$$

where $\sigma_1^2 = E[U_1^2]$ and $\sigma_2^2 = E[U_2^2]$. Note that U_1 and U_2 are zero mean here. To summarize, our proposed scheme is as follows,

$$Z = X + U_2, \quad \text{with} \quad U_2 \sim \mathcal{N}(0, \sigma_2^2) \tag{3.15}$$

and

$$\hat{x} = g(z, y) = aY + bZ, \tag{3.16}$$

where a and b are given in (3.14).

Now, let us evaluate the performance of the proposed scheme. With a little bit of algebra, one can show that the average distortion is

$$D = E[(\hat{X} - X)^2] = \frac{\sigma_1^2 \sigma_2^2}{\sigma_1^2 + \sigma_2^2}. \tag{3.17}$$

What will be the rate required to send Z? From the forward proof of the WZ Theorem, we have argued that

$$
\begin{aligned}
R &= I(X; Z) - I(Y; Z) \\
&= -h(Z|X) + h(Z|Y) \\
&= -h(X + U_1|X) + h(kY + U_1 + U_2|Y) \\
&= -h(U_1|X) + h(U_1 + U_2|Y) \\
&= -h(U_1) + h(U_1 + U_2) = \frac{1}{2}\log_2 \frac{\sigma_1^2 + \sigma_2^2}{\sigma_2^2}.
\end{aligned}
$$

If we substitute (3.17) into the last expression, we have

$$R(D) = \frac{1}{2}\log_2 \frac{\sigma_1^2}{D}. \tag{3.18}$$

Verify the Optimality of the Proposed Scheme

Let us digress a little and consider the closely related problem when side information Y is also available at the encoder. This turns out to be a much simpler extension of the classic rate-distortion problem and the theoretical limit is given by

$$R_{X|Y}(D) = \min_{p(\hat{x}|x,y)} I(X; \hat{X}|Y)$$

$$s.t. \quad E[d(X, \hat{X})] \le D \tag{3.19}$$

For the quadratic Gaussian case,

$$I(X; \hat{X}|Y) = H(X|Y) - H(X|\hat{X}, Y)$$

$$= H(X|Y) - H(\hat{X} + V|\hat{X}, Y)$$

$$\overset{(a)}{=} H(X|Y) - H(V|\hat{X}, Y)$$

$$\overset{(b)}{\geq} H(X|Y) - H(V)$$

$$\overset{(c)}{=} \frac{1}{2}\log_2 2\pi e\sigma_1^2 - \frac{1}{2}\log_2 2\pi eD$$

$$= \frac{1}{2}\log_2 \frac{\sigma_1^2}{D}, \tag{3.20}$$

where $V = X - \hat{X}$ is the reconstruction error of X and thus the variance of V is at most D. (a) is due to the fact that $\hat{X}$ reveals no information when it is given and thus essentially behaves like a constant, (b) comes from the fact that conditioning reduces entropy, and (c) is from the differential entropy equation of Gaussian sources and $X = Y + kU_1$ with the variance of U_1 equal to σ_1^2.

Since (3.20) holds for any $p(\hat{x}|x, y)$ satisfying the distortion constraint $E[d(X, \hat{X})] \leq D$, we have $R_{X|Y}(D) \geq \frac{1}{2}\log_2 \frac{\sigma_1^2}{D}$, which is a lower bound of the required rate even when side information is also available at the encoder. It is not immediately obvious if such a lower bound can be achieved or not. However, note that in (3.18) we have exactly the same rate-distortion performance as for our proposed scheme for the WZ problem. Given that, we can immediately conclude that the lower bound can indeed be achieved (i.e., by our proposed scheme). Moreover, as our proposed scheme achieves the lower bound, it is indeed optimum.

3.4 Rate Loss of the WZ Problem

As we can see from (3.20) and (3.1), the rate-distortion function is generally different when side information Y is available or not at the encoder. Of course, the former is always a lower bound of the latter since the encoder can always choose to ignore Y at the encoder even when it is available. Therefore, WZ is strictly a more demanding setup. We typically call the rate difference between the two setups the rate loss.

However, as we can see from the quadratic Gaussian case, it is possible that the required rate with an equal distortion constraint D is the same regardless of the availability of Y at the encoder. This *no-rate loss* scenario is actually very rare. Besides the joint Gaussian case mentioned above, another possible setup is a simple extension of the Gaussian case, namely, Y is a random variable with arbitrary distribution but X is equal to Y added with an independent Gaussian noise [16].

Binary Source Case

How about the doubly symmetric binary source discussed earlier? When Y is also available at the encoder, the rate required is simply

$$I(X; \hat{X}|Y) = H(X|Y) - H(X|\hat{X}, Y)$$
$$= H(X|Y) - H(\hat{X} + V|\hat{X}, Y)$$
$$= H(X|Y) - H(V|\hat{X}, Y)$$
$$\overset{(a)}{\geq} H(X|Y) - H(V)$$
$$= H(p) - H(D), \tag{3.21}$$

where V is the difference between X and $\hat{X}$, and the inequality (a) is achievable (and becomes equality) as long as V is chosen to be independent of $\hat{X}$ and Y. Thus, (3.21) describes the actual rate-distortion limit, which is illustrated in pale solid curve in Figure 3.5. It is clear from the figure that there is rate loss for this case.

While there is generally rate loss for the binary source case, it has been shown recently that if X and Y are related by a *Z-channel*, there is actually no rate loss [17]. This is one of the few known scenarios other than Gaussian cases where we have no rate loss.

Rate loss of General Cases

It is difficult to characterize the rate loss for general cases. However, it can be shown that in general the WZ rate loss is upper bounded by 0.5 bits/sample for generic sources with quadratic distortion measure and by 0.22 bits/sample for binary sources with Hamming-distance distortion measure [18].

4

Lossy Distributed Source Coding

In the last chapter we considered the WZ problem, where only one of two sources is compressed in a lossy manner. In this chapter we look at the most general case, lossy distributed source coding or multiterminal source coding, where two or more sources are compressed at the same time. Just as in the WZ problem, lossy distributed source coding is a rate-distortion problem. However, since each compressed source input can have different rates, we have a rate-tuple instead of a single rate value for each setup. Given a distortion measure $d(\cdot, \cdot)$ and distortion constraint $\mathcal{D}_j$ for each length-n source input X_j^n, a rate-tuple $(R_1, R_2, \cdots)$ is achievable if we can find a scheme such that the distortion constraint is satisfied, that is

$$D_j = E[d(X_j^n, \hat{X}_j^n)] \leq \mathcal{D}_j$$

and

$$R_j = \frac{H(M_j)}{n}$$

with M_j as the encoded output of X_j^n.

A closely related problem is the CEO problem, where the CEO of a company is interested in a piece of information and a number of managers help gather this information for him. In this case, we assume each manager (encoder) observes a noisy copy of the target information. Since only the target information is interesting, the decoder does not need to recover each observation obtained by a manager. The problem is simplified somewhat since we only have a single distortion constraint involved. Since the encoders only observe the actual source indirectly, the CEO problem is also known as the *indirect* multiterminal source coding problem. In contrast, the original setup in which the decoder attempts to recover all encoder inputs is sometimes known as the direct multiterminal source coding problem.

The theoretical limits of both these problems are still unknown in general. However, for some special cases, such as joint Gaussian inputs with the quadratic distortion measure, the theoretical limits are partially known and we will focus on those cases in this chapter.

Distributed Source Coding: Theory and Practice, First Edition. Shuang Wang, Yong Fang, and Samuel Cheng.
© 2017 John Wiley & Sons Ltd. Published 2017 by John Wiley & Sons Ltd.
Companion Website: www.wiley.com/go/cheng/dsctheoryandpractice

For general setup, an important well-known result is the Berger–Tung inner bound, which is useful to give the achievable proofs for both the direct and indirect multiterminal source coding problem. In the following, we will first describe the Berger–Tung inner bound and then discuss the indirect multiterminal source coding problem followed by the direct multiterminal source coding problem.

4.1 Berger–Tung Inner Bound

Consider the direct multiterminal source coding problem with two sources and distortion constraints

$$E[d(X_j, \hat{X}_j)] \leq D_j, j = 1, 2.$$

If we can find auxiliary random variables U_1 and U_2 such that

$$U_1, U_2 | X_1, X_2, \cdots, \sim p(u_1 | x_1) p(u_2 | x_2)$$

and decoding function $\hat{x}_1(U_1, U_2)$ and $\hat{x}_2(U_1, U_2)$ such that

$$E[d(X_j, \hat{x}_j(U_1, U_2))] \leq D_j, j = 1, 2,$$

then the following rate region is achievable

$$R_1 > I(X_1; U_1 | U_2)$$
$$R_2 > I(X_2; U_2 | U_1)$$
$$R_1 + R_2 > I(X_1, X_2; U_1, U_2).$$

4.1.1 Berger–Tung Scheme

The achievable proof can be shown by construction through the following setup, which will be referred to as the Berger–Tung Scheme.

Codebook Preparation
- Prepare random variables $U_1 \sim p(u_1 | X_1)$ and $U_2 \sim p(u_2 | X_2)$, and decoding functions $\hat{x}_1(U_1, U_2)$ and $\hat{x}_2(U_1, U_2)$ such that $E[d(\hat{x}_1(U_1, U_2), X_1)] \leq D_1$ and $E[d(\hat{x}_2(U_1, U_2), X_2)] \leq D_2$.
- Generate codewords $U_1^n(l_1)$, $l_1 = 1, \cdots, \lceil 2^{n\tilde{R}_1} \rceil$ and codewords $U_2^n(l_2)$, $l_2 = 1, \cdots, \lceil 2^{n\tilde{R}_2} \rceil$ by sampling U_1 and U_2.
- Randomly partition l_1 and l_2 into 2^{nR_1} and 2^{nR_2} bins, respectively. Denote $b_1(l_1)$ and $b_2(l_2)$ as the corresponding bin indices.

Encoding
- Given input sequences (x_1^n, x_2^n), find l_1 and l_2 such that $(U_1^n(l_1), x_1^n) \in \mathcal{T}_\epsilon^{(n)}(U_1, X_1)$ and $(U_2^n(l_2), x_2^n) \in \mathcal{T}_\epsilon^{(n)}(U_2, X_2)$.
- Send bin indices $b_1(l_1)$ and $b_2(l_2)$ to the decoder.

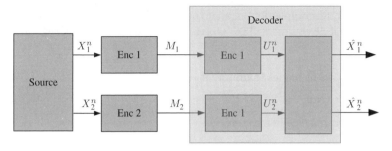

Figure 4.1 The Berger–Tung scheme.

Decoding

- Upon receiving $b_1(l_1)$ and $b_2(l_2)$, find $\tilde{l}_1$ and $\tilde{l}_2$ such that $(U_1^n(l_1), U_2^n(l_2)) \in T_\epsilon^{(n)}(U_1, U_2)$ and $b_1(\tilde{l}_1) = b_1(l_1), b_2(\tilde{l}_2) = b_2(l_2)$.
- Output $(\hat{x}_1^n(U_1^n(\tilde{l}_1), U_2^n(\tilde{l}_2)), \hat{x}_2^n(U_1^n(\tilde{l}_1), U_2^n(\tilde{l}_2)))$.

4.1.2 Distortion Analysis

Note that since the decoder picks $\tilde{l}_1 = l_1$ and $\tilde{l}_2 = l_2$, and $(U_1^n(l_1), U_2^n(l_2), x_1^n, x_2^n) \in T_\epsilon^{(n)}(U_1, U_2, X_1, X_2)$, by the definition of joint typicality and the Law of Large Number, we have $\frac{1}{n}d(x_1^n, \hat{x}_1^n) \to E[d(X_1, \hat{x}_1(U_1, U_2))] \leq D_1$ and $d(x_2^n, \hat{x}_2^n) \to E[d(X_2, \hat{x}_2(U_1, U_2))] \leq D_2$ as $n \to \infty$. Therefore, distortion constraints will be automatically satisfied if no encoding or decoding failures occur. Let us consider the following failure events[1] (with $\epsilon > \epsilon' > \epsilon''$).

$$\mathcal{E}_1 = \{\nexists l_1 \text{ or} \nexists l_2 \text{ such that } (U_1^n(l_1), X_1^n) \in T_{\epsilon''}^{(n)} \text{ and } (U_2^n(l_2), X_2^n) \in T_{\epsilon''}^{(n)}\},$$

$$\mathcal{E}_2 = \{(U_1^n(L_1), X_1^n, X_2^n) \notin T_{\epsilon'}^{(n)}\},$$

$$\mathcal{E}_3 = \{(U_1^n(L_1), U_2^n(L_2), X_1^n, X_2^n) \notin T_\epsilon^{(n)}\},$$

$$\mathcal{E}_4 = \{\exists \tilde{l}_1, \tilde{l}_2 \text{ such that } (U_1^n(\tilde{l}_1), U_2^n(\tilde{l}_2)) \in T_\epsilon^{(n)} \text{ and } b_i(\tilde{l}_i) = b_i(L_i)$$
$$\text{but } \tilde{l}_i \neq l_i, i = 1, 2\},$$

$$\mathcal{E}_5 = \{\exists \tilde{l}_1 \text{ such that } (U_1^n(\tilde{l}_1), U_2^n(l_2)) \in T_\epsilon^{(n)} \text{ and } b_1(\tilde{l}_1) = b_1(L_1)$$
$$\text{but } \tilde{l}_1 \neq L_1\},$$

$$\mathcal{E}_6 = \{\exists \tilde{l}_2 \text{ such that } (U_1^n(l_1), U_2^n(\tilde{l}_2)) \in T_\epsilon^{(n)} \text{ and } b_2(\tilde{l}_2) = b_2(L_2)$$
$$\text{but } \tilde{l}_2 \neq L_2\}.$$

We have

$$Pr(\text{Coding failure})$$
$$\leq Pr(\mathcal{E}_1) + Pr(\mathcal{E}_1^C \cap \mathcal{E}_2) + Pr(\mathcal{E}_1^C \cap \mathcal{E}_2^C \cap \mathcal{E}_3) + Pr(\mathcal{E}_4)$$
$$+ Pr(\mathcal{E}_5) + Pr(\mathcal{E}_6),$$

---

1 To slightly simplify the notation, we will omit the arguments of $T_\epsilon^{(n)}(X, Y)$ as $T_\epsilon^{(n)}$ for the set of jointly typical sequences of X^n and Y^n.

where $\mathcal{E}^C$ denotes the complement of an error event $\mathcal{E}$. We will show that each of the above probability terms will go to zero within the Berger–Tung region.

$Pr(\mathcal{E}_1) \to 0$:
Note that $Pr(\mathcal{E}_1) \to 0$ as n goes to infinity if

$$\tilde{R}_i > I(U_i; X_i) \quad \text{for } i = 1, 2 \tag{4.1}$$

by the covering lemma (see Lemma A.29).

$Pr(\mathcal{E}_1^C \cap \mathcal{E}_2) \to 0$:
If $(u_1^n(l_1), x_1^n) \in \mathcal{T}_{\epsilon''}^{(n)}$, as $X_2 \sim p(x_2|x_1)$, $Pr((u_1^n(l_1), x_1^n, X_2^n) \in \mathcal{T}_{\epsilon'}^{(n)}) \to 1$ as n goes to infinity by the conditional typical lemma (see Lemma A.24). Therefore $Pr(\mathcal{E}_1^C \cap \mathcal{E}_2) \to 0$.

$Pr(\mathcal{E}_1^C \cap \mathcal{E}_2^C \cap \mathcal{E}_3) \to 0$:
If $(u_1^n(l_1), x_1^n, x_2^n) \in \mathcal{T}_{\epsilon'}^{(n)}$ and $(u_2^n(l_2), x_2^n) \in \mathcal{T}_{\epsilon''}^{(n)}$, we have

$$2^{-n(H(U_2|X_2)+\delta(\epsilon))} \le p(u_2^n(l_2)|x_2^n) \le 2^{-n(H(U_2|X_2)-\delta(\epsilon))}$$

due to Lemmas A.21. By the Markov Lemma (Lemma A.26), we have $(u_1^n(l_1), u_2^n(l_2), x_1^n, x_2^n) \in \mathcal{T}_\epsilon^{(n)}$ and thus $Pr(\mathcal{E}_1^C \cap \mathcal{E}_2^C \cap \mathcal{E}_3) \to 0$ as $n \to \infty$.

$Pr(\mathcal{E}_4) \to 0$
Within any bin with the same $b_1(l_1)$ and $b_2(l_2)$ there are $2^{n(\tilde{R}_1 - R_1)}$ u_1-sequences and $2^{n(\tilde{R}_2 - R_2)}$ u_2-sequences. Therefore, by the mutual packing lemma (see Lemma A.28), $Pr(\mathcal{E}_4) \to 0$ if

$$\tilde{R}_1 - R_1 + \tilde{R}_2 - R_2 < I(U_1; U_2) \tag{4.2}$$

as $n \to \infty$.

$Pr(\mathcal{E}_5), Pr(\mathcal{E}_6) \to 0$
By the packing lemma (see Lemma A.27), $Pr(\mathcal{E}_5) \to 0$ and if

$$\tilde{R}_1 - R_1 < I(U_1, U_2)$$
$$\Rightarrow R_1 > \tilde{R}_1 - I(U_1; U_2) \overset{(a)}{>} I(U_1; X_1) - I(U_1; U_2) = I(U_1; X_1|U_2), \tag{4.3}$$

where (a) is due to (4.1). Similarly, $Pr(\mathcal{E}_6) \to 0$ if

$$R_2 > I(U_2; X_2|U_1). \tag{4.4}$$

When the constraints in (4.1), (4.2), (4.3), and (4.4) are satisfied, we have the code failure probabilities go to zero and thus the distortion constraints are satisfied. From (4.1) and (4.2) we have

$$I(U_1; X_1) + I(U_2; X_2) - R_1 - R_2 < I(U_1; U_2)$$
$$\Rightarrow R_1 + R_2 > I(U_1; X_1) + I(U_2; X_2) - I(U_1; U_2) = I(U_1, U_2; X_1, X_2),$$

where the last inequality is due to the Markov chain $U_1 \leftrightarrow X_1 \leftrightarrow X_2 \leftrightarrow U_2$. Along with (4.3) and (4.4), we see that the distortion constraints are satisfied when the rates lie within the Berger–Tung inner bound.

4.2 Indirect Multiterminal Source Coding

Consider a CEO with several managers each conducting some investigations for him. Assume that all managers are investigating the same event because of unintentional or intentional bias, the observations of the managers are slightly different. We can idealize it to the problem shown in Figure 4.2, which is known as the CEO problem, or sometimes the indirect multiterminal source coding problem.

Like other distributed source coding problems, this problem is very difficult in general and the achievable rate region is still unknown for the general statistics of X. However, for the quadratic Gaussian case, the problem has been solved in recent decades. In this chapter, we will only consider the simplest case with two encoders (managers). However, the result can be extended to cases with more managers [19, 20].

4.2.1 Quadratic Gaussian CEO Problem with Two Encoders

For the quadratic Gaussian case, the original source $X \sim \mathcal{N}(0, \sigma_X^2)$ is Gaussian and the observations are original sources added with an independent Gaussian noise, that is, $Y_1 = X + Z_1$ and $Y_2 = X + Z_2$ with $Z_1 \sim \mathcal{N}(0, \sigma_{Z_1}^2)$ and $Z_2 \sim \mathcal{N}(0, \sigma_{Z_2}^2)$. Note that without loss of generality we assume the source X and the observation noises are all zero-mean. Moreover, we assume Z_1 and Z_2 are independent.

The mean square distortion is used here, so the distortion D of the reconstructed source $\hat{X}$ is given by $D = E[(X - \hat{X})^2]$. For such a distortion D, the achievable rate region $\mathcal{R}(D)$ can be described by the following inequalities:

$$R_1 \geq r_1 + \frac{1}{2} \log \left(\frac{1}{D} \left(\frac{1}{\sigma_X^2} + \frac{1 - 2^{-2r_2}}{\sigma_{Z_2}^2} \right)^{-1} \right) \tag{4.5}$$

$$R_2 \geq r_2 + \frac{1}{2} \log \left(\frac{1}{D} \left(\frac{1}{\sigma_X^2} + \frac{1 - 2^{-2r_1}}{\sigma_{Z_1}^2} \right)^{-1} \right) \tag{4.6}$$

$$R_1 + R_2 \geq r_1 + r_2 + \frac{1}{2} \log \frac{\sigma_X^2}{D} \tag{4.7}$$

$$D \geq \left(\frac{1}{\sigma_X^2} + \frac{1 - 2^{-2r_1}}{\sigma_{Z_1}^2} + \frac{1 - 2^{-2r_2}}{\sigma_{Z_2}^2} \right)^{-1}, \tag{4.8}$$

where r_1 and r_2 are larger than or equal to 0.

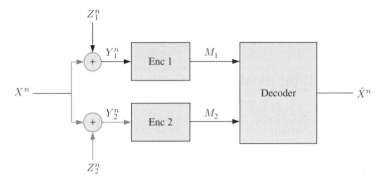

Figure 4.2 The CEO problem (indirect multiterminal source coding coding problem) with two managers (encoders).

Both the forward and converse proof will be given in the following. Throughout the proofs, the long Markov chain $M_1 \leftrightarrow Y_1^n \leftrightarrow X^n \leftrightarrow Y_2^n \leftrightarrow M_2$ is heavily used. The chain is actually a rather strong condition and implies various other subchains such as $X^n \leftrightarrow (Y_1^n, Y_2^n) \leftrightarrow (M_1, M_2)$ and $Y_2^n \leftrightarrow (X^n, Y_1^n, M_2) \leftrightarrow M_1$. Most of the time, we will specify which subchain is used in a proof step, but the reader should be aware that most of the subchains mentioned in the proof step originate from the aforementioned long Markov chain.

Forward Proof of Quadratic Gaussian CEO Problem with Two Terminals

The achievable proof is based on the Berger–Tung inner bound. First, it is convenient to realize that for $D \leq \sigma_X^2$, the specified rate region will be the same even when the inequality in (4.8) becomes equality. This can be shown by reducing r_1 and r_2 to increase the left-hand side of (4.8) to match D. Moreover, the new r_1 and r_2 can be selected in such a way that the left-hand sides of (4.5)–(4.7) do not increase. Since the expected distortion should be no larger than σ_X^2 even when nothing is sent, we can assume $D \leq \sigma_X^2$ and consider the inequality in (4.8) as equality from now on.

Now, let us consider the Berger–Tung scheme and have the encoder define

$$U_j = Y_j + V_j, \text{ where } V_j \sim \mathcal{N}(0, \sigma_{V_j}^2) \text{ with } \sigma_{V_j}^2 = \left(\frac{1}{\sigma_X^2} + \frac{1}{\sigma_{Z_j}^2} \right)^{-1}. \text{ Then, given } \mathbf{U} =$$

$(U_1 U_2)^T$, the decoder recovers X as $\hat{X} = E[X|\mathbf{U}]$. Therefore,

$$
\begin{aligned}
D &= E[(\hat{X} - X)^2] = \Sigma_{XX} - \Sigma_{X\mathbf{U}} \Sigma_{\mathbf{U}\mathbf{U}}^{-1} \Sigma_{\mathbf{U}X} \\
&= \sigma_X^2 - \left(E[XU_1] \; E[XU_2] \right) \begin{pmatrix} E[U_1^2] & E[U_1 U_2] \\ E[U_2 U_1] & E[U_2^2] \end{pmatrix}^{-1} \begin{pmatrix} E[XU_1] \\ E[XU_2] \end{pmatrix} \\
&= \sigma_X^2 + \left(\sigma_X^2 \; \sigma_X^2 \right) \begin{pmatrix} \sigma_X^2 + \sigma_{Z_1}^2 + \sigma_{V_1}^2 & \sigma_X^2 \\ \sigma_X^2 & \sigma_X^2 + \sigma_{Z_2}^2 + \sigma_{V_2}^2 \end{pmatrix}^{-1} \begin{pmatrix} \sigma_X^2 \\ \sigma_X^2 \end{pmatrix} \\
&= \left(\frac{1}{\sigma_{X_2}^2} + \frac{1}{\sigma_{Z_1}^2 + \sigma_{V_1}^2} + \frac{1}{\sigma_{Z_2}^2 + \sigma_{V_2}^2} \right)^{-1}
\end{aligned}
$$

Now, let $r_1 = \frac{1}{2}\log\left(1 + \frac{\sigma_{Z_1}^2}{\sigma_{V_1}^2}\right)$. Then $1 - 2^{2r_1} = 1 - \left(1 + \frac{\sigma_{Z_1}^2}{\sigma_{V_1}^2}\right)^{-1} = \frac{\sigma_{Z_1}^2}{\sigma_{V_1}^2 + \sigma_{Z_1}^2}$.

Similarly, we may let $r_2 = \frac{1}{2}\log\left(1 + \frac{\sigma_{Z_2}^2}{\sigma_{V_2}^2}\right)$ and have $1 - 2^{2r_2} = 1 - \left(1 + \frac{\sigma_{Z_2}^2}{\sigma_{V_2}^2}\right)^{-1}$

$= \frac{\sigma_{Z_2}^2}{\sigma_{V_2}^2 + \sigma_{Z_2}^2}$. Therefore,

$$D = \left(\frac{1}{\sigma_{X_2}^2} + \frac{1 - 2^{2r_1}}{\sigma_{Z_1}^2} + \frac{1 - 2^{2r_2}}{\sigma_{Z_2}^2}\right)^{-1}$$

as desired.

From the Berger–Tung inner bound, (R_1, R_2) are achievable if

$$R_1 \geq I(Y_1; U_1|U_2)$$
$$R_2 \geq I(Y_2; U_2|U_1)$$
$$R_1 + R_2 \geq I(U_1, U_2; Y_1, Y_2).$$

For the first bound,

$$R_1 \geq I(Y_1; U_1|U_2) = h(U_1|U_2) - h(U_1|Y_1, U_2)$$
$$= h(U_1|U_2) - h(U_1|Y_1)$$
$$= h(U_1|U_2) - \frac{1}{2}\log 2\pi e \sigma_{V_1}^2.$$

However,

$$\sigma_{U_1|U_2}^2 = \sigma_{U_1}^2 - \sigma_{U_1 U_2} \sigma_{U_2}^{-2} \sigma_{U_2 U_1}$$

$$= \sigma_{X_2}^2 + \sigma_{Z_1}^2 + \sigma_{V_1}^2 - \frac{(\sigma_{X_2}^2)^2}{\sigma_{X_2}^2 + \sigma_{Z_2}^2 + \sigma_{V_2}^2}$$

$$= \sigma_{Z_1}^2 + \sigma_{V_1}^2 + \frac{(\sigma_{Z_2}^2 + \sigma_{V_2}^2)\sigma_{X_2}^2}{\sigma_{X_2}^2 + \sigma_{Z_2}^2 + \sigma_{V_2}^2}$$

$$= (\sigma_{Z_1}^2 + \sigma_{V_1}^2)\frac{\sigma_{X_2}^2(\sigma_{Z_2}^2 + \sigma_{V_2}^2)}{\sigma_{X_2}^2 + \sigma_{Z_2}^2 + \sigma_{V_2}^2}\left(\frac{\sigma_{X_2}^2 + \sigma_{Z_2}^2 + \sigma_{V_2}^2}{\sigma_{X_2}^2(\sigma_{Z_2}^2 + \sigma_{V_2}^2)} + \frac{1}{\sigma_{Z_1}^2 + \sigma_{V_1}^2}\right)$$

$$= (\sigma_{Z_1}^2 + \sigma_{V_1}^2)\left(\frac{1}{\sigma_{X_2}^2} + \frac{1}{\sigma_{Z_2}^2 + \sigma_{V_2}^2}\right)^{-1}\left(\frac{1}{\sigma_{X_2}^2} + \frac{1}{\sigma_{Z_2}^2 + \sigma_{V_2}^2} + \frac{1}{\sigma_{Z_1}^2 + \sigma_{V_1}^2}\right)$$

$$= (\sigma_{Z_1}^2 + \sigma_{V_1}^2)\left(\frac{1}{\sigma_{X_2}^2} + \frac{1 - 2^{-2r_2}}{\sigma_{Z_2}^2}\right)^{-1}\frac{1}{D}.$$

Therefore, we can obtain (4.5) from

$$R_1 \geq h(U_1 | U_2) - \frac{1}{2}\log 2\pi e \sigma_{V_1}^2$$

$$= \frac{1}{2}\log \frac{\sigma_{U_1 | U_2}^2}{\sigma_{V_1}^2}$$

$$= \frac{1}{2}\log \frac{\sigma_{Z_1}^2 + \sigma_{V_1}^2}{\sigma_{V_1}^2} \left(\frac{1}{\sigma_{X_2}^2} + \frac{1 - 2^{-2r_2}}{\sigma_{Z_2}^2} \right)^{-1} \frac{1}{D}$$

$$= \frac{1}{2}\log 2^{2r_1} \left(\frac{1}{\sigma_{X_2}^2} + \frac{1 - 2^{-2r_2}}{\sigma_{Z_2}^2} \right)^{-1} \frac{1}{D}$$

$$= r_1 + \frac{1}{2}\log \left(\frac{1}{\sigma_{X_2}^2} + \frac{1 - 2^{-2r_2}}{\sigma_{Z_2}^2} \right)^{-1} \frac{1}{D}.$$

We can obtain (4.6) in a similar manner. Finally, for the sum bound in (4.7),

$$R_1 + R_2 \geq I(Y_1, Y_2; U_1, U_2) = h(U_1, U_2) - h(U_1, U_2 | Y_1, Y_2)$$

$$= \frac{1}{2}\log \frac{\left| \begin{pmatrix} \sigma_{U_1}^2 & \sigma_{U_1 U_2} \\ \sigma_{U_2 U_1} & \sigma_{U_2}^2 \end{pmatrix} \right|}{\sigma_{V_1}^2 \sigma_{V_2}^2}$$

$$= \frac{1}{2}\log \frac{1}{\sigma_{V_1}^2 \sigma_{V_2}^2} \left(\left(\sigma_{X_2}^2 + \sigma_{Z_1}^2 + \sigma_{V_1}^2 \right) \left(\sigma_{X_2}^2 + \sigma_{Z_2}^2 + \sigma_{V_2}^2 \right) - \sigma_X^4 \right)$$

$$= \frac{1}{2}\log \frac{1}{\sigma_{V_1}^2 \sigma_{V_2}^2} \left(\sigma_{X_2}^2 \left(\sigma_{Z_1}^2 + \sigma_{V_1}^2 \right) + \left(\sigma_{Z_2}^2 + \sigma_{V_2}^2 \right) \left(\sigma_{X_2}^2 + \sigma_{Z_1}^2 + \sigma_{V_1}^2 \right) \right)$$

$$= \frac{1}{2}\log \left(\frac{\sigma_{V_1}^2 + \sigma_{Z_1}^2}{\sigma_{V_1}^2} \right) \left(\frac{\sigma_{X_2}^2}{\sigma_{V_2}^2} + \frac{\sigma_{X_2}^2 (\sigma_{Z_2}^2 + \sigma_{V_2}^2)}{\sigma_{V_2}^2 (\sigma_{V_1}^2 + \sigma_{Z_1}^2)} + \frac{\sigma_{Z_2}^2 + \sigma_{V_2}^2}{\sigma_{V_2}^2} \right)$$

$$= \frac{1}{2}\log \left(\frac{\sigma_{V_1}^2 + \sigma_{Z_1}^2}{\sigma_{V_1}^2} \right) \left(\frac{\sigma_{V_2}^2 + \sigma_{Z_2}^2}{\sigma_{V_2}^2} \right) \left(\frac{\sigma_{X_2}^2}{\sigma_{V_2}^2 + \sigma_{Z_2}^2} + \frac{\sigma_{X_2}^2}{\sigma_{V_1}^2 + \sigma_{Z_1}^2} + 1 \right)$$

$$= \frac{1}{2}\log 2^{2r_1} 2^{2r_2} \frac{\sigma_{X_2}^2}{D} = r_1 + r_2 + \log \frac{\sigma_{X_2}^2}{D}$$

and this concludes the achievable proof.

Converse Proof of Quadratic Gaussian CEO Problem with Two Terminals
First, we will introduce a simple lemma.

Lemma 4.1 For an i.i.d. Gaussian input $X^n \sim \mathcal{N}(\mathbf{0}, \sigma_X^2 I)$ and a reconstructed $\hat{X}^n$ with a distortion constraint $D = \frac{1}{n}\sum_{i=1}^{n} E[(X_i - \hat{X}_i)^2]$

$$I(X^n; \hat{X}^n) \geq \frac{n}{2}\log \frac{\sigma_X^2}{D}.$$

Proof.

$$I(X^n; \hat{X}^n) = h(X^n) - h(\hat{X}^n | X^n)$$

$$\stackrel{(a)}{=} \frac{n}{2} \log(2\pi e \sigma_X^2) - h(\hat{X}^n - X^n | X^n)$$

$$\stackrel{(b)}{\geq} \frac{n}{2} \log(2\pi e \sigma_X^2) - h(\hat{X}^n - X^n)$$

$$\stackrel{(c)}{\geq} \frac{n}{2} \log \frac{\sigma_X^2}{D},$$

where (a) is due to (A.22), (b) is due to conditioning reducing entropy, and (c) is due to the fact that the Gaussian source has the largest entropy with fixed variance. $\qquad\square$

Let us define $r_1 \triangleq I(Y_1^n; M_1 | X^n)$ and $r_2 \triangleq I(Y_2^n; M_2 | X^n)$. Then, we have

$$n(R_1 + R_2)$$

$$\geq H(M_1, M_2)$$

$$\geq I(Y_1^n, Y_2^n; M_1, M_2)$$

$$\stackrel{(a)}{=} I(Y_1^n, Y_2^n; M_1, M_2) + I(X^n; M_1, M_2 | Y_1^n, Y_2^n)$$

$$= I(X^n, Y_1^n, Y_2^n; M_1, M_2)$$

$$= I(X^n; M_1, M_2) + I(Y_1^n, Y_2^n; M_1, M_2 | X^n)$$

$$= I(X^n; M_1, M_2) + I(Y_1^n; M_1, M_2 | X^n) + I(Y_2^n; M_1, M_2 | X^n, Y_1^n)$$

$$\stackrel{(b)}{=} I(X^n; M_1, M_2) + I(Y_1^n; M_1 | X^n) + I(Y_1^n; M_2 | X^n, M_1)^{\nearrow 0}$$

$$\qquad + I(Y_2^n; M_2 | X^n, Y_1^n) + I(Y_2^n; M_1 | X^n, Y_1^n, M_2)^{\nearrow 0}$$

$$\stackrel{(c)}{=} I(X^n; M_1, M_2) + I(Y_1^n; M_1 | X^n) + I(Y_2^n; M_2 | X^n, Y_1^n)$$

$$\stackrel{(d)}{\geq} I(X^n; \hat{X}^n) + \underbrace{I(Y_1^n; M_1 | X^n)}_{nr_1} + \underbrace{I(Y_2^n; M_2 | X^n)}_{nr_2}$$

$$\stackrel{(e)}{=} \frac{n}{2} \log \frac{\sigma_X^2}{D} + nr_1 + nr_2,$$

where in (a) $I(X^n; M_1, M_2 | Y_1^n, Y_2^n) = 0$ is due to $X^n \leftrightarrow (Y_1^n, Y_2^n) \leftrightarrow (M_1, M_2)$, (b) is due to $Y_1^n \leftrightarrow (X^n, M_1) \leftrightarrow M_2$, and $Y_2^n \leftrightarrow (X^n, Y_1^n, M_2) \leftrightarrow M_1$, (c) is due to the long Markov chain $M_1 \leftrightarrow Y_1^n \leftrightarrow X^n \leftrightarrow Y_2^n \leftrightarrow M_2$, (d) is due to data processing inequality and $X^n \leftrightarrow (M_1, M_2) \leftrightarrow \hat{X}^n$, and (e) is due to Lemma 4.1.

Now, for the bound of R_1,

$$nR_1 \geq H(M_1) \geq H(M_1 | M_2) \geq I(M_1; Y_1^n | M_2)$$

$$\stackrel{(a)}{=} I(M_1; Y_1^n | M_2) + I(M_1; X^n | Y_1^n, M_2)$$

$$= I(M_1; X^n, Y_1^n | M_2)$$

$$= I(X^n; M_1 | M_2) + I(Y_1^n; M_1 | X^n, M_2)$$

$$= I(M_1; X^n, Y_1^n | M_2)$$

$$\overset{(b)}{=} I(X^n; M_1 | M_2) + I(Y_1^n; M_1 | X^n, \cancel{M_2})$$

$$= I(X^n; M_1, M_2) - I(X^n; M_2) + I(Y_1^n; M_1 | X^n)$$

$$\geq [I(X^n; \hat{X}^n) - I(X^n; M_2)]^+ + \underbrace{I(Y_1^n; M_1 | X^n)}_{nr_1}$$

$$\overset{(c)}{\geq} \frac{n}{2} \log \frac{\sigma_X^2}{D} - I(X^n; M_2)]^+ + nr_1, \tag{4.9}$$

where $[x]^+ = \max\{x, 0\}$, (a) is due to $M_1 \leftrightarrow (Y_1^n, M_2) \leftrightarrow X^n$, (b) is due to the long Markov chain $M_1 \leftrightarrow Y_1^n \leftrightarrow X^n \leftrightarrow Y_2^n \leftrightarrow M_2$, and (c) is due to Lemma 4.1.

To bound $I(X^n; M_2)$, consider the best linear estimate of X_i given $Y_{2,i}$. From (D.20), we know that

$$X | Y_2 \sim \mathcal{N}(\Sigma_{XY} \Sigma_{YY}^{-1} Y_2, \Sigma_{XX} - \Sigma_{XY} \Sigma_{YY}^{-1} \Sigma_{YX}) = \mathcal{N}\left(\frac{\sigma_X^2}{\sigma_X^2 + \sigma_{Z_2}^2} Y_2, \frac{\sigma_X^2 \sigma_{Z_2}^2}{\sigma_X^2 + \sigma_{Z_2}^2} \right).$$

Therefore, the best linear estimate of X_i given $Y_{2,i}$, which should be just the mean of $X | Y_{2,i}$, is equal to $\tilde{X}_i \triangleq \frac{\sigma_X^2}{\sigma_X^2 + \sigma_{Z_2}^2} Y_{2,i}$ and write $X_i = \tilde{X}_i + W_i$, where $W_i \sim \mathcal{N}\left(0, \frac{\sigma_X^2 \sigma_{Z_2}^2}{\sigma_X^2 + \sigma_{Z_2}^2} \right)$ is the estimation error of $\tilde{X}_i$. By the orthogonality principle, we have $W_i [\perp\!\!\!\perp] Y_{2,i}$, where $\perp\!\!\!\perp$ denotes that W_i and $Y_{2,i}$ are independent.[2] Since the system is memoryless, we have $W^n \perp\!\!\!\perp Y_2^n$, and as both M_2 and $\tilde{X}^n$ are just functions of Y_2^n, we have $W^n \perp\!\!\!\perp (M_2, \tilde{X}^n)$. By Fact A.2, $W^n \perp\!\!\!\perp \tilde{X}^n | M_2$. Thus, we can apply the conditional entropy power inequality (A.27),

$$2^{\frac{2h(X^n | M_2)}{n}} \geq 2^{\frac{2h(\tilde{X}^n | M_2)}{n}} + 2^{\frac{2h(W^n | M_2)}{n}}$$

$$\overset{(a)}{=} 2^{\frac{2h(X^n | M_2)}{n}} + 2^{\frac{2h(W^n)}{n}}$$

$$= 2^{\frac{2}{n}[h(\tilde{X}^n | X^n, M_2) + I(\tilde{X}^n; X^n | M_2)]} + 2\pi e \frac{\sigma_X^2 \sigma_{Z_2}^2}{\sigma_X^2 + \sigma_{Z_2}^2}, \tag{4.10}$$

where (a) is due to $W^n \perp\!\!\!\perp M_2$. We have

$$2^{\frac{2}{n} h(\tilde{X}^n | X^n, M_2)} = 2^{\frac{2}{n}[h(\tilde{X}^n) - I(\tilde{X}^n; M_2, X^n)]}$$

$$\overset{(a)}{\geq} 2^{\frac{2}{n}[h(\tilde{X}^n) - I(Y_2^n; M_2, X^n)]}$$

$$\overset{(b)}{=} \left(\frac{\sigma_X^2}{\sigma_X^2 + \sigma_{Z_2}^2} \right)^2 2^{\frac{2}{n}[h(Y_2^n) - I(Y_2^n; M_2, X^n)]}$$

---

2 Since W_i and $Y_{2,i}$ are Gaussian, uncorrelatedness and independence are equivalent.

$$= \left(\frac{\sigma_X^2}{\sigma_X^2 + \sigma_{Z_2}^2} \right)^2 2^{\frac{2}{n} h(Y_2^n | M_2, X^n)}$$

$$= \left(\frac{\sigma_X^2}{\sigma_X^2 + \sigma_{Z_2}^2} \right)^2 2^{\frac{2}{n} [h(Y_2^n | X^n) - \underbrace{I(Y_2^n, M_2 | X^n)]}_{nr_2}}$$

$$= \left(\frac{\sigma_X^2}{\sigma_X^2 + \sigma_{Z_2}^2} \right)^2 \sigma_{Z_2}^2 2\pi e 2^{-2r_2},$$

where (*a*) is due to $(M_2, X^n) \leftrightarrow Y_2^n \leftrightarrow \tilde{X}^n$ and data processing inequality, (*b*) is due to $\tilde{X}^n = \frac{\sigma_X^2}{\sigma_X^2 + \sigma_{Z_2}^2} Y_2^n$, and

$$2^{\frac{2}{n} I(\tilde{X}^n; X^n | M_2)} = 2^{\frac{2}{n} [h(X^n | M_2) - h(X^n | \tilde{X}^n, M_2)]}$$

$$\overset{(a)}{=} 2^{\frac{2}{n} [h(X^n | M_2) - h(X^n | \tilde{X}^n, M_2, Y_2^n)]}$$

$$\overset{(b)}{=} 2^{\frac{2}{n} [h(X^n | M_2) - h(X^n | \tilde{X}^n, Y_2^n)]}$$

$$\overset{(c)}{=} 2^{\frac{2}{n} [h(X^n | M_2) - h(X^n | \tilde{X}^n)]}$$

$$= \frac{2^{\frac{2}{n} h(X^n | M_2)}}{2\pi e \frac{\sigma_X^2 \sigma_{Z_2}^2}{\sigma_X^2 + \sigma_{Z_2}^2}},$$

where (*a*) and (*c*) are due to the fact that Y_2^n is a function of $\tilde{X}^n$, and (*b*) is due to the fact that M_2 is a function of Y_2^n. Now, we can substitute $2^{\frac{2}{n} I(\tilde{X}^n; X^n | M_2)}$ and $2^{\frac{2}{n} h(\tilde{X}^n | X^n, M_2)}$ back into (4.10). Thus, we have

$$2^{\frac{2h(X^n | M_2)}{n}} \geq \left(\frac{\sigma_X^2}{\sigma_X^2 + \sigma_{Z_2}^2} \right) \left(2^{-2r_2} 2^{\frac{2}{n} h(X^n | M_2)} + 2\pi e \sigma_{Z_2}^2 \right)$$

and this gives us

$$2^{-\frac{2h(X^n | M_2)}{n}} \leq \frac{1 - \left(\frac{\sigma_X^2}{\sigma_X^2 + \sigma_{Z_2}^2} \right) 2^{-2r_2}}{2\pi e \left(\frac{\sigma_X^2 \sigma_{Z_2}^2}{\sigma_X^2 + \sigma_{Z_2}^2} \right)}.$$

Multiplying both sides by $2^{\frac{2}{n} h(X^n)} = 2\pi e \sigma_X^2$, we have

$$2^{\frac{2}{n} I(X^n; M_2)} \leq \frac{\sigma_X^2}{\left(\frac{\sigma_X^2 \sigma_{Z_2}^2}{\sigma_X^2 + \sigma_{Z_2}^2} \right)} - \frac{\sigma_X^2}{\sigma_{Z_2}^2} 2^{-2r_2}$$

$$
= \frac{\sigma_X^2 \left(\left(\dfrac{\sigma_X^2}{\sigma_X^2 + \sigma_{Z_2}^2} \right) + \left(\dfrac{\sigma_{Z_2}^2}{\sigma_X^2 + \sigma_{Z_2}^2} \right) \right)}{\left(\dfrac{\sigma_X^2 \sigma_{Z_2}^2}{\sigma_X^2 + \sigma_{Z_2}^2} \right)} - \frac{\sigma_X^2}{\sigma_{Z_2}^2} 2^{-2r_2}
$$

$$
= 1 + \frac{\sigma_X^2}{\sigma_{Z_2}^2} (1 - 2^{-2r_2}).
$$

Therefore, we have

$$
I(X^n; M_2) \leq \frac{n}{2} \log \left(1 + \frac{\sigma_X^2}{\sigma_{Z_2}^2} (1 - 2^{-2r_2}) \right) \tag{4.11}
$$

and substitute it back to (4.9) and we obtain the bound for R_1 in (4.5). By symmetry, we can swap r_1 and r_2, and replace $\sigma_{Z_2}^2$ with $\sigma_{Z_1}^2$ in (4.5) to obtain the bound for R_2.

For the final inequality, let us consider the best estimator of X given $\mathbf{Y} = (Y_1, Y_2)^T$. The conditional distribution is given by

$$
X | Y_1, Y_2 \sim \mathcal{N}(\Sigma_{XY} \Sigma_{YY}^{-1} \begin{pmatrix} Y_1 \\ Y_2 \end{pmatrix}, \Sigma_{XX} - \Sigma_{XY} \Sigma_{YY}^{-1} \Sigma_{YX})
$$

$$
= \mathcal{N} \left((\sigma_X^2 \sigma_X^2) \begin{pmatrix} \sigma_X^2 + \sigma_{Z_1}^2 & \sigma_X^2 \\ \sigma_X^2 & \sigma_X^2 \sigma_{Z_2}^2 \end{pmatrix}^{-1} \begin{pmatrix} Y_1 \\ Y_2 \end{pmatrix}, \right.
$$

$$
\left. \sigma_X^2 - (\sigma_X^2 \sigma_X^2) \begin{pmatrix} \sigma_X^2 + \sigma_{Z_1}^2 & \sigma_X^2 \\ \sigma_X^2 & \sigma_X^2 \sigma_{Z_2}^2 \end{pmatrix} \begin{pmatrix} \sigma_X^2 \\ \sigma_X^2 \end{pmatrix} \right)
$$

$$
= \mathcal{N} \left(\frac{\sigma_X^2 \sigma_{Z_2}^2 Y_1 + \sigma_X^2 \sigma_{Z_1}^2 Y_2}{\sigma_{Z_1}^2 \sigma_{Z_2}^2 + \sigma_X^2 \sigma_{Z_1}^2 + \sigma_X^2 \sigma_{Z_2}^2}, \frac{\sigma_X^2 \sigma_{Z_1}^2 \sigma_{Z_2}^2}{\sigma_{Z_1}^2 \sigma_{Z_2}^2 + \sigma_X^2 \sigma_{Z_1}^2 + \sigma_X^2 \sigma_{Z_2}^2} \right).
$$

Therefore, the best linear estimate $\tilde{X}_i$ given $Y_{1,i}$ and $Y_{2,i}$ is $\tilde{X}_i = \frac{\sigma_X^2 \sigma_{Z_2}^2 Y_{1,i} + \sigma_X^2 \sigma_{Z_1}^2 Y_{2,i}}{\sigma_{Z_1}^2 \sigma_{Z_2}^2 + \sigma_X^2 \sigma_{Z_1}^2 + \sigma_X^2 \sigma_{Z_2}^2}$.
As in the previous case, write $X_i = \tilde{X}_i + W_i$, where W_i is the error of the estimate. Then

$$
\mathbf{W}_i \sim \mathcal{N} \left(0, \frac{\sigma_X^2 \sigma_{Z_1}^2 \sigma_{Z_2}^2}{\sigma_{Z_1}^2 \sigma_{Z_2}^2 + \sigma_X^2 \sigma_{Z_1}^2 + \sigma_X^2 \sigma_{Z_2}^2} \right)
$$

$$
= \mathcal{N} \left(0, \left(\frac{1}{\sigma_{Z_1}^2} + \frac{1}{\sigma_{Z_2}^2} + \frac{1}{\sigma_X^2} \right)^{-1} \right).
$$

Define $\sigma_W^2 \triangleq \left(\frac{1}{\sigma_{Z_1}^2} + \frac{1}{\sigma_{Z_2}^2} + \frac{1}{\sigma_X^2} \right)^{-1}$. Note that we may rewrite $\tilde{X}_i = \frac{\sigma_W^2}{\sigma_{Z_1}^2} Y_{1,i} + \frac{\sigma_W^2}{\sigma_{Z_2}^2} Y_{2,i}$. Since the system is memoryless, we have

$$\tilde{X}^n = \frac{\sigma_W^2}{\sigma_{Z_1}^2} Y_1^n + \frac{\sigma_W^2}{\sigma_{Z_2}^2} Y_2^n \qquad (4.12)$$

and $X^n = \tilde{X}^n + W^n$. Moreover, since $W^n \perp\!\!\!\perp (Y_1^n, Y_2^n)$, we have $W^n \perp\!\!\!\perp (M_1, M_2, \tilde{X}^n)$. From Fact A.2, $W^n \perp\!\!\!\perp \tilde{X}^n | (M_1, M_2)$, therefore we can apply conditional entropy power inequality on $X^n = \tilde{X}^n + W^n$ and obtain

$$2^{\frac{2}{n}h(X^n|M_1,M_2)} \geq 2^{\frac{2}{n}h(\tilde{X}^n|M_1,M_2)} + 2^{\frac{2}{n}h(W^n|M_1,M_2)}$$

$$\overset{(a)}{=} 2^{\frac{2}{n}h(\tilde{X}^n|M_1,M_2)} + 2^{\frac{2}{n}h(W^n)}$$

$$= 2^{\frac{2}{n}h(\tilde{X}^n|X^n,M_1,M_2)+I(\tilde{X}^n;X^n|M_1,M_2)} + 2\pi e \sigma_W^2, \qquad (4.13)$$

where (a) is due to $W^n \perp\!\!\!\perp (M_1, M_2)$.

Since $Y_1^n \leftrightarrow (X^n, M_1, M_2) \leftrightarrow Y_2^n$, we can apply conditional entropy power inequality on (4.12) and obtain

$$2^{\frac{2}{n}h(\tilde{X}^n|X^n,M_1,M_2)} \geq \left(\frac{\sigma_W^2}{\sigma_{Z_1}^2} \right)^2 2^{\frac{2}{n}h(Y_1^n|X^n,M_1,M_2)} + \left(\frac{\sigma_W^2}{\sigma_{Z_2}^2} \right)^2 2^{\frac{2}{n}h(Y_2^n|X^n,M_1,M_2)}$$

$$\overset{(a)}{=} \left(\frac{\sigma_W^2}{\sigma_{Z_1}^2} \right)^2 2^{\frac{2}{n}h(Y_1^n|X^n)-I(Y_1^n;M_1,M_2|X^n)}$$

$$+ \left(\frac{\sigma_W^2}{\sigma_{Z_2}^2} \right)^2 2^{\frac{2}{n}h(Y_2^n|X^n)-I(Y_2^n;M_1,M_2|X^n)}$$

$$= 2\pi e \sigma_W^4 \left(\frac{2^{2r_1}}{\sigma_{Z_1}^2} + \frac{2^{2r_2}}{\sigma_{Z_2}^2} \right),$$

where in (a) we can remove M_1 and M_2 from the mutual information because of the long Markov chain $M_1 \leftrightarrow Y_1^n \leftrightarrow X^n \leftrightarrow Y_2^n \leftrightarrow M_2$, and

$$2^{\frac{2}{n}I(\tilde{X}^n;X^n|M_1,M_2)} = 2^{\frac{2}{n}[h(X^n|M_1,M_2)-h(X^n|\tilde{X}^n,M_1,M_2)]}$$

$$= 2^{\frac{2}{n}[h(X^n|M_1,M_2)-h(X^n|\tilde{X}^n)]}$$

$$= \frac{2^{\frac{2}{n}h(X^n|M_1,M_2)}}{2\pi e \sigma_W^2}.$$

Now, if we substitute $2^{\frac{2}{n}h(\tilde{X}^n|X^n,M_1,M_2)}$ and $2^{\frac{2}{n}I(\tilde{X}^n;X^n|M_1,M_2)}$ back into (4.13), we have

$$2^{\frac{2}{n}h(X^n|M_1,M_2)} \geq \sigma_W^2 2^{\frac{2}{n}h(X^n|M_1,M_2)} \left(\frac{2^{-2r_1}}{\sigma_{Z_1}^2} + \frac{2^{-2r_2}}{\sigma_{Z_2}^2} \right) + 2\pi e \sigma_W^2.$$

This gives us

$$2^{-\frac{2}{n}h(X^n|M_1,M_2)} \leq \frac{1 - \frac{\sigma_W^2}{\sigma_{Z_1}^2}2^{-2r_1} - \frac{\sigma_W^2}{\sigma_{Z_2}^2}2^{-2r_2}}{2\pi e \sigma_W^2}.$$

Multiplying $2^{\frac{2}{n}h(X^n)} = 2\pi e \sigma_X^2$ on both sides, we have

$$2^{\frac{2}{n}I(X^n;M_1,M_2)} \leq \frac{\sigma_X^2}{\sigma_W^2} - \frac{\sigma_W^2}{\sigma_{Z_1}^2}2^{-2r_1} - \frac{\sigma_W^2}{\sigma_{Z_2}^2}2^{-2r_2}$$

$$= \frac{\sigma_X^2\sigma_{Z_1}^2 + \sigma_X^2\sigma_{Z_2}^2 + \sigma_{Z_1}^2\sigma_{Z_2}^2}{\sigma_{Z_1}^2\sigma_{Z_2}^2} - \frac{\sigma_W^2}{\sigma_{Z_1}^2}2^{-2r_1} - \frac{\sigma_W^2}{\sigma_{Z_2}^2}2^{-2r_2}$$

$$= 1 + \frac{\sigma_W^2}{\sigma_{Z_1}^2}(1 - 2^{-2r_1}) + \frac{\sigma_W^2}{\sigma_{Z_2}^2}(1 - 2^{-2r_2})$$

therefore,

$$\frac{1}{2}\log\left(1 + \frac{\sigma_W^2}{\sigma_{Z_1}^2}(1 - 2^{-2r_1}) + \frac{\sigma_W^2}{\sigma_{Z_2}^2}(1 - 2^{-2r_2})\right)$$

$$\geq \frac{I(X^n;M_1,M_2)}{n}$$

$$\overset{(a)}{\geq} \frac{I(X^n;\hat{X}^n)}{n}$$

$$\overset{(b)}{\geq} \frac{1}{2}\log\frac{\sigma_X^2}{D},$$

where (a) is from data processing inequality and (b) is from Lemma 4.1. Rearranging the terms will give us the final inequality in (4.8).

4.3 Direct Multiterminal Source Coding

Now, let us consider the direct multiterminal problem with two joint Gaussian sources. As in the CEO case, we consider the quadratic distortion measure and so we can write

$$D_{1,i} = E[d(X_{1,i}, \hat{X}_{1,i})] = E[(X_{1,i} - \hat{X}_{1,i})^2]$$

and

$$D_{2,i} = E[d(X_{2,i}, \hat{X}_{2,i})] = E[(X_{2,i} - \hat{X}_{2,i})^2].$$

Given the distortion constraint D_1 and D_2, we need

$$D_1 = \frac{1}{n}\sum_{i=1}^{n}D_{1,i} \leq D_1$$

and

$$D_2 = \frac{1}{n} \sum_{i=1}^{n} D_{2,i} \leq D_2.$$

In the remaining section, we will show that the achievable rate region is given by

$$R_1(D_1) = \{(R_1, R_2) | R_1 \geq \frac{1}{2} \log \frac{1 - \rho^2 + \rho^2 - 2^{-2R_2}}{D_1}, R_2 \geq 0\}$$

$$R_2(D_2) = \{(R_1, R_2) | R_1 \geq 0, R_2 \geq \frac{1}{2} \log \frac{1 - \rho^2 + \rho^2 - 2^{-2R_1}}{D_2}\}$$

$$R_s(D_1, D_2) = \{(R_1, R_2) | R_1 + R_2 \geq \frac{1}{2} \log \frac{(1 - \rho^2)\beta(D_1, D_2)}{2D_1 D_2}\}, \quad (4.14)$$

where $\beta(D_1, D_2) = 1 + \sqrt{1 + \frac{4\rho^2}{(1-\rho^2)^2} D_1 D_2}$. The proof follows closely to that in [21] with some additional intermediate steps given.

4.3.1 Forward Proof of Gaussian Multiterminal Source Coding Problem with Two Sources

Without loss of generality, let us assume $D_1 \leq D_2$ and the source inputs are zero-mean and unit variance. Therefore, we can write $\mathbf{X} = (X_1, X_2)^T \sim \mathcal{N}(0, \Sigma_{XX})$ with $\Sigma_{XX} = \begin{pmatrix} 1 & \rho \\ \rho & 1 \end{pmatrix}$. Let us consider the Berger–Tung scheme and construct the auxiliary random variable $\mathbf{U} = (U_1, U_2)^T = \mathbf{X} + \mathbf{V}$, where $\mathbf{V} = (V_1, V_2)^T \sim \mathcal{N}(0, \Sigma_{VV})$ with $\Sigma_{VV} = \begin{pmatrix} N_1 & 0 \\ 0 & N_2 \end{pmatrix}$. Since Σ_{VV} is diagonal, we have $U_1 \leftrightarrow X_1 \leftrightarrow X_2 \leftrightarrow U_2$ as required by the Berger–Tung scheme.

Since

$$p(\mathbf{x}|\mathbf{u}) \propto p(\mathbf{u}|\mathbf{x})p(\mathbf{x})$$
$$= \mathcal{N}(\mathbf{u}; \mathbf{x}, \Sigma_{VV})\mathcal{N}(\mathbf{x}; 0, \Sigma_{XX})$$
$$= \mathcal{N}(\mathbf{x}; \mathbf{u}, \Sigma_{VV})\mathcal{N}(\mathbf{x}; 0, \Sigma_{XX})$$
$$\propto \mathcal{N}(\mathbf{x}; (\Sigma_{VV}^{-1} + \Sigma_{XX}^{-1})\Sigma_{VV}^{-1}\mathbf{u}, (\Sigma_{VV}^{-1} + \Sigma_{XX}^{-1})^{-1}),$$

$$\Sigma_{X|U} = (\Sigma_{XX}^{-1} + \Sigma_{VV}^{-1})^{-1}$$
$$= \left(\begin{pmatrix} 1 & \rho \\ \rho & 1 \end{pmatrix}^{-1} + \begin{pmatrix} N_1 & 0 \\ 0 & N_2 \end{pmatrix}^{-1} \right)^{-1}$$
$$= \left(\frac{1}{1 - \rho^2} \begin{pmatrix} 1 & -\rho \\ -\rho & 1 \end{pmatrix} + \begin{pmatrix} 1/N_1 & 0 \\ 0 & 1/N_2 \end{pmatrix} \right)^{-1} \quad (4.15)$$

$$= \frac{1}{(1 + N_1)(1 + N_2) - \rho^2}$$
$$\begin{pmatrix} (1 + N_2)N_1 - \rho^2 N_1 & \rho N_1 N_2 \\ \rho N_1 N_2 & (1 + N_1)N_2 - \rho^2 N_2 \end{pmatrix}.$$

Assuming that the expected reconstructed distortions are D_1 and D_2, we may then also write

$$\Sigma_{X|U} = \begin{pmatrix} D_1 & \theta \sqrt{D_1 D_2} \\ \theta \sqrt{D_1 D_2} & D_2 \end{pmatrix}$$

with some $\theta \in [-1, 1]$. Therefore, by comparing the upper left and lower right elements we have

$$D_1 = \frac{N_1(1 + N_2 - \rho^2)}{(1 + N_1)(1 + N_2) - \rho^2} \tag{4.16}$$

and

$$D_2 = \frac{N_2(1 + N_1 - \rho^2)}{(1 + N_1)(1 + N_2) - \rho^2}. \tag{4.17}$$

Note that $N_2 \geq N_1$ as $N_2 - N_1 = (D_2 - D_1)((1 + N_1)(1 + N_2) - \rho^2)/\rho^2 \geq 0$ since $(1 + N_1)(1 + N_2) \geq 1 \geq \rho^2$ and $D_2 \geq D_1$ as we assumed earlier.

Now, if we compare ρ^2 with $\frac{1 - D_2}{1 - D_1}$,

$$\rho^2 \overset{?}{\approx} \frac{1 - D_2}{1 - D_1} = \frac{(1 + N_1) - (1 - N_2)\rho^2}{(1 + N_2) - (1 - N_1)\rho^2}$$

$$\Leftrightarrow (1 - N_1)\rho^4 - 2\rho^2 + (1 + N_1) \overset{?}{\approx} 0$$

Define $f(\rho^2, N_1) = (1 - N_1)(\rho^2)^2 - 2(\rho^2) + (1 + N_1)$, where the only supposed valid ranges for ρ^2 and N_1 are $\rho^2 \in [0, 1]$ and $N_1 \in [0, \infty)$. As shown in Figure 4.3, this region appears to overlap with the region of $f(\rho^2, N_1) < 0$ and equivalently $\rho^2 < \frac{1 - D_2}{1 - D_1}$. This strongly hints that $\rho^2 = \frac{1 - D_2}{1 - D_1}$ is a watershed case and we should consider the cases $\rho^2 < \frac{1 - D_2}{1 - D_1}$ and $\rho^2 \geq \frac{1 - D_2}{1 - D_1}$ separately, as we discuss later. Before that, let us also compute the required rate constraint for the Berger–Tung scheme.

From (4.15) again,

$$\Sigma_{X|U}^{-1} = \Sigma_{VV}^{-1} + \Sigma_{XX}^{-1}$$

$$\Rightarrow \frac{1}{D_1 D_2 (1 - \theta^2)} \begin{pmatrix} D_2 & -\theta \sqrt{D_1 D_2} \\ -\theta \sqrt{D_1 D_2} & D_1 \end{pmatrix}$$

$$= \frac{1}{1 - \rho^2} \begin{pmatrix} 1 & -\rho \\ -\rho & 1 \end{pmatrix} + \begin{pmatrix} 1/N_1 & 0 \\ 0 & 1/N_2 \end{pmatrix}$$

$$\Rightarrow -\frac{\theta \sqrt{D_1 D_2}}{D_1 D_2 - \theta^2 D_1 D_2} = -\frac{\rho}{1 - \rho^2}$$

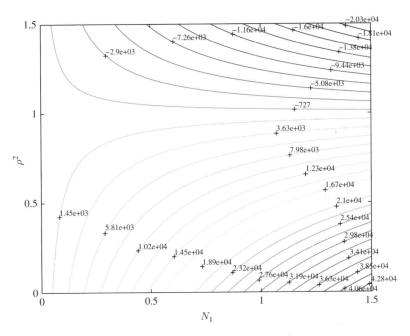

Figure 4.3 $f(\rho^2, N_1) = (1 - N_1)(\rho^2)^2 - 2(\rho^2) + (1 + N_1)$. Note that the region $(\rho^2, N_1)|(0 \leq \rho^2 \leq 1, N_1 \geq 0)$ appears to overlap with that of $f(\rho^2, N_1) > 0$ (or equivalently $\rho^2 < \frac{1-D_2}{1-D_1}$). This strongly suggests that we should consider the case $\rho^2 < \frac{1-D_2}{1-D_1}$ and $\rho^2 \geq \frac{1-D_2}{1-D_1}$ separately.

$$\Rightarrow \theta = -\frac{(1 - \rho^2) \pm \sqrt{(1 - \rho^2)^2 + 4\rho^2 D_1 D_2}}{2\rho \sqrt{D_1 D_2}}.$$

Since $|\theta| \leq 1$, the only possible solution is

$$\theta = \frac{\sqrt{(1 - \rho^2)^2 + 4\rho^2 D_1 D_2} - (1 - \rho^2)}{2\rho \sqrt{D_1 D_2}}. \tag{4.18}$$

Therefore, the minimum sum rate to achieve the required distortion is

$$\begin{aligned}
I(U_1, U_2; X_1, X_2) &= h(X_1, X_2) - h(X_1, X_2 | U_1, U_2) \\
&= \frac{1}{2} \log \frac{|\Sigma_{XX}|}{|\Sigma_{X|U}|} \\
&= \frac{1}{2} \log \frac{1 - \rho^2}{D_1 D_2 (1 - \theta^2)} \\
&= \frac{1}{2} \log \frac{(1 - \rho^2)\beta(D_1, D_2)}{2D_1 D_2},
\end{aligned}$$

where $\beta(D_1, D_2) = 1 + \sqrt{1 + \frac{4\rho^2}{(1-\rho^2)^2 D_1 D_2}}$ as defined earlier.

The minimum R_1 required by the Berger–Tung scheme is

$$
\begin{aligned}
I(X_1; U_1 | U_2) &= h(U_1 | U_2) - h(U_1 | X_1, U_2) \\
&= h(U_1 | U_2) - h(U_1 | X_1) \\
&= \frac{1}{2} \log \frac{\sigma_{U_1 | U_2}}{\sigma_{U_1 | X_1}} \\
&\overset{(a)}{=} \frac{1}{2} \log \frac{E[U_1^2] - E[U_1 U_2] E[U_2^2]^{-1} E[U_2 U_1]}{E[U_1^2] - E[U_1 X_1] E[X_1^2]^{-1} E[X_1 U_1]} \\
&= \frac{1}{2} \log \frac{1 + N_1 - \rho^2 / (1 + N_2)}{1 + N_1 - 1} \\
&= \frac{1}{2} \log \frac{(1 + N_1)(1 + N_2) - \rho^2}{N_1 (1 + N_2)},
\end{aligned}
\tag{4.19}
$$

where (a) is due to (D.20). Similarly,

$$
I(X_2; U_2 | U_1) = \frac{1}{2} \log \frac{(1 + N_1)(1 + N_2) - \rho^2}{N_1 (1 + N_2)}.
\tag{4.20}
$$

Since

$$
\begin{aligned}
I(U_1, U_2 | X_1, X_2) &= I(U_1; X_1, X_2) + I(U_2; X_1, X_2 | U_1) \\
&= I(U_1; X_1) + I(X_2; U_1 | X_1)^{\;0} + I(U_2, X_2 | U_1) \\
&\quad + I(U_2; X_1 | X_2, U_1)^{\;0}
\end{aligned}
$$

$(R_1, R_2) = (I(U_1; X_1), I(U_2; X_2 | U_1))$ is the lower right corner point of the Berger–Tung region with the given N_1 and N_2. Moreover, we have

$$
R_1 = I(U_1; X_1) = \frac{1}{2} \log \frac{1 + N_1}{N_1} \Leftrightarrow N_1 = \frac{1}{2^{2R_1} - 1}
$$

and so we can further represent R_2 as

$$
\begin{aligned}
R_2 &= \frac{1}{2} \log \frac{(1 + N_1)(1 + N_2) - \rho^2}{N_2 (1 + N_1)} \\
&= \frac{1}{2} \log \underbrace{\frac{(1 + N_1)(1 + N_2) - \rho^2}{N_2 (1 + N_1 - \rho^2)}}_{1/D_2} \frac{N_2 (1 + N_1 - \rho^2)}{N_2 (1 + N_1)} \\
&= \frac{1}{2} \log \frac{1 - \rho^2 + \rho^2 2^{-2R_1}}{D_2}.
\end{aligned}
\tag{4.21}
$$

Therefore, this corner point falls right on the boundary of $\mathcal{R}_2(D_2)$. Similarly, the rate pair $(I(U_1; X_1 | U_2), I(U_2; X_2))$, which corresponds to the upper left corner point of the Berger–Tung region for the given N_1 and N_2, falls right on the boundary of $\mathcal{R}_1(D_1)$, as shown in Figure 4.4.

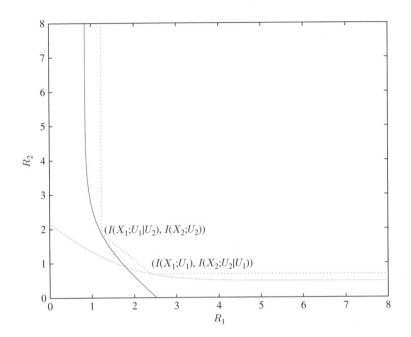

Figure 4.4 Corner points of the Berger–Tung region of a fixed N_1 and N_2 fall right on the boundary of $R_1(D_1)$ and $R_2(D_2)$.

Case 1: $(1 - D_2) \leq \rho^2(1 - D_1)$
For this case, the region $R_s(D_1, D_2) \cap R_1(D_1, D_2) \cap R_2(D_1, D_2)$ simplifies to $R_1(D_1, D_2)$ as we have $R_1(D_1, D_2) \subseteq R_s(D_1, D_2)$ and $R_1(D_1, D_2) \subseteq R_2(D_1, D_2)$ as shown below.

Proof of $R_1(D_1, D_2) \subseteq R_s(D_1, D_2)$ Consider D_2' satisfies that $R_s(D_1, D_2') = \{(R_1, R_2) | R_1 + R_2 \leq \frac{1}{2} \log \frac{1}{D_1}\}$. Then, we have

$$\frac{(1 - \rho^2)\beta(D_1, D_2)}{2D_1 D_2'} = \frac{1}{D_1}$$
$$\Rightarrow D_2' = 1 - \rho^2 + \rho^2 D_1.$$

Now, if we assume $(1 - D_2) \leq \rho^2(1 - D_1)$, we have $D_2 \geq 1 - \rho^2 + \rho^2 D_1 = D_2'$. Then, consider $(R_1, R_2) \in R_1(D_1)$,

$$R_1 + R_2 \geq \frac{1}{2} \log \frac{1 - \rho^2 + \rho^2 2^{-2R_2}}{D_1} + R_2$$
$$= \frac{1}{2} \log \frac{2^{2R_2}(1 - \rho^2) + \rho^2}{D_1}$$

$$\overset{(a)}{\geq} \frac{1}{2} \log \frac{1}{D_1},$$

where (a) is due to the fact that $2^{2R_2}(1 - \rho^2) + \rho^2$ is an increasing function of R_2 and equal to 1 when R_2 is 0. Therefore, we have

$$\mathcal{R}_1(D_1) \subseteq \mathcal{R}_s(D_1, D_2')$$

$$\overset{(a)}{\subseteq} \mathcal{R}_s(D_1, D_2),$$

where (a) is because $D_2 \geq D_2'$ given $(1 - D_2) \leq \rho^2(1 - D_1)$.

Proof of $\mathcal{R}_1(D_1, D_2) \subseteq \mathcal{R}_2(D_1, D_2)$ Note that we can represent $\mathcal{R}_2(D_2)$ alternatively as

$$\mathcal{R}_2(D_2) = \{(R_1, R_2)|R_1 \geq \frac{1}{2} \log \frac{\rho^2}{D_2 2^{2R_2} + \rho^2 - 1}, R_2 \geq 0\},$$

and $(1 - D_2) \leq \rho^2(1 - D_1) \Rightarrow D_1 \leq \frac{D_2 + \rho^2 - 1}{\rho^2}$. If we consider any $(R_1, R_2) \in \mathcal{R}_1(D_1)$, we have

$$R_1 \geq \frac{1}{2} \log \frac{1 - \rho^2 + \rho^2 2^{-2R_2}}{D_1}$$

$$\geq \frac{1}{2} \log \frac{\rho^2(1 - \rho^2 + \rho^2 2^{-2R_2})}{D_2 + \rho^2 - 1}$$

$$\overset{(a)}{\geq} \frac{1}{2} \log \frac{\rho^2}{D_2 2^{2R_2} + \rho^2 - 1},$$

where (a) is due to Lemma 4.2. Therefore, we have $\mathcal{R}_1(D_1) \subseteq \mathcal{R}_2(D_2)$ also when $(1 - D_2) \leq \rho^2(1 - D_1)$.

Lemma 4.2

$$\frac{1 - \rho^2 + \rho^2 2^{-2R_2}}{D_2 + \rho^2 - 1} \geq \frac{1}{D_2 2^{2R_2} + \rho^2 - 1}$$

for non-negative R_2.

Proof. The inequality is equivalent to $(1 - \rho^2 + \rho^2 2^{-2R_2})(D_2 2^{2R_2} + \rho^2 - 1) \geq D_2 + \rho^2 - 1$. Define $f(R_2) \triangleq (1 - \rho^2 + \rho^2 2^{-2R_2})(D_2 2^{2R_2} + \rho^2 - 1) - (D_2 + \rho^2 - 1)$ and then we just need to show $f(R_2) \geq 0$ for non-negative R_2. First, note that $f(0) = 0$, and we can write

$$f(R_2) = (1 - \rho^2 + \rho^2 2^{-2R_2})(D_2 2^{2R_2} + \rho^2 - 1) - (D_2 + \rho^2 - 1)$$

$$= \underbrace{D_2(1 - \rho^2)2^{2R_2} + (\rho^2 - 1)\rho^2 2^{-2R_2} +}_{\text{independent of } R_2} \cdots$$

$$= (1 - \rho^2)\underbrace{(D_2 2^{2R_2} - \rho^2 2^{-2R_2})}_{\text{increasing function of } R_2} + \underbrace{\cdots}_{\text{independent of } R_2} \cdot$$

Therefore $f(R_2)$ is a monotonically increasing function of R_2 and thus $f(R_2) \geq f(0) = 0$ for non-negative R_2. $\qquad \square$

Let us now focus on the lower corner point of $\mathcal{R}_1(D_1)$, which is $(\frac{1}{2} \log \frac{1}{D_1})$ from (4.14). Note that we cannot find N_1 and N_2 by solving (4.16) and (4.17) as no finite valid solution exists. However, we can pick $N_2 = \infty$, which gives us $I(U_2; X_2) = I(U_2; X_2 | U_1) = 0$ and $I(U_1; X_1) = I(U_1; X_1 | U_2) = \frac{1}{2} \log \frac{1+N_1}{N_1}$. Note that the two corner points of the Berger–Tung region overlap in this case. Now, pick $R_2 = I(U_2; X_2 | U_1) = 0$ and $R_1 = I(U_1; X_1) = \frac{1}{2} \log \frac{1+N_1}{N_1} = \frac{1}{2} \log \frac{1}{D_1}$, this requires us to pick $N_1 = \frac{D_1}{1-D_1}$. For this pair of (R_1, R_2), the Berger–Tung scheme guarantees that

$$D_1 = \frac{N_1(1 + N_2 - \rho^2)}{(1 + N_1)(1 + N_2) - \rho^2} = \frac{N_1}{1 + N_1} = D_1$$

and

$$D_2 = \frac{N_2(1 + N_1 - \rho^2)}{(1 + N_1)(1 + N_2) - \rho^2} = \frac{1 + N_1 - \rho^2}{1 + N_1}$$
$$= 1 - \rho^2(1 - D_1) \leq D_2,$$

where the last inequality is from our assumption that $(1 - D_2) \leq \rho^2(1 - D_1)$. Therefore, by selecting the suggested N_1 and N_2, the corner point $(\frac{1}{2} \log \frac{1}{D_1}, 0)$ is achievable.

To achieve other boundary points of $\mathcal{R}_1(D_1)$, let us consider $(\tilde{N}_1, \tilde{N}_2)$ with an arbitrary $\tilde{N}_2 < \infty$ but choose $\tilde{N}_1$ such that the resulting $\tilde{D}_1$ is the same as before. This choice ensures that the distortion constraint of X_1 is satisfied. Moreover, it guarantees that the upper left corner of the Berger–Tung region for the given $(\tilde{N}_1, \tilde{N}_2)$ falls right on the boundary of $\mathcal{R}_1(D_1)$, as shown in (4.21). It remains to show that the distortion constraint for X_2 is not violated to complete the proof for this case. Since

$$\tilde{D}_1 = \frac{\tilde{N}_1(1 + \tilde{N}_2 - \rho^2)}{(1 + \tilde{N}_1)(1 + \tilde{N}_2) - \rho^2} = \frac{\tilde{N}_1(1 + \tilde{N}_2 - \rho^2)}{(1 + \tilde{N}_1)(1 + \tilde{N}_2 - \rho^2) + \tilde{N}_1 \rho} = \frac{N_1}{1 + N_1}$$
$$= D_1,$$

we have $\tilde{N}_1 \geq N_1$ as $\tilde{N}_1 \rho \geq 0$ and $1 + \tilde{N}_2 - \rho^2 \geq 0$. Now, consider

$$\frac{\tilde{D}_2}{\tilde{D}_1} = \frac{\tilde{N}_2(1 + \tilde{N}_1 - \rho^2)}{\tilde{N}_1(1 + \tilde{N}_2 - \rho^2)} = \frac{\frac{1}{\tilde{N}_1} + \frac{1}{1-\rho^2}}{\frac{1}{\tilde{N}_2} + \frac{1}{1-\rho^2}} \overset{(a)}{\leq} \frac{\frac{1}{N_1} + \frac{1}{1-\rho^2}}{\frac{1}{N_2} + \frac{1}{1-\rho^2}} = \frac{D_2}{D_1},$$

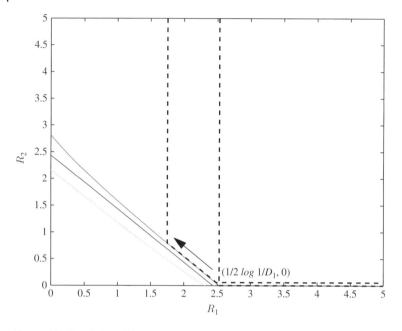

Figure 4.5 $(1 - D_2) \le \rho^2(1 - D_1)$: $R_1(D_1) \cap R_2(D_2) \cap R_s(D_1, D_2) = R_1(D_1)$. Corner point of the boundary is achieved by picking $N_1 = \frac{D_1}{1-D_1}$ and $N_2 = \infty$. Other boundary points can be achieved by picking $(\tilde{N}_1, \tilde{N}_2)$ such that $\tilde{D}_1 = \frac{\tilde{N}_1(1+\tilde{N}_2-\rho^2)}{(1+\tilde{N}_1)(1+\tilde{N}_2)-\rho^2} = \frac{N_1}{1+N_1} = D_1$.

where (a) is due to $\tilde{N}_1 \ge N_1$ and $\tilde{N}_2 \le N_2$. Since $\tilde{D}_1 = D_1$, $\tilde{D}_2 \le D_2$ and the distortion constraint for X_2 is also satisfied. In summary, all boundary points of $R_1(D_1)$ can be achieved through selecting some $\tilde{N}_2 \in [0, \infty)$ and repeating the above procedure. Finally, the rate pairs in the interior of $R_1(D_1)$ can be achieved by simply padding R_1 to the corresponding boundary point to the left of the target point.

Case 2: $(1 - D_2) > \rho^2(1 - D_1)$

For this case, the three regions are not subsets of one another. Moreover, there is a valid solution of (N_1, N_2) for (4.16) and (4.17). The corner points $((I(X_1; U_1|U_2), I(X_2; U_2))$ and $(I(X_1; U_1), I(X_2; U_2|U_1))$ corresponding to this (N_1, N_2) achieve the sum rate bound $R_1 + R_2 = I(X_1, X_2; U_1, U_2)$. Moreover, the upper left corner point falls on the boundary of $R_1(D_1)$ and the lower right corner point falls on the boundary of $R_2(D_2)$. Therefore, we can immediately show that the corner points and convex combination between the two points can be achieved by the chosen N_1 and N_2. Now, for other boundary points of $R_1(D_1)$ we can choose $\tilde{N}_1 \ge N_1$ and $\tilde{N}_2 \le N_2$ such that

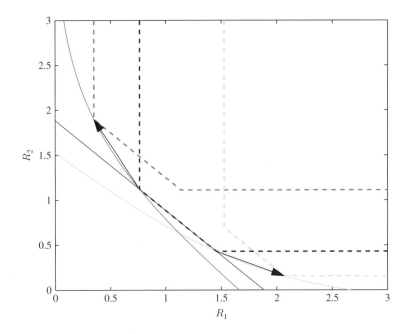

Figure 4.6 $(1 - D_2) > \rho^2(1 - D_1)$. Corner points of the boundary are achieved by picking (N_1, N_2) satisfying (4.16) and (4.17) (black dashed line boundary in the middle). Other boundary points can be achieved by picking some $(\tilde{N}_1, \tilde{N}_2)$ such that $D_1 = \frac{\tilde{N}_1(1+\tilde{N}_2-\rho^2)}{(1+\tilde{N}_1)(1+\tilde{N}_2)-\rho^2} = D_1$ (darker gray dashed line boundary on the left) and some other $(\tilde{N}_1, \tilde{N}_2)$ such that $D_2 = \frac{\tilde{N}_2(1+\tilde{N}_1-\rho^2)}{(1+\tilde{N}_1)(1+\tilde{N}_2)-\rho^2} = D_2$ (lighter gray dashed line boundary on the right).

$\tilde{D}_1 = \frac{\tilde{N}_1(1+\tilde{N}_1-\rho^2)}{(1+\tilde{N}_1)(1+\tilde{N}_2)-\rho^2} = \frac{N_1(1+N_2-\rho^2)}{(1+N_1)(1+N_2)-\rho^2} = D_1$ is invariant. We can also show that $\tilde{D}_2 \leq D_2$ just as in Case 1. Moreover, the upper left corner of the corresponding achievable region will fall on the boundary of $\mathcal{R}_1(D_1)$, as shown in Figure 4.6. Therefore, all boundary points of $\mathcal{R}_1(D_1)$ can be achieved by repeating the above argument and selecting different $(\tilde{N}_1, \tilde{N}_2)$.

By the same token, we can show that the boundary points of $\mathcal{R}_2(D_2)$ can be achieved by the lower left corner of an achievable region for some adjusted (N_1, N_2) pair with D_2 kept constant. Finally, the interior rate pairs can be achieved by rate padding, as in Case 1.

4.3.2 Converse Proof of Gaussian Multiterminal Source Coding Problem with Two Sources

We will use the matrix inversion formula several times in this proof, therefore we will include it as a lemma below.

Lemma 4.3 Matrix inversion formula

$$(A - F_{AC}CF_{CA})^{-1} = A^{-1} + A^{-1}F_{AC}(C^{-1} - F_{CA}A^{-1}F_{AC})^{-1}F_{CA}A^{-1}$$

Proof. Applying row operations on $\begin{pmatrix} A & F_{AC} & I & 0 \\ F_{CA} & C & 0 & I \end{pmatrix}$, we have

$$\begin{pmatrix} A & F_{AC} & I & 0 \\ F_{CA} & C & 0 & I \end{pmatrix} \to \begin{pmatrix} I & A^{-1}F_{AC} & A^{-1} & 0 \\ F_{CA} & C & 0 & I \end{pmatrix}$$

$$\to \begin{pmatrix} I & A^{-1}F_{AC} & A^{T} & 0 \\ 0 & C - F_{CA}A^{-1}F_{AC} & -F_{CA}A^{-1} & I \end{pmatrix}$$

$$\to \begin{pmatrix} I & A^{-1}F_{AC} & A^{-1} & 0 \\ 0 & I & -(C - F_{CA}A^{-1}F_{AC})^{-1}F_{CA}A^{-1} & (C - F_{CA}A^{-1}F_{AC})^{-1} \end{pmatrix}$$

$$\to \begin{pmatrix} I & 0 & A^{-1} + A^{-1}F_{AC}(C - F_{CA}A^{-1}F_{AC})^{-1}F_{CA}A^{-1} & -A^{-1}F_{AC}(C - F_{CA}A^{-1}F_{AC})^{-1} \\ 0 & I & -(C - F_{CA}A^{-1}F_{AC})^{-1}F_{CA}A^{-1} & (C - F_{CA}A^{-1}F_{AC})^{-1} \end{pmatrix}.$$

On the other hand,

$$\begin{pmatrix} A & F_{AC} & I & 0 \\ F_{CA} & C & 0 & I \end{pmatrix} \to \begin{pmatrix} A & F_{AC} & I & 0 \\ C^{-1}F_{CA} & I & 0 & C^{-1} \end{pmatrix}$$

$$\to \begin{pmatrix} A - F_{AC}C^{-1}F_{CA} & 0 & I & -F_{AC}C^{-1} \\ C^{-1}F_{CA} & I & 0 & C^{-1} \end{pmatrix}$$

$$\to \begin{pmatrix} I & 0 & (A - F_{AC}C^{-1}F_{CA})^{-1} & -(A - F_{AC}C^{-1}F_{CA})F_{AC}C^{-1} \\ C^{-1}F_{CA} & I & 0 & C^{-1} \end{pmatrix}$$

$$\to \begin{pmatrix} I & 0 & (A - F_{AC}C^{-1}F_{CA})^{-1} & -(A - F_{AC}C^{-1}F_{CA})F_{AC}C^{-1} \\ 0 & I & -C^{-1}F_{CA}(A - F_{AC}C^{-1}F_{CA})^{-1} & C^{-1} + C^{-1}F_{CA}(A - F_{AC}C^{-1}F_{CA})^{-1}F_{AC}C^{-1} \end{pmatrix}.$$

Therefore, $\begin{pmatrix} A & F_{AC} \\ F_{CA} & C \end{pmatrix}^{-1} =$

$$\begin{pmatrix} A^{-1} + A^{-1}F_{AC}(C - F_{CA}A^{-1}F_{AC})^{-1}F_{CA}A^{-1} & -A^{-1}F_{AC}(C - F_{CA}A^{-1}F_{AC})^{-1} \\ -(C - F_{CA}A^{-1}F_{AC})^{-1}F_{CA}A^{-1} & (C - F_{CA}A^{-1}F_{AC})^{-1} \end{pmatrix}$$

$$= \begin{pmatrix} (A - F_{AC}C^{-1}F_{CA})^{-1} & -(A - F_{AC}C^{-1}F_{CA})F_{AC}C^{-1} \\ -C^{-1}F_{CA}(A - F_{AC}C^{-1}F_{CA})^{-1} & C^{-1} + C^{-1}F_{CA}(A - F_{AC}C^{-1}F_{CA})^{-1}F_{AC}C^{-1} \end{pmatrix}.$$

Comparing the upper left element completes the proof. □

Bounds for R_1 and R_2

$$nR_1 \geq H(M_1) \geq H(M_1|M_2)$$

$$\overset{(a)}{=} I(M_1; X_1^n|M_2) + H(M_1|X_1^n, M_2)^{\to 0}$$

$$= h(X_1^n|M_2) - h(X_1^n|M_1, M_2)$$

$$\overset{(b)}{=} h(X_1^n|M_2) - h(X_1^n|M_1, M_2, \hat{X}_1^n)$$

$$\geq h(X_1^n|M_2) - h(X_1^n|\hat{X}_1^n)$$

$$= h(X_1^n|M_2) - \sum_{i=1}^{n} h(X_{1,i}|\hat{X}_1^n, X^{i-1})$$

$$= h(X_1^n|M_2) - \sum_{i=1}^{n} h(X_{1,i}|\hat{X}_{1,i})$$

$$= h(X_1^n|M_2) - \sum_{i=1}^{n} h(X_{1,i} - \hat{X}_{1,i}|\hat{X}_{1,i})$$

$$\geq h(X_1^n|M_2) - \sum_{i=1}^{n} h(X_{1,i} - \hat{X}_{1,i})$$

$$\overset{(c)}{\geq} h(X_1^n|M_2) - n\left(\frac{1}{n}\sum_{i=1}^{n} h_g(D_{1,i})\right)$$

$$\overset{(d)}{\geq} h(X_1^n|M_2) - nh_g\left(\frac{1}{n}\sum_{i=1}^{n} D_{1,i}\right)$$

$$= h(X_1^n|M_2) - \frac{n}{2}\log(2\pi e D_1),$$

where (a) is because M_1 is deterministic given X_1^n, (b) is because $\hat{X}_1^n$ is deterministic given M_1 and M_2, (c) is because the Gaussian entropy function $h_g(D_{1,i}) \triangleq \frac{1}{2}\log(2\pi e D_{1,i})$ results in the largest entropy for a fixed variance, and (d) is because $h_g(\cdot)$ is concave and because of Jensen's inequality. Therefore, we have

$$\frac{h(X_1^n|M_2)}{n} \leq R_1 + \frac{1}{2}\log 2\pi e D_1 \tag{4.22}$$

Moreover,

$$nR_2 \geq H(M_2) \geq I(M_2; X_2^n) = h(X_2^n) - h(X_2^n|M_2)$$

$$\geq \frac{n}{2}\log(2\pi e) - h(X_2^n|M_2). \tag{4.23}$$

For a pair of joint Gaussian variables X_1 and X_2, we can always write $X_1 = \rho X_2 + V$ with variable V independent of X_2. Note that $E[X_1 X_2] = E[\rho X_2 X_2] = \rho$ as desired. Since $E[X_1 X_1] = \rho^2 E[X_2 X_2] + E[V^2] = 1$, we need

$$E[V^2] = (1 - \rho^2). \tag{4.24}$$

Since the source is memoryless, $V_i \perp\!\!\!\perp X_{2,i}$ for all i implies $V^n \perp\!\!\!\perp X_2^n$, and since M_2 and ρX_2^n are just functions of X_2^n, we have $V^n \perp\!\!\!\perp (\rho X_2^n, M_2)$, which in turn implies $V^n \perp\!\!\!\perp \rho X_2^n|M_2$ from Fact A.2. Therefore we can apply conditional entropy power inequality on $X_1^n = \rho X_2^n + V^n$ and obtain

$$2^{\frac{2}{n}h(X_1^n|M_2)} \geq 2^{\frac{2}{n}h(\rho X_2^n|M_2)} + 2^{\frac{2}{n}h(V^n|M_2)}$$

$$=\rho^2 2^{\frac{2}{n}h(X_2^n|M_2)} + 2\pi e(1-\rho^2)$$

$$\overset{(a)}{\geq} \rho^2 2^{\frac{2}{n}(-nR_2+\frac{n}{2}\log(2\pi e))} + 2\pi e(1-\rho^2)$$

$$=2\pi e(\rho^2 2^{-2R_2} + 1 - \rho^2),$$

where (a) is due to (4.23).

Therefore,

$$\frac{h(X_1^n|M_2)}{n} \geq \frac{1}{2}\log 2\pi e(\rho^2 2^{-2R_2} + 1 - \rho^2). \tag{4.25}$$

Combining this with (4.22), we have

$$R_1 \geq \frac{1}{2}\log \frac{1 + \rho^2 + \rho^2 - 2^{-2R_1}}{D_1}. \tag{4.26}$$

This essentially ensures that an achievable (R_1, R_2) pair has to fall into $\mathcal{R}_1(D_1)$. Similarly, we can prove that an achievable pair has to fall into $\mathcal{R}_2(D_2)$ as well.

Collaborative Lower Bound

For the third region $\mathcal{R}_s(D_1, D_2)$, let us write the reconstruction error $\mathbf{W}_i = \mathbf{X}_i - \hat{\mathbf{X}}_i$ and the average covariance $\Sigma_D = \frac{1}{n}\sum_{i=1}^n \Sigma_{\mathbf{W}_i \mathbf{W}_i}$. Now, let us first consider the collaborative lower bound assuming the two encoders cooperate. We have

$$n(R_1 + R_2) \geq H(M_1, M_2)$$

$$\geq I(X_1^n, X_2^n; M_1, M_2)$$

$$= h(X_1^n, X_2^n) - h(X_1^n, X_2^n|M_1, M_2)$$

$$= \sum_{i=1}^n h(X_{1,i}, X_{2,i}) - h(X_1^n, X_2^n|M_1, M_2)$$

$$= \frac{n}{2}\log(2\pi e)^2|\Sigma_{\mathbf{XX}}| - h(X_1^n, X_2^n|M_1, M_2)$$

$$= \frac{n}{2}\log(2\pi e)^2|\Sigma_{\mathbf{XX}}| - \sum_{i=1}^n h(X_{1,i}, X_{2,i}|M_1, M_2, X_1^{i-1}, X_2^{i-1})$$

$$\geq \frac{n}{2}\log(2\pi e)^2|\Sigma_{\mathbf{XX}}| - \sum_{i=1}^n h(X_{1,i}, X_{2,i}|M_1, M_2)$$

$$\overset{(a)}{=} \frac{n}{2}\log(2\pi e)^2|\Sigma_{\mathbf{XX}}| - \sum_{i=1}^n h(X_{1,i}, X_{2,i}|M_1, M_2, \hat{X}_{1,i}, \hat{X}_{2,i})$$

$$\geq \frac{n}{2}\log(2\pi e)^2|\Sigma_{\mathbf{XX}}| - \sum_{i=1}^n h(X_{1,i}, X_{2,i}|\hat{X}_{1,i}, \hat{X}_{2,i})$$

$$= \frac{n}{2}\log(2\pi e)^2|\Sigma_{\mathbf{XX}}| - \sum_{i=1}^n h(\underbrace{X_{1,i} - \hat{X}_{1,i}, X_{2,i} - \hat{X}_{2,i}}_{W_i}|\hat{X}_{1,i}, \hat{X}_{2,i})$$

$$\geq \frac{n}{2} \log (2\pi e)^2 |\Sigma_{XX}| - \sum_{i=1}^{n} h(\mathbf{W}_i)$$

$$\overset{(b)}{\geq} \frac{n}{2} \log (2\pi e)^2 |\Sigma_{XX}| - \sum_{i=1}^{n} \frac{1}{2} \log (2\pi e)^2 |\Sigma_{\mathbf{W}_i \mathbf{W}_i}|$$

$$\overset{(c)}{\geq} \frac{n}{2} \log (2\pi e)^2 |\Sigma_{XX}| - \frac{n}{2} \log (2\pi e)^2 \left| \frac{1}{n} \sum_{i=1}^{n} \Sigma_{\mathbf{W}_i \mathbf{W}_i} \right|$$

$$= \frac{n}{2} \log \frac{|\Sigma_{XX}|}{|\Sigma_D|} \geq \frac{n}{2} \log \frac{|\Sigma_{XX}|}{|S(\theta)|}, \tag{4.27}$$

where (a) is because $\hat{X}_1^n$ and $\hat{X}_2^n$ are functions of M_1 and M_2, (b) is because Gaussian distribution has the highest entropy for a fixed covariance, and (c) is due to the concavity of $\log \det(\cdot)$ and Jensen's inequality.

Since $|\Sigma_{XX}| = \left| \begin{pmatrix} 1 & \rho \\ \rho & 1 \end{pmatrix} \right| = 1 - \rho^2$ and $|S(\theta)| = \left| \begin{pmatrix} D_1 & \theta \sqrt{D_1 D_2} \\ \theta \sqrt{D_1 D_2} & D_2 \end{pmatrix} \right| = D_1 D_2 (1 - \theta^2)$,

$$R_1 + R_2 \geq \frac{1}{2} \log \frac{1 - \rho^2}{D_1 D_2 (1 - \theta^2)}. \tag{4.28}$$

μ-sum Bound

To obtain a tight bound for the sum rate, we need to consider another bound besides the collaborative bound. Let the μ-sum Y be given by

$$Y = \mu_1 X_1 + \mu_2 X_2 + Z = \mu^T \mathbf{X} + Z,$$

where $Z \sim \mathcal{N}(0, N)$ is independent of X_1 and X_2. Then

$$n(R_1 + R_2) \geq H(M_1, M_2) \geq I(M_1, M_2; X^n, Y^n)$$

$$= h(\mathbf{X}^n, Y^n) - h(\mathbf{X}^n, Y^n | M_1, M_2)$$

$$= h(\mathbf{X}^n, Y^n) - h(Y^n | M_1, M_2) - h(\mathbf{X}^n | Y^n, M_1, M_2)$$

$$= \sum_{i=1}^{n} h(\mathbf{X}_i, Y_i) - \sum_{i=1}^{n} h(Y_i | M_1, M_2, Y^{i-1})$$

$$- \sum_{i=1}^{n} h(\mathbf{X}_i | Y^n, M_1, M_2, \mathbf{X}^{i-1})$$

$$\geq \sum_{i=1}^{n} h(\mathbf{X}_i, Y_i) - \sum_{i=1}^{n} h(Y_i | M_1, M_2) - \sum_{i=1}^{n} h(\mathbf{X}_i | Y^n, M_1, M_2) \tag{4.29}$$

For the first summation,

$$
\begin{aligned}
\sum_{i=1}^{n} h(\mathbf{X}_i, Y_i) &= \sum_{i=1}^{n} \frac{1}{2} \log{(2\pi e)^3} \left| E\left[\begin{pmatrix} \mathbf{X}_i \\ Y_i \end{pmatrix} \begin{pmatrix} \mathbf{X}_i^T & Y_i^T \end{pmatrix} \right] \right| \\
&= \sum_{i=1}^{n} \frac{1}{2} \log{(2\pi e)^3} \left| \begin{pmatrix} \Sigma_{\mathbf{XX}} & \Sigma_{\mathbf{XX}}^T \mu \\ \mu^T \Sigma_{\mathbf{XX}} & \mu^T \Sigma_{\mathbf{XX}} \mu + N \end{pmatrix} \right| \\
&= \frac{n}{2} \log{(2\pi e)^3} \left| \begin{pmatrix} \Sigma_{\mathbf{XX}} & \Sigma_{\mathbf{XX}}^T \mu \\ 0 & N \end{pmatrix} \right| \\
&= \frac{n}{2} \log((2\pi e)^3 N |\Sigma_{\mathbf{XX}}|) \quad\quad (4.30)
\end{aligned}
$$

For the second summation,

$$
\begin{aligned}
\sum_{i=1}^{n} h(Y_i | M_1, M_2) &\le \sum_{i=1}^{n} h(Y_i | M_1, M_2, \hat{\mathbf{X}}_i) \\
&\le \sum_{i=1}^{n} h(Y_i - \mu^T \hat{\mathbf{X}}_i | M_1, M_2, \hat{\mathbf{X}}_i) \\
&\le \sum_{i=1}^{n} h(Y_i - \mu^T \hat{\mathbf{X}}_i) \\
&\le \sum_{i=1}^{n} h(\mu^T(\mathbf{X}_i - \hat{\mathbf{X}}_i) + Z) \\
&\le n \left(\frac{1}{n} \sum_{i=1}^{n} \frac{1}{2} \log(2\pi e)(\mu^T \Sigma_{\mathbf{W}_i \mathbf{W}_i} \mu + N) \right) \\
&\overset{(a)}{\le} \frac{n}{2} \log \left((2\pi e) \left(\mu^T \left(\frac{1}{n} \sum_{i=1}^{n} \Sigma_{\mathbf{W}_i \mathbf{W}_i} \right) \mu + N \right) \right) \\
&= \frac{n}{2} \log((2\pi e)(\mu^T \Sigma_D \mu + N)) \\
&\le \frac{n}{2} \log((2\pi e)(\mu^T S(\theta) \mu + N)), \quad\quad (4.31)
\end{aligned}
$$

where (a) is due to the concavity[3] of $\log(\mu^T A \mu + N)$ as a function of A and Jensen's inequality.

For the last summation, first note that the equality in (4.27) is achieved when $\mathbf{W}_i \perp\!\!\!\perp \hat{\mathbf{X}}_i$, which in turn implies

$$
\Sigma_{\hat{\mathbf{X}}_i \hat{\mathbf{X}}_i} = \Sigma_{\mathbf{X}_i \mathbf{X}_i} - \Sigma_{\mathbf{W}_i \mathbf{W}_i}.
$$

---

3 Because $\mu^T A \mu + N$ is a linear function in terms of A
$(\mu^T (\alpha A + (1 - \alpha)B)\mu + N = \alpha \mu^T A \mu + N + (1 - \alpha)\mu^T B \mu + N)$ and log function is concave.

Now, let us define $\hat{\mathbf{Y}}_i = \begin{pmatrix} Y_i \\ \hat{\mathbf{X}}_i \end{pmatrix}$ and we have

$$\Sigma_{\mathbf{X}_i\hat{\mathbf{Y}}_i} = \begin{pmatrix} \Sigma_{\mathbf{X}_i\mathbf{X}_i}\mu & \Sigma_{\mathbf{X}_i\mathbf{X}_i} - \Sigma_{\mathbf{W}_i\mathbf{W}_i} \end{pmatrix}$$

and

$$\Sigma_{\hat{\mathbf{Y}}_i\hat{\mathbf{Y}}_i} = \begin{pmatrix} E[Y_iY_i] & E[Y_i\hat{\mathbf{X}}_i^T] \\ E[\hat{\mathbf{X}}_iY_i^T] & E[\hat{\mathbf{X}}_i\hat{\mathbf{X}}_i^T] \end{pmatrix}$$
$$= \begin{pmatrix} \mu^T\Sigma_{\mathbf{X}_i\mathbf{X}_i}\mu + N & \mu^T(\Sigma_{\mathbf{X}_i\mathbf{X}_i} - \Sigma_{\mathbf{W}_i\mathbf{W}_i}) \\ (\Sigma_{\mathbf{X}_i\mathbf{X}_i} - \Sigma_{\mathbf{W}_i\mathbf{W}_i})\mu & \Sigma_{\mathbf{X}_i\mathbf{X}_i} - \Sigma_{\mathbf{W}_i\mathbf{W}_i} \end{pmatrix}.$$

If we consider $\tilde{\mathbf{X}}_i = \underbrace{\left(\frac{1}{n}\sum_{i=1}^{n}\Sigma_{\mathbf{X}_i\hat{\mathbf{Y}}_i}\right)\left(\frac{1}{n}\sum_{i=1}^{n}\Sigma_{\hat{\mathbf{Y}}_i\hat{\mathbf{Y}}_i}\right)^{-1}}_{A}\hat{\mathbf{Y}}_i$ as an estimate[4] of $\mathbf{X}_i$

given $\hat{\mathbf{Y}}_i$, then we have

$$\frac{1}{n}\sum_{i=1}^{n}\Sigma_{\mathbf{X}_i|M_1,M_2,Y^n} \preceq \frac{1}{n}\sum_{i=1}^{n}\Sigma_{\mathbf{X}_i-\hat{\mathbf{X}}_i,\mathbf{X}_i-\hat{\mathbf{X}}_i}.$$

However,

$$\frac{1}{n}\sum_{i=1}^{n}\Sigma_{\mathbf{X}_i-\hat{\mathbf{X}}_i,\mathbf{X}_i-\hat{\mathbf{X}}_i}$$
$$=\frac{1}{n}\sum_{i=1}^{n}E[(\mathbf{X}_i - A\hat{\mathbf{Y}}_i)(\mathbf{X}_i - A\hat{\mathbf{Y}}_i)^T]$$
$$=\frac{1}{n}\sum_{i=1}^{n}\Sigma_{\mathbf{X}_i\mathbf{X}_i} - \frac{1}{n}\sum_{i=1}^{n}E[\mathbf{X}_i\hat{\mathbf{Y}}_i^T]A^T - A\frac{1}{n}\sum_{i=1}^{n}E[\hat{\mathbf{Y}}_i\mathbf{X}_i^T]$$
$$+A\frac{1}{n}\sum_{i=1}^{n}E[\hat{\mathbf{Y}}_i\hat{\mathbf{Y}}_i^T]A^T$$
$$=\Sigma_{\mathbf{XX}} - \left(\frac{1}{n}\sum_{i=1}^{n}\Sigma_{\mathbf{X}_i\hat{\mathbf{Y}}_i}\right)\left(\frac{1}{n}\sum_{i=1}^{n}\Sigma_{\hat{\mathbf{Y}}_i\hat{\mathbf{Y}}_i}\right)\left(\frac{1}{n}\sum_{i=1}^{n}\Sigma_{\hat{\mathbf{Y}}_i\mathbf{X}_i}\right)$$
$$=\Sigma_{\mathbf{XX}} - \Sigma_{\mathbf{X}\hat{\mathbf{Y}}}\Sigma_{\hat{\mathbf{Y}}\hat{\mathbf{Y}}}^{-1}\Sigma_{\hat{\mathbf{Y}}\mathbf{X}},$$

where $\Sigma_{\mathbf{X}\hat{\mathbf{Y}}} = \frac{1}{n}\sum_{i=1}^{n}\Sigma_{\mathbf{X}_i\hat{\mathbf{Y}}_i} = \left(\Sigma_{\mathbf{XX}}\mu \ \ \Sigma_{\mathbf{XX}} - \frac{1}{n}\sum_{i=1}^{n}\Sigma_{\mathbf{W}_i\mathbf{W}_i}\right) = \left(\Sigma_{\mathbf{XX}}\mu \ \ \Sigma_{\mathbf{XX}} - \Sigma_D\right)$

and $\Sigma_{\hat{\mathbf{Y}}\hat{\mathbf{Y}}} = \frac{1}{n}\sum_{i=1}^{n}\Sigma_{\hat{\mathbf{Y}}_i\hat{\mathbf{Y}}_i} = \begin{pmatrix} \mu^T\Sigma_{\mathbf{XX}}\mu + N & \mu^T\left(\Sigma_{\mathbf{XX}} - \frac{1}{n}\sum_{i=1}^{n}\Sigma_{\mathbf{W}_i\mathbf{W}_i}\right) \\ \left(\Sigma_{\mathbf{XX}} - \frac{1}{n}\sum_{i=1}^{n}\Sigma_{\mathbf{W}_i\mathbf{W}_i}\right)\mu & \Sigma_{\mathbf{XX}} - \frac{1}{n}\sum_{i=1}^{n}\Sigma_{\mathbf{W}_i\mathbf{W}_i} \end{pmatrix}$

---

4 Note that $\Sigma_{\mathbf{X}_i\hat{\mathbf{Y}}_i}\Sigma_{\hat{\mathbf{Y}}_i\hat{\mathbf{Y}}_i}^{-1}\hat{\mathbf{Y}}_i$ is the best linear estimate of $\mathbf{X}_i$. Virtually A is obtained from averaging the statistics over the n inputs of $\mathbf{X}$ and Y.

$$= \begin{pmatrix} \mu^T \Sigma_{\mathbf{XX}} \mu + N & \mu^T(\Sigma_{\mathbf{XX}} - \Sigma_D) \\ (\Sigma_{\mathbf{XX}} - \Sigma_D)\mu & \Sigma_{\mathbf{XX}} - \Sigma_D \end{pmatrix}. \text{ Thus,}$$

$$\frac{1}{n} \sum_{i=1}^{n} \Sigma_{\mathbf{X}_i - \hat{\mathbf{X}}_i, \mathbf{X}_i - \hat{\mathbf{X}}_i}$$

$$= \Sigma_{\mathbf{XX}} - (\Sigma_{\mathbf{XX}}\mu \; \Sigma_{\mathbf{XX}} - \Sigma_D) \begin{pmatrix} \mu^T \Sigma_{\mathbf{XX}} \mu + N & \mu^T(\Sigma_{\mathbf{XX}} - \Sigma_D) \\ (\Sigma_{\mathbf{XX}} - \Sigma_D)\mu & \Sigma_{\mathbf{XX}} - \Sigma_D \end{pmatrix} \begin{pmatrix} \mu^T \Sigma_{\mathbf{XX}} \\ \Sigma_{\mathbf{XX}} - \Sigma_D \end{pmatrix}$$

$$= \underbrace{\Sigma_{\mathbf{XX}}}_{A^{-1}} - \Sigma_{\mathbf{XX}} \underbrace{(\mu I - \Sigma_{\mathbf{XX}}^{-1}\Sigma_D)}_{F_{AC}} \begin{pmatrix} \mu^T \Sigma_{\mathbf{XX}} \mu + N & \mu^T(\Sigma_{\mathbf{XX}} - \Sigma_D) \\ (\Sigma_{\mathbf{XX}} - \Sigma_D)\mu & \Sigma_{\mathbf{XX}} - \Sigma_D \end{pmatrix}$$

$$\underbrace{\left(I - \Sigma_D \Sigma_{\mathbf{XX}}^{-1} \right) \Sigma_{\mathbf{XX}}}_{F_{CA}}$$

$$= A^{-1} - A^{-1} F_{AC} \left(\begin{pmatrix} N & 0 \\ 0 & \Sigma_D - \Sigma_D \Sigma_{\mathbf{XX}}^{-1} \Sigma_D \end{pmatrix} + F_{CA} A^{-1} F_{AC} \right) F_{CA} A^{-1}$$

$$\overset{(a)}{=} \left(A + F_{AC} \begin{pmatrix} N & 0 \\ 0 & \Sigma_D - \Sigma_D \Sigma_{\mathbf{XX}}^{-1} \Sigma_D \end{pmatrix}^{-1} F_{CA} \right)^{-1}$$

$$= \left(\Sigma_{\mathbf{XX}}^{-1} + (\mu \; I - \Sigma_{\mathbf{XX}}^{-1}\Sigma_D) \begin{pmatrix} N^{-1} & 0 \\ 0 & (\Sigma_{\mathbf{XX}} - \Sigma_D \Sigma_{\mathbf{XX}}^{-1} \Sigma_D)^{-1} \end{pmatrix} \begin{pmatrix} \mu^T \\ I - \Sigma_D \Sigma_{\mathbf{XX}}^{-1} \end{pmatrix} \right)^{-1}$$

$$= \left(\Sigma_D^{-1} + \frac{\mu\mu^T}{N} \right)^{-1}$$

$$\overset{(b)}{\preceq} \left(S(\theta)^{-1} + \frac{\mu\mu^T}{N} \right)^{-1}$$

$$\overset{(c)}{=} S(\theta) - S(\theta)\mu(N + \mu^T S(\theta)\mu)\mu^T S(\theta),$$

where (a) and (c) are due to the matrix inversion formula (Lemma 4.3), and (b) is due to $\Sigma_D \preceq S(\theta)$.

Now if we pick $\mu = \left(1/\sqrt{D_1} \; 1/\sqrt{D_2} \right)^T$, $N + \mu^T S(\theta)\mu$ simplifies to $N + 2(1 + \theta)$. After some simple algebra we obtain

$$\frac{1}{n} \sum_{i=1}^{n} \Sigma_{\mathbf{X}_i - \hat{\mathbf{X}}_i, \mathbf{X}_i - \hat{\mathbf{X}}_i} \preceq S(\theta) - \frac{S(\theta)\mu\mu^T S(\theta)}{N + 2(1 + \theta)}$$

$$= \begin{pmatrix} (1 - \alpha)D_1 & (\theta - \alpha)\sqrt{D_1 D_2} \\ (\theta - \alpha)\sqrt{D_1 D_2} & (1 - \alpha)D_2 \end{pmatrix},$$

where

$$\alpha = \frac{(1 + \theta)^2}{N + 2(1 + \theta)}. \tag{4.32}$$

Moreover, if we insist $X_{1,i} \leftrightarrow Yi \leftrightarrow X_{2,i}$, we have $E[(X_{1,i} - E[X_{1,i}])(X_{2,i} - E[X_{2,i}])|Y^n, M_1, M_2] = E[X_{1,i} - E[X_{1,i}]|Y^n, M_1, M_2]E[X_{2,i} - E[X_{2,i}]|Y^n, M_1, M_2]$

$= 0$. Therefore, $\frac{1}{n} \sum_{i=1}^{n} \Sigma_{X_i|M_1,M_2,Y^n}$ will be a diagonal matrix. From Corollary D.1, we need $\rho_{X_1X_2} = \rho_{X_1Y}\rho_{X_2Y}$, which implies $\rho = \frac{(\mu_1+\mu_2\rho)(\mu_2+\mu_1\rho)}{\mu_1^2+\mu_2^2+2\mu_1\mu_2\rho+N}$ and thus we have

$$N = \frac{1-\rho^2}{\rho\sqrt{D_1D_2}}. \tag{4.33}$$

Now, since

$$\frac{1}{n} \sum_{i=1}^{n} \Sigma_{X_i|M_1,M_2,Y^n} \preceq \frac{1}{n} \sum_{i=1}^{n} \Sigma_{X_i-\hat{X}_i,X_i-\hat{X}_i}$$

$$\preceq \begin{pmatrix} (1-\alpha)D_1 & (\theta-\alpha)\sqrt{D_1D_2} \\ (\theta-\alpha)\sqrt{D_1D_2} & (1-\alpha)D_2 \end{pmatrix}$$

and $\frac{1}{n} \sum_{i=1}^{n} \Sigma_{X_i|M_1,M_2,Y^n}$ is a diagonal matrix, we have

$$\left| \frac{1}{n} \sum_{i=1}^{n} \Sigma_{X_i|M_1,M_2,Y^n} \right| \leq D_1D_2(1+\theta-2\alpha)^2 = \left(\frac{N(1+\theta)}{2(1+\theta)+N} \right)^2 D_1D_2 \tag{4.34}$$

from the following lemma and (4.32).

Lemma 4.4 If a 2×2 diagonal matrix D satisfies $0 \preceq D \preceq \begin{pmatrix} a & c \\ c & b \end{pmatrix}$, then $|D| \leq (\sqrt{ab} + c)^2$.

Proof. Let $D = \begin{pmatrix} d_1 & 0 \\ 0 & d_2 \end{pmatrix}$. Then, we have

$$\begin{pmatrix} d_1 & 0 \\ 0 & d_2 \end{pmatrix} \preceq \begin{pmatrix} a & c \\ c & b \end{pmatrix}$$

$$\Rightarrow \begin{pmatrix} \sqrt{a} & 0 \\ 0 & -\sqrt{b} \end{pmatrix} \begin{pmatrix} d_1 & 0 \\ 0 & d_2 \end{pmatrix} \begin{pmatrix} \sqrt{a} & 0 \\ 0 & -\sqrt{b} \end{pmatrix}$$

$$\preceq \begin{pmatrix} \sqrt{a} & 0 \\ 0 & -\sqrt{b} \end{pmatrix} \begin{pmatrix} a & c \\ c & b \end{pmatrix} \begin{pmatrix} \sqrt{a} & 0 \\ 0 & -\sqrt{b} \end{pmatrix}$$

$$\Rightarrow \begin{pmatrix} \frac{d_1}{a} & 0 \\ 0 & \frac{d_2}{b} \end{pmatrix} \preceq \begin{pmatrix} 1 & -\frac{c}{\sqrt{ab}} \\ -\frac{c}{\sqrt{ab}} & 1 \end{pmatrix}.$$

We can complete the proof using the following lemma. $\square$

Lemma 4.5 If $0 \preceq \begin{pmatrix} d_1 & 0 \\ 0 & d_2 \end{pmatrix} \preceq \begin{pmatrix} a & c \\ c & a \end{pmatrix}$, then $(a-c) \geq \sqrt{d_1d_2}$.

Proof. From the given conditions, we have

$$\begin{cases} (a - d_1)(a - d_2) \geq c^2 \\ a \geq d_1 \\ a \geq d_2 \end{cases}$$

By the fact that arithmetic means are always larger than geometric means,

$$\left(a - \frac{d_1 + d_2}{2} \right)^2 \geq (a - d_1)(a - d_2) \geq c^2.$$

Since $a \geq (d_1 + d_2)/2$, we have $\left(a - \frac{d_1 + d_2}{2} \right) \geq c$ and use again the fact that arithmetic means are always larger than geometric means,

$$(a - c) \geq \frac{d_1 + d_2}{2} \geq \sqrt{d_1 d_2}.$$

$\square$

From (4.34), the final summation in (4.29) can be bounded by

$$
\begin{aligned}
\sum_{i=1}^{n} h(\mathbf{X}_i | Y^n, M_1, M_2) &= n \frac{1}{n} \sum_{i=1}^{n} \frac{1}{2} \log (2\pi e)^2 |\Sigma_{\mathbf{X}_i | Y_n, M_1, M_2}| \\
&\overset{(a)}{\leq} \frac{n}{2} \log (2\pi e)^2 \left| \frac{1}{n} \sum_{i=1}^{n} \Sigma_{\mathbf{X}_i | Y_n, M_1, M_2} \right| \\
&\leq \frac{n}{2} \log \left((2\pi e)^2 D_1 D_2 \left(\frac{N(1 + \theta)}{2(1 + \theta) + N} \right)^2 \right), \quad (4.35)
\end{aligned}
$$

where (a) is due to the concavity of the log det$(\cdot)$ function and Jensen's inequality.

Substituting (4.30), (4.31), (4.35), and (4.33) into (4.29), we finally obtain the μ-sum bound as

$$
\begin{aligned}
n(R_1 + R_2) &\geq \frac{n}{2} \frac{\log |\Sigma_{\mathbf{XX}}| (2(1 + \theta) + N)}{D_1 D_2 N (1 + \theta)^2} \\
&\geq \frac{n}{2} \log \frac{(1 - \rho^2) + 2\rho \sqrt{D_1 D_2}(1 + \theta)}{D_1 D_2 (1 + \theta)^2}. \quad (4.36)
\end{aligned}
$$

Combining this with the collaborative bound in (4.28), we have

$$
R_1 + R_2 \geq \min_{\theta \in [-1,1]} \max \left\{ \begin{array}{l} \frac{1}{2} \log \frac{1 - \rho^2}{D_1 D_2 (1 - \theta^2)}, \\ \frac{1}{2} \log \frac{(1 - \rho^2) + 2\rho \sqrt{D_1 D_2}(1 + \theta)}{D_1 D_2 (1 + \theta)^2} \end{array} \right\}. \quad (4.37)
$$

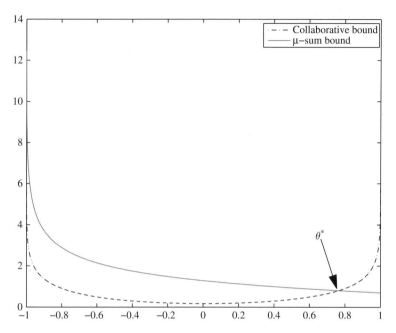

Figure 4.7 The sum rate bounds for the Gaussian multiterminal source coding problems $(D_1 = 0.3, D_2 = 0.5, \rho = 0.9)$.

Figure 4.7 shows the collaborative sum bound and μ-sum bound as functions of θ. One can see that the optimum θ^* simply occurs when the two bounds overlap. Solving

$$\frac{1}{2}\log\frac{1-\rho^2}{D_1D_2(1-\theta^2)} = \frac{1}{2}\log\frac{(1-\rho^2)+2\rho\sqrt{D_1D_2}(1+\theta)}{D_1D_2(1+\theta)^2}. \tag{4.38}$$

gives $\theta^* = -\frac{(1-\rho^2)\pm\sqrt{(1-\rho^2)^2+4\rho^2D_1D_2}}{2\rho\sqrt{D_1D_2}}$. As the solution with $\pm$ as minus sign has magnitude larger than 1, the only possible solution is

$$\theta^* = -\frac{(1-\rho^2)+\sqrt{(1-\rho^2)^2+4\rho^2D_1D_2}}{2\rho\sqrt{D_1D_2}}.$$

Substituting θ^* back to (4.28), we get $R_1+R_2 \geq \frac{1}{2}\log\frac{(1-\rho^2)\beta(D_1,D_2)}{2D_1D_2}$. This concludes the converse proof.

Part II

Implementation

5

Slepian–Wolf Code Designs Based on Channel Coding

In this chapter we will investigate the practical design of SW coding based on channel coding principles. While the possibility for the practical implementation of SW coding was suggested by Wyner in 1974 [22], it was not until the late 1990s that SW coding was actually implemented.

As described in Chapter 2, the proof of SW coding is based on random binning. While the concept of binning is simple, it is extremely powerful. In essence, the source space is partitioned into bins and the encoder identifies the bin that a input source belongs to. The label of the bin instead of the source itself is passed to the decoder, which will then estimate the original source based on the bin information and the correlation among sources.

Essentially all practical SW coding can be interpreted using binning. The main practical difficulty is how one should partition the input source space into bins. In theory, random binning is asymptotically optimal, but decoding becomes extremely difficult when bins are constructed purely randomly. In practice, error correction coding, or simply channel coding, provides a flexible solution to this problem. We will focus on SW coding based on channel coding in this chapter and describe implementation of both *asymmetric* and *non-asymmetric* SW coding. We will also briefly describe SW coding based on distributed arithmetic coding, a completely different paradigm from the channel coding approach. Finally, we look into adaptive SW coding and discuss the latest development and trend of SW coding design.

5.1 Asymmetric SW Coding

We refer to asymmetric SW coding as the case when only one of the joint sources is compressed whereas the rest are transmitted directly to the decoder. The sources that are transmitted directly are used as side information for the source to be compressed. Asymmetric SW coding is therefore essentially iden-

Distributed Source Coding: Theory and Practice, First Edition. Shuang Wang, Yong Fang, and Samuel Cheng.
© 2017 John Wiley & Sons Ltd. Published 2017 by John Wiley & Sons Ltd.
Companion Website: www.wiley.com/go/cheng/dsctheoryandpractice

tical to lossless source coding with side information given to the decoder (but not to the encoder).

Example 5.1 *(Temperatures of two cities).* The observatories at Glasgow and Edinburgh both want to send their temperature data to London. Due to the close vicinity of the two cities, the temperature data should be highly correlated to each other. Assuming that the observatory at Edinburgh has compressed the data independent of the data from Glasgow (e.g., using Huffmann coding) and transmitted them to London already, the observatory at Glasgow can save transmission rate by taking advantage of the data from Edinburgh as side information using asymmetric SW coding even though the people in Glasgow do not know the data from Edinburgh.

5.1.1 Binning Idea

The amazing outcome of SW Coding Theorem (see Chapter 2) is that there is no loss of asymmetric SW coding comparing to the case when side information is also given to the encoding. In other words, as long as side information is given to the decoder, it doesn't matter if the side information is given to the encoder or not.

This may seem mysterious at first but it turns out that the basic idea of asymmetric SW coding is very simple. Basically, the alphabet of the source to be compressed, $\mathcal{X}$, will be partitioned into a number of bins. During encoding, the index of the bin that the input lies in will be transmitted to the decoder instead of the input itself. For example, $|\mathcal{X}| = 8$ and we partition $|\mathcal{X}|$ into four bins each with two elements. Therefore, without SW coding, we generally need $\log_2 8 = 3$ bits to convey the input to the decoder. With SW coding, we only need $\log_2 4 = 2$ bits.

Of course the decoder can't tell what is the actual input with the bin index alone, but the side information now comes to the rescue. Essentially, the decoder should pick the element within the bin that is "best" matched with the side information. Whether the side information matches with an element in a bin is usually assessed statistically.

To consolidate this idea, let us consider a simple numerical example that was introduced by Pradhan *et al.* [5].

Example 5.2 *(Compression of a 3-bit sequence with side information).* Consider two binary sequences of length 3 where the two sequences are correlated in such a way that they can differ by no more than 1 bit. Denote the two sequences as $\mathbf{X}$ and $\mathbf{Y}$, respectively. Assuming that $\mathbf{Y}$ is given to both the encoder and the decoder, apparently we can compress $\mathbf{X}$ into 2 bits since we can represent and transmit $\mathbf{X} - \mathbf{Y} \in \{[000]^T, [001]^T, [010]^T, [100]^T\}$ with $\log_2 4 = 2$ bits instead of transmitting $\mathbf{X}$ directly. At the decoder, given $\mathbf{X} - \mathbf{Y}$ and the side information $\mathbf{Y}$, $\mathbf{X}$ can be recovered as $(\mathbf{X} - \mathbf{Y}) + \mathbf{Y}$.

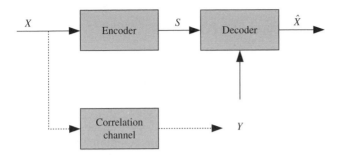

Figure 5.1 The block diagram of SW coding. Side information Y can be treated as the source X passing through a hypothetical channel.

Example 5.3 *(Asymmetric SW coding of a 3-bit sequence).* Continuing with Example 5.2, the side information $\mathbf{Y}$ is now given only to the decoder but not the encoder. So the encoder cannot compute $\mathbf{X} - \mathbf{Y}$ anymore. How can we still compress $\mathbf{X}$ into two bits?

Let us partition all possibilities of $\mathbf{X}$ into four bins as follows:

- Bin 0: $\{[000]^T, [111]^T\}$;
- Bin 1: $\{[001]^T, [110]^T\}$;
- Bin 2: $\{[010]^T, [101]^T\}$;
- Bin 3: $\{[100]^T, [011]^T\}$.

Now, let $\mathbf{X} = [001]^T$ and $\mathbf{Y} = [000]^T$. At the decoder, the index of the bin that $\mathbf{X} = [001]^T$ lies in, that is, 1, will be transmitted to the decoder. Since there are only four bins, we only need $\log_2 4 = 2$ bits to represent the bin index. From the bin index, the decoder knows that $\mathbf{X}$ is either $[001]^T$ or $[110]^T$. Since it is given that $\mathbf{X}$ and $\mathbf{Y}$ can differ no more than 1 bit we know that $\mathbf{X}$ is $[001]^T$.

5.1.2 Syndrome-based Approach

If we take a closer look at the binning scheme described in Example 5.3, we see that an element in each bin is assigned as the exact bit flip of the other. This is reasonable because it will be easier to identify the correct original input $\mathbf{X}$ from the side information $\mathbf{Y}$ if the elements in each bin are far away from each other. More precisely, we can consider each bin as a channel code and $\mathbf{X}$ passes through a hypothetical channel with channel output $\mathbf{Y}$ (see Figure 5.1). Then, asymmetric SW decoding will indeed closely resemble channel coding.

One remaining problem is how to assign the space of input into these channel-code-like bins efficiently. Wyner first suggested linear binning based on coset codes [22]. Basically, a linear block code is chosen and then elements with the same syndrome will be assigned to a same bin. More precisely, let H be a parity check matrix of any channel code. An input $\mathbf{x}$ will be assigned to the bin with index $\mathbf{s} = H\mathbf{x}$.

Hamming Binning

Let us now move to a more complicated example.

Example 5.4 *(Asymmetric SW coding of 7-bit sequences).* Assume that we have two binary sequences $\mathbf{X}$ and $\mathbf{Y}$ of length 7. As in Example 5.3, the two sequences can differ by no more than 1 bit. We assume that both $\mathbf{X}$ and $\mathbf{Y}$ are randomly and uniformly distributed. In other words, Pr $(X_i = 0) = Pr(X_i = 1) = Pr(Y_i = 0) = Pr(Y_i = 1) = 0.5, i = 1, 2, \cdots, 7$. Thus, $H(\mathbf{X}) = H(\mathbf{Y}) = 7$ bits, where $H(\cdot)$ denotes the entropy function. On the other hand, since $\mathbf{X}$ and $\mathbf{Y}$ differ by no more than 1 bit, $\mathbf{X} - \mathbf{Y} \in \{[0000000]^T, [0000001]^T, \cdots, [1000000]^T\}$ can only takes $8 = 2^3$ different values. Therefore, $H(\mathbf{X}|\mathbf{Y}) = 3$ bits and $H(\mathbf{X}, \mathbf{Y}) = H(\mathbf{Y}) + H(\mathbf{X}|\mathbf{Y}) = 10$ bits.

Similar to the previous examples, lossless compression is simple if joint encoding is allowed because we can compress $\mathbf{Y}$ with 7 bits first and then compress the difference $\mathbf{X} - \mathbf{Y}$ with 3 bits. At the decoder, both $\mathbf{Y}$ and $\mathbf{X} - \mathbf{Y}$ can be recovered and thus so as $\mathbf{X}$.

However, if the collaboration of encoders is not allowed, how can we compress $\mathbf{X}$ and $\mathbf{Y}$ into a total of 10 bits? As in the previous case, we would like to partition the space of $\mathbf{X}$ into bins. Since the target rate is 3 bits, ideally we only needs $2^3 = 8$ bins. But how?

Consider $H = \begin{bmatrix} 1001011 \\ 0101101 \\ 0010111 \end{bmatrix}$. We will partition the space of $\mathbf{X}$ by H such that $\mathbf{X}$ that belong to the same bin will have the same $H\mathbf{X}$. In other words, the bin index of $\mathbf{X}$ is equal to $H\mathbf{X}$.

A keen reader may notice that H is actually a possible parity check matrix of the (7,4)-Hamming code. From Appendix B we learn that any codeword of a Hamming code has a minimum Hamming distance of 3 from another codeword. Further, it is easy to verify that this distance property holds also for all other bins with nonzero syndrome. Therefore, any two elements in the same bin will be at least 3 bits away from each other. Thus, given the bin index, we expect perfect recovery of $\mathbf{x}$ since the side information $\mathbf{y}$ is no more than 1 Hamming distance away from $\mathbf{x}$. Let us summarize the encoding and decoding steps in the following.

SW Encoding

The encoder simply computes and passes the syndrome of the source $\mathbf{x}$ to the decoder.

SW Decoding

Unlike conventional channel decoding, SW decoding will recover the source as a code vector of the received syndrome (bin index) instead of a codeword. Let $\mathbf{s} = H\mathbf{x}$ be the syndrome of $\mathbf{x}$ and let $\mathbf{s}' = H\mathbf{y}$. Assuming that $\mathbf{s} - \mathbf{s}'$ is equivalent

$$\overbrace{}^{i-1}\ \overbrace{}^{7-i}$$

to the ith column of H, then $\mathbf{z} = [\,0 \cdots 0\,1\,0 \cdots 0\,]^T$ has syndrome $\mathbf{s} - \mathbf{s}'$. Thus $\hat{\mathbf{x}} = \mathbf{y} + \mathbf{z}$ provides a good estimate of $\mathbf{x}$ since it has syndrome $\mathbf{s}' + \mathbf{s} - \mathbf{s}' = \mathbf{s}$ and differs from $\mathbf{y}$ with no more than 1 bit.

Example 5.5 (Hamming binning). Let $\mathbf{x} = [1100010]^T$ and $\mathbf{y} = [1100011]^T$. $\mathbf{x}$ is compressed into the 3-bit syndrome $\mathbf{s} = H\mathbf{x} = [011]^T$. At the decoder, $\mathbf{y}$ is given and $\mathbf{s}' = H\mathbf{y} = [100]^T$. Hence, $\mathbf{s} - \mathbf{s}' = [111]^T$ and $\mathbf{z} = [0000001]^T$. Thus, $\hat{\mathbf{x}} = \mathbf{y} + \mathbf{z} = [1100010]^T = \mathbf{x}$ as desired.

It is interesting to point out that the code described above is *perfect*, which means that the total number of bits used to compress the source attains the possible minimum (i.e., 10 bits in this case). The notion of perfectness here is directly analogous to that in channel coding [23].

LDPC-based SW Coding

In practice, there is no need to find $\mathbf{z}$ first for SW decoding. It is much more efficient to obtain $\mathbf{x}$ directly. SW decoding will be almost identical to channel decoding with the exception that the decoder aims to find the most probable code vector with the received syndrome instead of a codeword (with all-zero syndrome).

Since we can treat any block code as a "low-density" parity check code, we can decode $\mathbf{x}$ just as in LDPC decoding. Let us illustrate this with the setting similar to that in Example 5.5, that is, both source $\mathbf{x}$ and size information $\mathbf{y}$ are 7 bits long. Unlike the previous example, we do not restrict $\mathbf{x}$ and $\mathbf{y}$ to differ by no more than 1 bit. Instead, we assume that the ith bits of $\mathbf{x}$ and $\mathbf{y}$ (x_i and y_i) are related by $x_i = y_i + z_i$, where z_i is a Bernoulli random variable with probability to be 1 equal to p.

Example 5.6 (LDPC-based SW decoding). The encoding step is identical to before. In Figure 5.2 we illustrate the encoding step using a Tanner graph (factor graph). Given the received syndrome $\mathbf{s}$ and the side information $\mathbf{y}$, we can decode the source $\mathbf{x}$ using belief propagation just as discussed in Appendix B. In Figure 5.2 the filled variable nodes corresponding to $\mathbf{y}$ and $\mathbf{s}$ have known values whereas the unfilled variable nodes corresponding to $\mathbf{x}$ are unknown. The square factor nodes specify the correlation among variables. The *correlation* factor nodes connecting $\mathbf{x}$ and $\mathbf{y}$ determine the correlation between the two. In practice, a correlation factor node can then be defined as

$$f_i(x_i, y_i) = \begin{cases} p, & \text{if } x_i \neq y_i, \\ 1 - p, & \text{otherwise.} \end{cases} \tag{5.1}$$

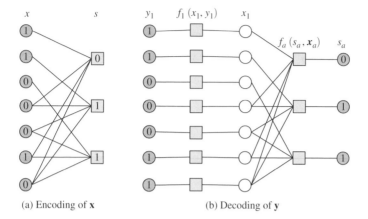

(a) Encoding of **x**   (b) Decoding of **y**

Figure 5.2 Factor graphs for encoding and decoding using LDPC-based SW coding.

The *check* factor nodes connecting **x** and **s** should return 1 only if the corresponding check is satisfied, that is, the bitwise sum of all variable nodes connecting to it is equal to 0.

Just as in LDPC decoding, since the variables are binary, it is much more efficient to pass likelihood values or log-likelihood values for belief propagation instead of probabilities of 1 and 0 separately. One can easily verify that the belief propagation decoding of **x** is almost identical to that in LDPC decoding for error correction. Algorithm 5.1 summarizes the decoding procedure with belief propagation using log-likelihoods as messages.

5.1.3 Parity-based Approach

A potential weakness of the syndrome-based approach is that no protection is provided to the compressed (syndrome) bits. If there is any transmission error of the syndrome bits, the decoded output will be completely wrong. Of course, one may cascade SW encoding with an additional channel coding step to correct any potential transmission errors.

Alternatively, SW encoding and channel encoding can be accomplished in one step by the parity-based approach. As the name suggests, parity bits instead of syndrome bits will be sent to the decoder. Consider the parity check matrix H of a channel code. An input **x** is compressed to the syndrome $s = H\mathbf{x}$ for the syndrome-based approach. On the other hand, **x** is compressed to **z** if $H \begin{bmatrix} x \\ z \end{bmatrix} = 0$ for the parity-based approach. At the decoder, given the side information **y** and the received parity bits **ẑ** (**ẑ** can be different from **z**), the decoder can estimate the original source **x** taking advantage of the fact that $\begin{bmatrix} x \\ z \end{bmatrix}$ has to be a codeword. More precisely, the decoder can find the closest codeword **w** from $\begin{bmatrix} y \\ \hat{z} \end{bmatrix}$ and then the first portion of w will be the desired estimate **x̂**.

Algorithm 5.1: Summary of LDPC-based SW decoding using BP

- **Initialize:**

$$L_{ai} = \begin{cases} 0, & \text{if } a \text{ is a check factor node,} \\ (-1)^{y_i} \log \frac{1-p}{p}, & \text{if } a \text{ is a channel factor node.} \end{cases} \quad (5.2)$$

- **Messages update:**
  - Variable nodes for **x** to check factor nodes:

$$L_{ia} \leftarrow \sum_{b \in N(i) \setminus i} L_{ia}, \quad (5.3)$$

  - Check factor nodes to variable nodes for **x**:

$$L_{ai} \leftarrow 2(-1)^{s_a} \tanh^{-1} \left(\prod_{j \in N(a) \setminus i} \tanh \left(\frac{L_{ja}}{2} \right) \right) \quad (5.4)$$

- **Belief update:** Denote the net belief in terms of log-likelihood ratio as β_i for variable node i, then

$$\beta_i \leftarrow \sum_{a \in N(i)} L_{ai} \quad (5.5)$$

- **Stopping criteria:** Repeat message update and belief update steps until maximum number of iterations is reached or all checks are satisfied. Output the hard threshold values of net beliefs as the estimated coded bits.

Example 5.7 (Parity-based SW coding of 4 bits). Consider two binary sequences **X** and **Y** of length 4 and, just as in previous examples, the two sequences can differ no more than 1 bit. Assuming that **Y** is given to the decoder, we may compress **X** into 3 bits using the $(7,4)$-Hamming code. We can write its generator matrix G as $\begin{bmatrix} 1000110 \\ 0100011 \\ 0010101 \\ 0001111 \end{bmatrix} = [I|P^T]$, where $P = \begin{bmatrix} 1011 \\ 1101 \\ 0111 \end{bmatrix}$ and the parity check matrix $H = [P|I] = \begin{bmatrix} 1011100 \\ 1101010 \\ 0111001 \end{bmatrix}$. To compress **x**, the encoder computes the parity bits $\mathbf{z} = P\mathbf{x}$ and transmits them to the decoders. For example, let $\mathbf{x} = [1000]^T$, then $\mathbf{z} = [110]^T$. Let us assume the channel for transmitting the parity bits is noisy and the $\hat{\mathbf{z}} = [100]^T$ is received instead, and also assume that the side information $\mathbf{y} = \mathbf{x} = [1000]^T$. We then have $\begin{bmatrix} \mathbf{y} \\ \hat{\mathbf{z}} \end{bmatrix} = [1000100]^T$. We will need to find the closest codeword to $\begin{bmatrix} \mathbf{y} \\ \hat{\mathbf{z}} \end{bmatrix}$. Since the code is very simple, we can achieve this using syndrome decoding. First, compute $H \begin{bmatrix} \mathbf{y} \\ \hat{\mathbf{z}} \end{bmatrix}$ to get $[010]^T$. By inspection, we can see that $[010]^T$ is also the syndrome of $[0000010]^T$. Thus, we expect that the second last bit has an error and hence the correct codeword should be $[1000100]^T + [0000010]^T = [1000110]^T$. Therefore, the estimate $\hat{\mathbf{x}}$ is the first 4 bits, $[1000]^T$, which is the same as the original $\mathbf{x}$.

In the above example, since the $(7, 4)$-Hamming code can correct 1 bit of error, the decoder can reconstruct the source correctly as long as the sum of the Hamming distance between $\hat{\mathbf{x}}$ and $\hat{\mathbf{y}}$ and the number of bit errors in the received parity bit is less than 1. Comparing to Example 5.3, we have much lower compression efficiency. This is mainly due to the extra redundancy needed to rectify any potential errors of the compressed bits.

The encoding of the parity-based approach requires a systematic channel code where a codeword is always composed of a cascade of input information bits with computed parity bits. Actually, any systematic code will work well with the parity-based approach, for example irregular repeat accumulated (IRA) codes can be used to implement it. Note that unlike the LDPC-based approach for syndrome-based SW coding, the decoding procedure is identical to that in channel decoding. No modification is needed in the belief propagation step.

5.1.4 Syndrome-based Versus Parity-based Approach

Theoretically, the potential of a binning approach is only restricted by how flexibly one can group the elements of the source space into bins. Thus, it may be reasonable to interpret the "power" of a scheme as the ability to partition the input space into bins. It turns out that both syndrome-based and parity-based approaches are equally "powerful" in this regard.

Theorem 5.1 The theoretical performances of the syndrome-based and the parity-based approaches are equivalent in terms of their abilities in partitioning the input space.

Proof. The proof is based on simple observation. Consider any parity-based scheme with parity check matrice H of size $m \times n$, where we split H into $[H^s H^p]$, where H^s and H^p consist of the first $n - m$ columns and the last m columns, respectively. Then, for any input $\mathbf{x}$ of length $(n - m)$, we have a parity vector $\mathbf{y}$ such that $H^s\mathbf{x} = H^p\mathbf{y}$. Therefore, if we establish a syndrome-based scheme with parity check matrices H^s, we have the bin specified by syndrome $\mathbf{s} = H^p\mathbf{y}$ in the syndrome scheme equivalent to the bin specified by parity $\mathbf{y}$ in the parity scheme. Therefore, any partition that can be established by the parity approach can also be constructed using the syndrome approach.

On the other hand, it is apparent that the syndrome scheme is a special case of the parity-based approach for $H^p = I$. Thus, any partition that can be constructed by the syndrome scheme is representable by the parity scheme as well. $\square$

The decision of which of the two approaches to use is usually based on practical concerns. If we expect absolutely no error will occur during transmission of the compressed bits, then it is always reasonable to use the syndrome-based approach since there is an overhead in terms of lower compression efficiency for the parity-based approach. However, the decision is less obvious if transmission loss can occur. Compared to applying an additional protection layer on top of the syndrome-based approach, the parity-based approach has the advantage of being more compact and thus easier to implement. Moreover, since each coding layer has an intrinsic loss that is difficult to eliminate completely, higher compression efficiency is possible for the parity-based approach since it only has one coding layer instead of two. On the other hand, the code design of the parity-based approach can be more difficult since the correlation channel and the lossy channel of the compressed bits generally do not match precisely. Thus, unequal error protection should be applied differently to the parity bits and the information bits. While such a requirement can easily be achieved for the two-stage design in the syndrome-based approach, it is less obvious for the parity-based approach.

5.2 Non-asymmetric SW Coding

For non-asymmetric SW coding, multiple sources rather than only one source will be compressed. We will focus on the simplest case with only two sources in this section. Consider two sources X and Y. SW Coding Theorem suggests that one can compress the two sources independently as long as the joint rate is larger than $H(X, Y)$ and the individual rates are larger than $H(X|Y)$ and $H(Y|X)$. However, the asymmetric SW coding described in the last section only allows us to approach the rate-tuples at the "corners" of the SW region (also see Figure 2.3 in Chapter 2), that is, $(H(X|Y), H(Y))$ and $(H(X), H(Y|X))$.

Probably the "simplest" way to approach any rate-tuple is by time sharing. Assume that we have two asymmetric SW coding schemes A and B operating close to rate-tuples $(H(X|Y), H(Y))$ and $(H(X), H(Y|X))$. We can consider a combined scheme by using scheme A with r fraction of time and scheme B with $1 - r$ fraction of time. By varying r within the interval $[0, 1]$, we can approach any rate-tuple (R_X, R_Y) with $R_X + R_Y = H(X, Y)$.

A more involved approach is so-called source splitting [24, 25]. In a nutshell, one of the sources, say X, is split into two sources U and V, where only U is coded directly and transmitted to the decoder. Equipped with U, the second encoder can compressed Y at $H(Y|U)$ with asymmetric SW coding (note that $H(Y|U) > H(Y|X)$, why?). Finally, V is transmitted with the help of both U and Y at rate $H(V|U, Y)$. Then, the overall transmitted rate will be $(H(U) + H(V|U, Y), H(Y|U))$. By adjusting how X is split, one may also achieve all possible rate-tuples in the SW region.

In the rest of this section we will focus instead on another approach based on a technique known as code partitioning, which is likely to be the state-of-the-art approach in terms of overall performance [26, 27].

5.2.1 Generalized Syndrome-based Approach

When the source is binary, it turns out that non-asymmetric SW decoding can be attained quite efficiently by using conventional channel decoding. The major trick is the idea of code partitioning, where one main code is designed and used for decoding whereas at the same time this main code is partitioned into channel subcodes to be used at each encoder.

Let us revisit the 7-bit sequence SW coding example using Hamming code (Example 5.5). Note that the parity check matrix of the Hamming code has the form $[P|I_3]$, where $P = \begin{bmatrix} 1011 \\ 1101 \\ 0111 \end{bmatrix}$. In Example 5.5, $\mathbf{y}$ is transmitted directly to the decoder with 7 bits, but only the 3-bit syndrome of $\mathbf{x}$, $H\mathbf{x}$, is passed to the decoder.

Now, instead of asymmetric SW coding, let us assume we want to compress both $\mathbf{x}$ and $\mathbf{y}$ into the same number of bits. Since the total entropy is 10 bits, we expect 5 bits for both sides. To achieve this, we will split the columns of P equally among two matrices P_1 and P_2, that is, $P = [P_1|P_2]$, and $P_1 = \begin{bmatrix} 10 \\ 11 \\ 01 \end{bmatrix}$ and $P_2 = \begin{bmatrix} 11 \\ 01 \\ 11 \end{bmatrix}$.

Define the "coding" matrices of the first and second encoders, H_1 and H_2, as

$$H_1 = \begin{bmatrix} I_2 & 0_{2\times2} & 0_{2\times3} \\ 0_{3\times2} & P_2 & I_3 \end{bmatrix} \tag{5.6}$$

and

$$H_2 = \begin{bmatrix} 0_{2\times2} & I_2 & 0_{2\times3} \\ P_1 & 0_{3\times2} & I_3 \end{bmatrix}, \tag{5.7}$$

where $0_{m\times n}$ is an all-zero matrix of size $m \times n$ and I_n is an identity matrix of size $n \times n$. As both H_1 and H_2 have five rows, the two encoders will symmetrically compress incoming 7-bit inputs into five bits.

We call this a code partitioning approach because of the following.

Lemma 5.1 *(Code partitioning).* The codes defined by H_1 and H_2 partition the code defined by H.

Proof. Let

$$G_1 = [0_{2\times2}|I_2|P_2^T] \tag{5.8}$$

and

$$G_2 = [I_2|0_{2\times2}|P_1^T]. \tag{5.9}$$

We have $H_1 G_1^T = H_2 G_2^T = 0$ and thus G_1 and G_2 are possible generator matrices of H_1 and H_2. Note that

$$H \begin{bmatrix} G_2 \\ G_1 \end{bmatrix}^T = 0, \tag{5.10}$$

and so $G \triangleq \begin{bmatrix} G_2 \\ G_1 \end{bmatrix}$ is a generator matrix of H. Moreover, since the row spaces of G_1 and G_2 are linearly independent, the codes defined by H_1 and H_2 partition the code defined by H. □

To explain the decoding procedure, let us denote the input source pair as $\mathbf{x} = [x_1, x_2, \cdots, x_7]^T$ and $\mathbf{y} = [y_1, y_2, \cdots, y_7]^T$. Then, encoder 1 outputs syndrome $\mathbf{u} = H_1 \mathbf{x} = \begin{bmatrix} x_1^2 \\ x_5^7 + P_2 x_3^4 \end{bmatrix}$ and encoder 2 outputs syndrome $\mathbf{v} = H_2 \mathbf{y} = \begin{bmatrix} y_3^4 \\ y_5^7 + P_1 y_1^2 \end{bmatrix}$.

On receiving the syndromes $\mathbf{u}$ and $\mathbf{v}$, the joint decoder rearranges the bits of $\mathbf{u}$ and $\mathbf{v}$ and pad zeros to form $\mathbf{t}_1$ and $\mathbf{t}_2$ as follows,

$$\mathbf{t}_1 = \begin{bmatrix} u_1^2 \\ 0_{2\times1} \\ u_3^5 \end{bmatrix} = \begin{bmatrix} x_1^2 \\ 0_{2\times1} \\ x_5^7 + P_2 x_3^4 \end{bmatrix} \tag{5.11}$$

and

$$\mathbf{t}_2 = \begin{bmatrix} 0_{2\times1} \\ v_1^2 \\ v_3^5 \end{bmatrix} = \begin{bmatrix} 0_{2\times1} \\ y_3^4 \\ y_5^7 + P_1 y_1^2 \end{bmatrix}. \tag{5.12}$$

Interestingly, note that

$$\mathbf{t}_1 + \mathbf{t}_2 + \mathbf{x} + \mathbf{y} = \begin{bmatrix} y_1^2 \\ x_3^4 \\ P_1 y_1^2 + P_2 x_3^4 \end{bmatrix} = \begin{bmatrix} y_1^2 \\ x_3^4 \\ P \begin{bmatrix} y_1^2 \\ x_3^4 \end{bmatrix} \end{bmatrix} = \begin{bmatrix} I_4 \\ P \end{bmatrix} \begin{bmatrix} y_1^2 \\ x_3^4 \end{bmatrix} \tag{5.13}$$

and hence $\mathbf{t}_1 + \mathbf{t}_2 + \mathbf{x} + \mathbf{y}$ is a codeword of the large code C as

$$H(\mathbf{t}_1 + \mathbf{t}_2 + \mathbf{x} + \mathbf{y}) = [P|I_3] \begin{bmatrix} I_4 \\ P \end{bmatrix} \begin{bmatrix} y_1^2 \\ x_3^4 \end{bmatrix} = (P + P) \begin{bmatrix} y_1^2 \\ x_3^4 \end{bmatrix} = 0_{3\times1}. \tag{5.14}$$

Therefore, we can consider $\mathbf{t}_1 + \mathbf{t}_2$ as a corrupted code vector of $\mathbf{t}_1 + \mathbf{t}_2 + \mathbf{x} + \mathbf{y}$ passing through a channel with $\mathbf{x} + \mathbf{y}$ as channel noise. When $\mathbf{x}$ and $\mathbf{y}$ are highly correlated, the code word $\mathbf{t}_1 + \mathbf{t}_2 + \mathbf{x} + \mathbf{y}$ can be recovered by channel decoding with a high probability. Under the correlation model of this problem, $\mathbf{x}$ and $\mathbf{y}$ differ by no more than 1 bit and thus $\mathbf{x} + \mathbf{y}$ corresponds to random error vectors with at most 1 bit error. As Hamming code has 1 bit error correcting capability, all the "error" introduced by $\mathbf{x} + \mathbf{y}$ will be fixed and $\mathbf{t}_1 + \mathbf{t}_2 + \mathbf{x} + \mathbf{y}$ will always be recovered. Given that $\mathbf{t}_1 + \mathbf{t}_2 + \mathbf{x} + \mathbf{y}$ is known, we can obtain y_1^2 and x_3^4 from its first 4 bits. Moreover, we can subtract the last 3 bits of $\mathbf{u}$ by $P_2 x_3^4$

to obtain x_5^7. Similarly, subtracting the last 3 bits of $\mathbf{v}$ by $P_1 y_1^2$ gives us y_5^7. Finally, x_1^2 and y_3^4 can be read directly from the first 2 bits of $\mathbf{u}$ and $\mathbf{v}$ and this gives us the entire $\mathbf{x}$ and $\mathbf{y}$.

Let us take another example and compress $\mathbf{x}$ and $\mathbf{y}$ into 4 bits and 6 bits, respectively. Write the parity check matrices of the two encoders, H_1 and H_2, into

$$H_1 = \begin{bmatrix} 1 & 0_{1\times2} & 0_{1\times3} \\ 0_{3\times1} & P_2 & I_3 \end{bmatrix} \tag{5.15}$$

and

$$H_2 = \begin{bmatrix} 0_{3\times1} & I_3 & 0_{3\times3} \\ P_1 & 0_{3\times3} & I_3 \end{bmatrix} \tag{5.16}$$

instead, where $P_1 = \begin{bmatrix} 1 \\ 1 \\ 0 \end{bmatrix}$ and $P_2 = \begin{bmatrix} 011 \\ 101 \\ 111 \end{bmatrix}$ are the first column and the last three columns of P. Then, the compressed inputs $\mathbf{u} = H_1\mathbf{x}$ and $\mathbf{v} = H_2\mathbf{y}$ have 4 and 6 bits, respectively. At the decoder, rearrange $\mathbf{u}$ and $\mathbf{v}$ to form $t_1 = \begin{bmatrix} u_1 \\ 0_{3\times1} \\ u_2^4 \end{bmatrix}$ and $t_2 = \begin{bmatrix} 0 \\ v_1^3 \\ v_4^6 \end{bmatrix}$. Again, it is easy to verify that $\mathbf{t}_1 + \mathbf{t}_2 + \mathbf{x} + \mathbf{y} = \begin{bmatrix} I_4 \\ P \end{bmatrix} \begin{bmatrix} y_1 \\ x_2^4 \end{bmatrix}$ is a codeword. On receiving $\mathbf{t}_1 + \mathbf{t}_2$, the decoder can recover $\mathbf{t}_1 + \mathbf{t}_2 + \mathbf{x} + \mathbf{y}$ using channel decoding. y_1 and x_2^4 can be read out from the first 4 bits of $\mathbf{t}_1 + \mathbf{t}_2 + \mathbf{x} + \mathbf{y}$. By subtracting u_2^4 by $P_2 x_2^4$ and v_4^6 by $P_1 y_1$, we get x_5^7 and y_5^7. Finally, x_1 and y_2^4 can be read out directly from u_1 and v_1^3.

5.2.2 Implementation using IRA Codes

Theoretically, one can use any channel code, including LDPC codes, to implement the generalized syndrome-based approach. In practice, we see that the main coding matrix is restricted to the form of $[P|I]$, that is, the parity check matrix of some systematic code. One may want to connect the variable and check nodes using H directly (as if it is a LDPC code). However, the identity part in the parity check matrix introduces a large number of variable nodes of degree 1, which makes the estimation of these nodes very difficult because of the lack of diversified information. This makes the direct implementation perform very poorly.

It turns out that this problem can be solved by considering H as the parity check matrix of an IRA code instead. This avoids the problem of having variable nodes of degree 1 and thus can result in much better performance than the direct implementation. We will elaborate our discussion with a concrete example as follows.

Example 5.8 *(Non-asymmetric SW coding using a systematic IRA code).* Consider two correlated length 12 binary inputs $\mathbf{x}$ and $\mathbf{y}$ compressed

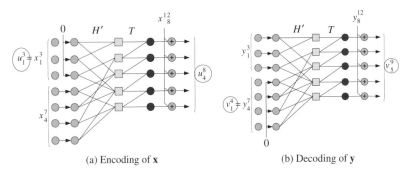

(a) Encoding of **x**   (b) Decoding of **y**

Figure 5.3 Non-asymmetric SW for two length-12 binary sources (a) **x** and (b) **y** using an IRA code, where the compressed bits, **u** and **v**, are circled.

non-asymmetrically into 8 and 9 bits, respectively. Let the parity check matrix H of the main IRA code be $H = [P|I] = [TH'|I]$, where $T = \begin{bmatrix} 10000 \\ 11000 \\ 11100 \\ 11110 \\ 11111 \end{bmatrix}$ and

$H' = \begin{bmatrix} 1100000 \\ 0010010 \\ 1001100 \\ 0100101 \\ 0001011 \end{bmatrix}$, and thus $P = \begin{bmatrix} 1100000 \\ 1110010 \\ 0111110 \\ 0011011 \\ 0010000 \end{bmatrix}$. Then for a length 12 vector **w**, H**w** can be constructed as multiplying the first 7 bits of **w** to H' followed by T and adding the result with the last 5 bits of **w**.

Now, to construct a non-asymmetric scheme, we split P horizontally into 5×3 and 5×4 matrices, P_1 and P_2. Thus $P_1 = \begin{bmatrix} 110 \\ 111 \\ 011 \\ 001 \\ 001 \end{bmatrix}$ and $P_2 = \begin{bmatrix} 0000 \\ 0010 \\ 1110 \\ 1011 \\ 0000 \end{bmatrix}$. So from the

previous section, we have $H_1 = \begin{bmatrix} I_3 & 0_{3\times4} & 0_{3\times5} \\ 0_{5\times3} & P_2 & I_5 \end{bmatrix}$ and $H_2 = \begin{bmatrix} 0_{4\times3} & I_4 & 0_{4\times5} \\ P_1 & 0_{5\times4} & I_5 \end{bmatrix}$.

Note that the first 3 bits of $\mathbf{u} = H_1\mathbf{x}$ are just the first 3 bits of **x** and the last 5 bits are generated in a similar manner to H**w** except that x_1^3 is first set to 0. This is illustrated in Figure 5.3(a). Similarly, v_1^4 is the same as y_4^7 and v_5^9 are constructed as H**w** except that y_4^7 is set to 0, as in Figure 5.3(b). In summary, the compressed outputs **u** and **v** are given by $\begin{bmatrix} x_1^3 \\ P_2 x_4^7 + x_8^{12} \end{bmatrix}$ and $\begin{bmatrix} y_4^7 \\ P_1 y_1^3 + y_8^{12} \end{bmatrix}$.

Note that the encoding steps ensure $H \begin{bmatrix} 0_{3\times1} \\ x_4^7 \\ x_8^{12}+u_4^8 \end{bmatrix} = 0_{5\times1}$ and $H \begin{bmatrix} y_1^3 \\ 0_{4\times1} \\ y_8^{12}+v_5^9 \end{bmatrix} = 0_{5\times1}$.

Thus $H \left(\begin{bmatrix} 0_{3\times1} \\ x_4^7 \\ x_8^{12}+u_4^8 \end{bmatrix} + \begin{bmatrix} y_1^3 \\ 0_{4\times1} \\ y_8^{12}+v_5^9 \end{bmatrix} \right) = H \begin{bmatrix} y_1^3 \\ x_4^7 \\ u_4^8+v_5^9+(x_8^{12}+y_8^{12}) \end{bmatrix} = 0_{5\times1}$ and hence

$\begin{bmatrix} y_1^3 \\ x_4^7 \\ u_4^8+v_5^9+(x_8^{12}+y_8^{12}) \end{bmatrix}$ is a codeword. If we consider $X + Y$ as a binary noise, then $u_1^3 = x_1^3 = y_1^3 + (x_1^3 + y_1^3)$ can be treated as a noisy version of y_1^3. Similarly, v_1^4 and $u_4^8 + v_5^9$ can be treated as noisy versions of x_4^7 and $u_4^8 + v_5^9 + (x_8^{12} + y_8^{12})$,

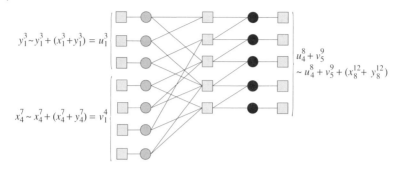

Figure 5.4 Factor graph for non-asymmetric SW decoding.

respectively. Therefore, we have the noisy estimate of the entire codeword $\begin{bmatrix} y_1^3 \\ x_4^7 \\ u_4^8+v_5^9+(x_8^{12}+y_8^{12}) \end{bmatrix}$. If the correlation statistics $p_{X,Y}(x,y)$ are known, the sum statistic $p_{X+Y}(z)$ will be known as well so we can estimate the prior probability of y_1^3 as $p(y_1^3|u_1^3) = p_{X+Y}(u_1^3 - y_1^3)$. The prior probabilities for the remaining portions of the codeword can also be estimated in a similar manner and thus we can decode the codeword using belief propagation just as regular LDPC or IRA codes (see also Figure 5.4). Once we decode y_1^3 and x_4^7, we can recover x_8^{12} and y_8^{12} as $u_4^8 - P_2 x_4^7$ and $v_5^9 - P_1 y_1^3$. Of course, we can read out x_1^3 and y_4^7 directly from u_1^3 and v_4^7.

5.3   Adaptive Slepian–Wolf Coding

In the previous sections we discussed several SW coding implementations, including both asymmetric and non-asymmetric setups based on channel coding principles and a distributed arithmetic coding technique that is based on source coding principles. However, for all the above approaches we have assumed that the perfect knowledge on correlation statistics among the sources is known. In practical application such as sensor networks, the correlation between sensors is usually unknown and can even vary over time. Apparently, encoders, which only have access to one of many sources, can never estimate the correlation among sources. Therefore, the decoder has to take responsibility for estimating the correlation besides decoding the sources.

We refer to adaptive SW coding here as approaches that can handle sources whose correlation is not fixed and may change over time. As a consequence, another advantage of such adaptive schemes is that they don't need any prior knowledge of the correlation between sources since this knowledge can be estimated directly. A naïve adaptive approach is to simply transmit the uncompressed sources to the decoder and have the decoder estimate the correlation directly. The disadvantage of such a scheme is that an extra over-head is introduced since the sources cannot be compressed during correlation estimation. Moreover, the correlation estimation phrase and decoding phrase must be interchanged periodically to ensure any variation in correlation can be captured and updated. Coordination of the transition of these phrases can introduce additional implementation complexity.

We will instead describe the more elegant asymmetric SW coding approach proposed in [28], where the decoder adaptively estimates the correlation and reconstructs the source at the same time. The asymmetric approach was fur-ther extended into the non-asymmetric case in [29]. However, the extension is rather straightforward and thus we will only focus on the discussion of the asymmetric case. We refer interested readers to [29] for the detail of the adaptive non-asymmetric scheme.

5.3.1 Particle-based Belief Propagation for SW Coding

Let us illustrate the adaptive approach by continuing with Example 5.6. However, we should not assume that the correlation between the source and side information is fixed or known. In other words, $x_i = y_i + z_i$, where z_i is the Bernoulli variable such that the probability of z_i equal to 1 is p_i. The encoding procedure is equivalent to the previous case. On the other hand, since the cor-relation parameters are unknown, they should be treated as random variables as well and this can be incorporated in the factor graph, as shown in Figure 5.5. The ith correlation factor node, $f_i(x_i, y_i, p_i)$, should now be defined as

$$f_i(x_i, y_i, p_i) = \begin{cases} p_i, & \text{if } x_i \neq y_i, \\ 1 - p_i, & \text{otherwise.} \end{cases} \tag{5.17}$$

If we assume that the correlation does not change rapidly, we may impose additional constraints on neighboring correlation variables. For example, we can link the correlation variables p_1 and p_2 by another factor node $f_{1,2}(p_1, p_2)$, where the correlation function can be described by a Gaussian kernel, that is, $f_{1,2}(p_1, p_2) = \exp((p_1 - p_2)^2 / \lambda)$. Note that λ here is a hyperprior which should roughly match with the rate of change of the correlation statistics. For example, if one expects the correlation statistics to change rather rapidly, a small λ should be used. In practice, the performance of the scheme is not very sensitive to the choice of λ [28].

With the factor graph as constructed in Figure 5.5, one can in theory recon-struct both the correlation statistics and the source **x** simultaneously using the

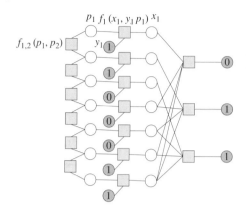

Figure 5.5 Factor graph for adaptive SW decoding. Note that the filled variable nodes (circles) are known but the blank variable nodes are unknown and need to be estimated.

belief propagation (sum-product) algorithm. However, note that the correlation variables p_i are continuous but belief propagation can only handle discrete variables, therefore some modification of the original belief propagation algorithm is needed to handle the continuous variables.

The most straightforward approach, as proposed in [28], is to employ Monte Carlo simulation. Here, we will just refer to it as particle-based belief propagation (PBP). The key idea is to model the distribution of a parameter variable p_i by a number of particles, for example five. The locations of the particles are denoted $p_i^{(1)}, p_i^{(2)}, p_i^{(3)}, p_i^{(4)}$, and $p_i^{(5)}$. The *weights* of the particles are given by $w_i^{(1)}, w_i^{(2)}, w_i^{(3)}, w_i^{(4)}$, and $w_i^{(5)}$, where without loss of generality, we assume that the weights are positive and $\sum_{j=1}^{5} w_i^{(j)} = 1$. Then the probability density function of p_i can be approximated as $\sum_{j=1}^{5} w_i^{(j)} \delta(p_i - p_i^{(j)})$, where $\delta(\cdot)$ is the Dirac-delta function. Moreover, we can approximate p_i as $p_i \approx \sum_{j=1}^{5} w_i^{(j)} p_i^{(j)}$.

Initially, the decoder will set $p_i^{(j)}$ randomly and $w_i^{(j)} = 1/5$ for all i and j. Now, since each p_i is only represented by a finite number of five possible values, conventional belief propagation can be applied and the resulting probability for particle j will be assigned to $w_i^{(j)}$. To ensure numerical stability (i.e., to avoid improbable particles having extremely small weights), a resampling step is applied. In general, a resampling method will convert a set of weighted particles into a set of unweighted ones. Moreover, it requires that the distribution of the resulting particles should be almost the same as the original one. There are many different resampling methods, but they often have similar steps. Let us denote N as the sample size ($N = 5$ in this case). First, normalize the original sample weights to have a sum of 1 and partition the interval $[0, 1)$ according to the normalized weights. Second, generate a set of N almost uniform numbers u_k in $[0, 1)$. Finally, a new particle at location $p^{(j)}$ (corresponding to the jth weight) is generated if a u_k falls in the jth interval. The major difference among different resampling methods often lies in how u_k are generated. For systematic resampling [29],

$$u_k = \frac{(k-1) + \tilde{u}}{N}, \tag{5.18}$$

where $\tilde{u}$ is sampled uniformly from $[0, 1)$.

In [28], the systematic resampling step as described above is used, which is also commonly used in particle filtering. Figure 5.6 provides concrete examples. For simplicity and without any confusion, we omit the index i in $p_i^{(j)}$ and $w_i^{(j)}$ for the following description. The original locations and weights of the five particles are shown in the topmost figure. We can see that the fourth particle has the highest weight ($w^{(4)}$) and thus is the most probable. To perform systematic resampling, the interval $[0, 1]$ is partitioned into five segments proportional to the initial weights $w^{(1)}, \cdots, w^{(5)}$ as shown in the middle figure. Five samples are generated according to (5.18). If a sample falls in the jth interval, then a new particle at location $p^{(j)}$ is generated. Note that the resampling step ensures that the new particle is generated at the location of an original particle with probability proportional to the corresponding weight. As shown in the middle figure, $p^{(2)}$ is chosen once, whereas $p^{(3)}$ and $p^{(4)}$ are chosen twice each. Therefore, the locations of the new particles are $\tilde{p}^{(1)} = p^{(2)}$, $\tilde{p}^{(2)} = \tilde{p}^{(3)} = p^{(3)}$ and $\tilde{p}^{(4)} = \tilde{p}^{(5)} = p^{(4)}$. Note that all weights should also be readjusted to $1/5$, as shown in the bottom figure.

After systematic resampling, multiple particles will concentrate at the same location. To ensure diversity of particles, the particles will perform a "random walk" by simply adding the locations of the particle with a random Gaussian noise. The conventional BP will be applied again after the random walk and the overall PBP algorithm is summarized in Algorithm 5.2.

Algorithm 5.2: Summary of PBP for adaptive LDPC decoding

---

1. Initialize variables and particles (weights and locations).
2. Perform BP. Exit if solution is found (i.e., checks are satisfied).
3. Apply systematic resampling to particles.
4. Apply random walk to particle locations.
5. Go to 2.

5.4 Latest Developments and Trends

Much recent attention has been given to the design of adaptive SW coding for sources with varying correlation. Besides non-asymmetric SW coding,

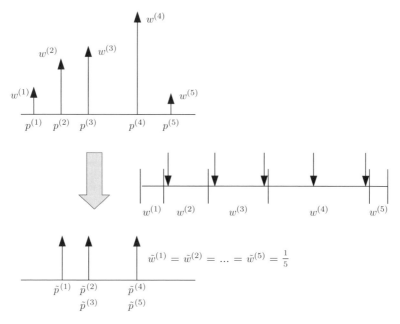

Figure 5.6 An example of systematic resampling with five samples.

the approach described in Section 5.3 has also been extended to lossy distributed source coding (WZ coding and multiterminal source coding, as will be described in Chapters 7 and 8). Unlike the non-adaptive approaches described in Sections 5.1 and 5.2, the particle-based approaches described in last section have deviated quite drastically from conventional channel coding. In particular, it is probably more insightful to interpret the decoding step as an inference problem directly. This is also suggested in [23] when SW coding for more than two sources is considered.

Note that even though the adaptive approach estimates the correlation between the source and the side information there is still one remaining problem of determining the compression rates. One possibility is to use rateless codes, as in [30, 31].

Despite their similarity, the setups in Examples 5.5 and 5.6 are not strictly the same. The former is real lossless (assuming that there is only one difference between the two sources) but the later is near-lossless (as it is possible to have more than one difference between the two sources and thus the decoder may fail). For a true lossless setup, one may describe perfect code, as in Example 5.5. Non-asymmetric perfect SW codes with two sources can be generated out of Hamming code using the code-partitioning approach. This is possible since code partitioning does not change the sum rate and it is trivial to verify that Hamming code is perfect for the asymmetric case. The existence of perfect SW codes for more than two sources is less obvious. Ma and Cheng showed that "perfect" SW codes actually exist for more than two sources [32], where a perfect SW coding with three sources of length 21 are constructed to compress the sources into a total of 27 bits. However, the correlation setup is the simplest nontrivial one. As in Example 5.5, only one out of the 21 bits can be different among the three sources. Results for more complicated correlations exist and we direct readers to the paper itself [33, 34].

6

Distributed Arithmetic Coding

In Chapter 5 we described how to implement *Slepian–Wolf coding* (SW coding) with channel codes, while in this chapter we will show how to implement SW coding with source codes. *Distributed arithmetic coding* (DAC) is developed on the basis of the source coding principle, which can be realized either with interval overlapping [35–37] or with bitstream puncturing [38]. This chapter will introduce only the DAC scheme with interval overlapping. Before we get into detailed discussions on this approach, we will give a brief overview on *arithmetic coding* (AC), which is essential to the understanding of DAC [39].

Although DAC is an effective implementation of SW coding, its properties have not yet received a thorough analysis. In this chapter, after a description of DAC, we will introduce the concept of the *DAC spectrum*, which is a promising tool to answer a series of theoretical problems about DAC. Finally, we will present two effective techniques to improve the coding efficiency of DAC.

6.1   Arithmetic Coding

Targeted to conventional source coding, AC is a type of entropy coding, where the symbol x is compressed on average approaching its minimum required length $-\log p(x)$. The key idea of AC is illustrated in Figure 6.1(a), where the interval $[0, 1)$ is partitioned into intervals with lengths proportional to the probabilities of the symbols of a source. Let us call each of these intervals the *probability interval* of the corresponding symbol. The main trick is that a symbol can be uniquely represented by a real number within the probability interval of the symbol. Moreover, the longer the probability interval, the fewer number of bits needed to represent the number.

As in Figure 6.1(a), 0.001b (0.125 in decimal representation) lies inside the probability interval of a and thus can be used to represent a uniquely with its bits after the decimal point (i.e., 001). Similarly, b can be identified by the number 0.1b and thus can be coded as 1. Note that we are really referring to the interval $[0.a_1 a_2 \cdots a_n, 0.a_1 a_2 \cdots a_n \dot{1}]$ when we consider the real number

Distributed Source Coding: Theory and Practice, First Edition. Shuang Wang, Yong Fang, and Samuel Cheng.
© 2017 John Wiley & Sons Ltd. Published 2017 by John Wiley & Sons Ltd.
Companion Website: www.wiley.com/go/cheng/dsctheoryandpractice

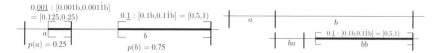

Figure 6.1 Representing a symbol using an interval. $[0, 1)$ is partitioned into regions with lengths proportional to the probabilities of the symbols. Any interval (which in turn can be represented by a number) that lies within the region corresponding to a symbol can be used to generate the codeword of the symbol.

$0.a_1 a_2 \cdots a_n$. Therefore, it is important to require this region to lie completely inside the probability interval of the corresponding symbol.

Apparently, as the probability interval gets shorter, a real number with higher precision is needed to represent the interval. It is easy to verify that it needs approximately $-\log p(x)$ bits to represent a probability interval of x. The length of the codeword approximates to the optimum code length.

We can easily see why this coding approach allows unique decoding. By construction, the regions that identify any two symbols cannot overlap since both regions lie completely inside the non-overlapping probability intervals of the two symbols. As an immediate consequence, the following lemma provides a sufficient condition for unique decoding.

Lemma 6.1 *(Prefix condition).* Any codeword of the code constructed above cannot be a prefix of another codeword.

Proof. Assuming that the Lemma is false, there exists two codewords $\mathbf{a} = a_1 a_2 \cdots a_n$ and $\mathbf{b} = b_1 b_2 \cdots b_m$ such that $\mathbf{a}$ is a prefix of $\mathbf{b}$. This means that $m \geq n$ and $a_i = b_i$, for $i = 1, \cdots n$. Then, we have $0.b_1 b_2 \cdots b_m$ lying inside both intervals $([0.a_1 a_2 \cdots a_n, 0.a_1 a_2 \cdots a_n \dot{1}]$ and $[0.b_1 b_2 \cdots b_n, 0.b_1 b_2 \cdots b_n \dot{1}]$ specified by $\mathbf{a}$ and $\mathbf{b}$. This contradicts the assumption that the corresponding intervals for each symbols cannot overlap. □

Actually, the prefix condition not only guarantees the uniqueness of the codewords. With little inspection, the condition ensures that each symbol can be decoded instantaneously once the codeword of the symbol is received by the decoder.

If the coding trick described above is only used for one symbol at a time, we will be compressing the source rather inefficiently. (In fact, we are "expanding" rather than "compressing" in our example since the source bit rate is 1 bit/sample but the average "compressed" bit rate is $0.25 \times 3 + 0.75 \times 1 = 1.5$ bit/sample.) The real power of AC is that it can handle a group of symbols with arbitrary size easily. Let us say we are going to encode bb. Instead of sending a codeword right away for the first symbol (b), we can first further partition the interval of b into intervals for ba ($[0.25, 0.4375)$) and bb ($[0.4375, 1)$). On receiving the second b, the encoder realizes that *code interval* is $[0.4375, 1)$

and can represent this code interval as described before. In this case, 0.1b and its corresponding region lie inside the probability interval of *bb* and thus *bb* can be compressed as a single bit 1. Of course, we do not need to stop at the second symbol, we can continue to partition the code interval into finer and finer regions and ultimately the interval will become very short and converge to a real number. This real number corresponds precisely to the compressed symbols.

Due to the finite precision of any computing device, scaling is a required step to avoid underflow problems. Let LOW and HIGH be the lower and upper bounds of the code interval, respectively. During encoding, the encoder knows that the next bit has to be 0 whenever a) HIGH < 0.5. Therefore, the encoder can safely output a bit 0 and scales up the region by two, that is, LOW ← 2LOW and HIGH ← 2HIGH. On the other hand, if b) LOW > 0.5, the encoder knows that next output bit has to be 1 and thus it can output 1 and scales up the region by setting LOW ← 2(LOW − 0.5) and HIGH ← 2(HIGH − 0.5).

The scaling steps described above are not complete. With only these two scaling rules, it may be that LOW = $0.5 - \epsilon_1$ and HIGH = $0.5 + \epsilon_2$ with very small ϵ_1 and ϵ_2. To avoid this kind of underflow, the code interval should also be scaled by two when c) HIGH < 0.75 and LOW > 0.25. In that case, LOW ← 2(LOW − 0.25) and HIGH ← 2(HIGH − 0.25). Note that the encoder cannot tell whether the next output bit should be 1 or 0 at this stage. However, for each scaling-up, the encoder expects the code interval to be closer to 0.5. Actually, one can easily verify that if the code interval turns out to be less than 0.5 after one scaling, the next two output bits should be 01 as the code interval will lie in the second quadrant if no scaling has been applied. Similarly, if the code interval turns out to be larger than 0.5, the next two output bits should be 10.

In general, the encoder will store the number of times that Case c) happens and let the number be BITS-TO-FOLLOW. When finally the encoder realizes the code interval is smaller than 0.5 (i.e., Case a), the next BITS-TO-FOLLOW +1 output bits should be 011 · · · 1 and BITS-TO-FOLLOW should be reset to 0. If the opposite (Case b) happens, the next BITS-TO-FOLLOW+1 output bits will be 100 · · · 0 and again BITS-TO-FOLLOW should be reset to 0.

Decoding is very similar to encoding except that an extra variable VALUE comes into the picture. VALUE can be interpreted as the location of the code interval up to the maximum available precision. For example, let the bit precision of the coders be 6. At the beginning of decoding, VALUE is constructed from the first 6 bits of the incoming bit stream. From VALUE, the decoder will know precisely which symbol the code interval lies on and thus this symbol will be the next output symbol. As the output symbol is known, the code interval will be partitioned and shrunk, as in encoding. The scaling procedure is almost identical except that the decoder needs to update VALUE as well during scaling. Moreover, VALUE is scaled up by two, and an incoming bit is read as the

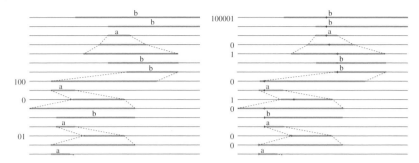

Figure 6.2 An example of encoding (left) and decoding (right) of AC. A sequence of symbols *bbabbabaa* · · · is compressed into a bitstream 1000010101000 · · ·.

least significant bit of VALUE. This step is important to maintain the precision of VALUE.

Example 6.1 *(Arithmetic coding).* A detailed coding example is shown in Figure 6.2. A sequence of symbols *bbabbabaa* · · · is compressed into 1000010101000 · · ·.

The left- and right-hand figures in Figure 6.2 describe the status of encoding and decoding, respectively. The thick horizontal line is the code interval. In the left figure, we can see that the code interval shrinks to lie within the range [0.25, 0.75) as symbols *b, b, a* are coded. The code interval scales up twice afterward and thus BITS-TO-FOLLOW becomes two. The scaling-up steps are highlighted by the dashed lines connecting the corresponding LOWs and HIGHs of code intervals of adjacent time steps. As two symbols *b, b* are coded afterward, the code interval now falls in the range [0.5, 1), thus the code interval expands and 100 are output. The explanation of the rest of the left-hand figure is similar and thus is omitted.

The right-hand figure looks very similar to the left-hand figure except that there is a cross mark within the code interval for each iteration step. The cross mark corresponds to VALUE, as described previously. Since 6 bit precision is used in this example, 6 bits 100001 are read by the decoder to generate VALUE at the beginning of decoding. As VALUE falls inside the probability interval of *b*, the decoder knows that the first output symbol is *b*. Similarly, it can tell from the next two iteration steps that *b* and *a* are the next two symbols. Case c scaling occurs in the fourth and fifth iterations and two bits are read by the decoder to keep VALUE to maintain the precision of VALUE. The explanation for the rest of the figure is similar and will be left as an exercise for readers.

Due to space limitation, we have only described some fundamental elements of AC. For detailed implementation issues, we recommend the classic reference by Witten *et al.* [40].

6.2 Distributed Arithmetic Coding

AC can be extended rather intuitively to the case of asymmetric SW coding. Since the average code rate decreases as the average code interval increases, one natural way to achieve a lower code rate is simply to increase the probability intervals for all symbols. Of course, this unavoidably introduces overlapping among the probability intervals. For example, we can extend the intervals for symbols a and b to $[0, 0.3)$ and $[0.2, 1)$, respectively. The encoding process is essentially identical to the original arithmetic coding.

Example 6.2 *(Encoding of distributed AC).* As shown in the left-hand figure of Figure 6.3, a sequence of symbols $bbbbabbabba \cdots$ is compressed (to $100111110110 \cdots$).

Decoding is performed in a similar way to conventional AC. When VALUE (as defined in last subsection) lies exclusively in the probability interval of a or b (i.e., VALUE > 0.3 or VALUE < 0.2), then the symbol can be unambiguously determined. Then, the decoding step is precisely the same as before. However, if VALUE lies inside the intersection of probability intervals of a and b (i.e., 0.2 < VALUE < 0.3), the decoder cannot be certain which symbol is actually coded. In this case, the decoder has to accept that both possibilities may happen and needs to consider both cases. Note that all state information (such as VALUE, LOW, HIGH, BITS-TO-FOLLOW) needs to be saved separately for both cases. Essentially, every time an ambiguity occurs, the decoder splits itself into two and each decoder considers one of the two cases and continues independent of the other.

The splitting process described above will result in exponentially many decoding possibilities and form a tree, as shown in Figure 6.4. It is important to find the most probable decoding path. Unlike conventional arithmetic

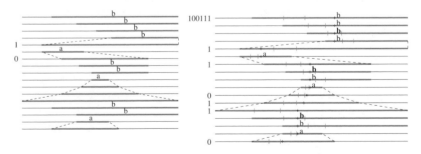

Figure 6.3 An example of encoding (left) and decoding (right) of distributed AC. A sequence of symbols $bbbbabbabba \cdots$ is compressed into the bitstream $100111110110 \cdots$. The symbols that cannot be determined immediately during the decoding process are shown in bold. Only the correct decoding "route" is shown.

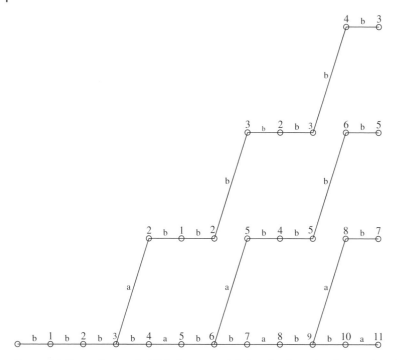

Figure 6.4 Scores (log-probabilities) computed during distributed arithmetic decoding. The higher scores correspond to higher probabilities. A node is split when the symbol cannot be determined unambiguously.

decoding, the side information Y is provided. Given the realization of Y at the same time step, the decoder can compute the likelihoods of X to be a and b. The overall likelihood for a sequence x^n can then be computed as the product of the likelihood of x_i at each time step. In practice, it is more convenient to use the log-likelihood function. Moreover, without loss of generality we can scale the log-likelihood function arbitrarily without affecting the decoding result. If the correlation between X and Y does not change over time and the correlation is symmetric (i.e., $Pr\{X = a|Y = b\} = Pr\{X = b|Y = a\}$ and $Pr\{X = a|Y = a\} = Pr\{X = b|Y = b\}$), we can simply scale the log-likelihood to unity. In essence, the log-likelihood (score) is 1 when $X = Y$ and -1 otherwise (assuming that $Pr\{X = Y\} > Pr\{X \neq Y\}$). Therefore, ultimately we just need to search through all the decoding "paths" and find the one that results in the maximum score. After the optimum path is found, the decoding symbols can be traced back by backtracking.

One way of avoiding an exponential increase in possible paths is to keep only the K leave nodes that have the highest log-likelihoods (scores) after each processed symbol.

Example 6.3 *(Decoding of distributed AC).* Continuing with the previous example, decoding of the compressed bit stream is very similar to conventional arithmetic decoding. As shown in right-hand figure of Figure 6.3, VALUE is computed and a symbol is decoded unambiguously if VALUE falls exclusively in one probability interval. This happens for the first two symbols *bb*. However, for the third symbol, VALUE lies in the intersection of both probability intervals of *a* and *b*, as shown in the figure. We highlight the symbol *b* in bold to indicate that the decoded symbol is ambiguous (the alternative case with *a* as the decoded symbol is not shown). This is also illustrated in Figure 6.4 as the decoder "splits" into two cases. As we can see from the figure, splitting also occurs after decoding the sixth and ninth symbols (corresponding symbols are highlighted in Figure 6.3 as well). In Figure 6.4, the cumulative scores that are based on the side information are also shown. For clarity purposes, the scores are sorted vertically with the maximum scores shown at the bottom. Therefore, the lowest path (*x*-axis) is the most probable and thus the decoded symbols are *bbbbabbabba* · · ·.

One of the potential advantage of distributed AC is that the probability model can be adjusted very flexibly for each coded symbol. However, this needs to be done at both encoder and decoder. Therefore, such adjustment may not be possible unless some feedback of correlation statistics is allowed.

6.3 Definition of the DAC Spectrum

6.3.1 Motivations

Since the emergence of DAC, a thorough analysis of its properties has become an important and interesting topic for the research community. Although it is rather intuitive that the classic AC may achieve the Shannon limit as code length increases, few systematic analyses on DAC have been found in the literature up to now. It has been found by experiments that the residual errors of DAC codes cannot be removed by increasing the rate [35, 36]. Therefore, instead of the achievable rate, we may quantitatively characterize the coding efficiency of DAC codes by the residual *frame-error rate* (FER) and *symbol-error rate* (SER) at a given rate.

As the DAC decoding proceeds, more and more branches will be created and all decoding branches will form an incomplete tree. Potentially, the number of decoding branches will increase exponentially as the DAC decoding proceeds, while a feasible decoder can afford only linear or near-linear complexity, implying that the decoding tree must be pruned in practice. Hence, the complexity of a practical DAC decoder may be quantitatively characterized by the maximum number of allowed decoding branches.

We define the breadth-K DAC decoder as the DAC decoder allowing no more than K branches during the decoding. In particular, the breadth-∞ DAC decoder is abbreviated as the *ideal DAC decoder*, which means no branch will be pruned during the decoding. In practice, the selection of K is a trade-off: the smaller the K value, the lower the decoding complexity, but the poorer the coding efficiency, and vice versa. Therefore, the coding efficiency of the DAC is closely linked with the decoding complexity and the best coding efficiency is achieved by the ideal decoder. Now the analyses on DAC codes can be finished by answering the following three questions:

- *Complexity of the ideal DAC decoder*: How many branches will be created as the DAC decoding proceeds, if no branch is pruned?
- *Performance of the ideal DAC decoder*: How about the residual FER and SER if DAC codes are parsed by the ideal decoder?
- *Performance of a practical DAC decoder*: How does K impact the residual FER and SER?

A systematic solution to these questions requires a powerful analysis tool. The first tentative work on this topic appeared in [41], where a theoretical tool named *spectrum* is proposed. On the basis of [41], some further advances on this topic were achieved in [42]. Due to the difficulty, all related works currently available are restricted to equiprobable binary sources [41, 42]. Nevertheless, the spectrum is a promising tool to answer a series of theoretical problems of DAC codes for general source models.

The rest of this chapter is arranged as below. After an in-depth analysis on the encoding and decoding processes of DAC codes, we will give the definition of the DAC spectrum. Then we will formulate the DAC spectrum and show how to calculate it by theoretical or numerical means. Afterwards, some questions regarding the decoding complexity and coding efficiency of DAC codes are solved by exploiting the DAC spectrum as a tool. Finally, two techniques are presented to improve the coding efficiency of DAC codes.

6.3.2 Initial DAC Spectrum

Let random variable $X \sim p_X(x)$ for $x \in \mathbb{B} \triangleq \{0, 1\}$ be the equiprobable binary source to be compressed at the sender, whose *probability mass function* (pmf) is $p_X(1) = p_X(0) = 0.5$. Let random variable $Y \sim p_Y(y)$ for $y \in \mathbb{B}$ be the binary side information available only at the receiver, which is correlated to X by the conditional pmf $p_{Y|X}(1|0) = p_{Y|X}(0|1) = p$ and $p_{Y|X}(0|0) = p_{Y|X}(1|1) = (1 - p)$. As conventions, we call p the *crossover probability* between X and Y. Further, let $X^n \triangleq (X_0, \cdots, X_{n-1})$ be the tuple of n independent and identically distributed (i.i.d.) random variables with $X_i \sim p_X(x)$ for all $i \in [0 : n) \triangleq \{0, \cdots, (n-1)\}$ and $Y^n \triangleq (Y_0, \cdots, Y_{n-1})$ be the tuple of n i.i.d. random variables with $Y_i|\{X_i = x\} \sim p_{Y|X}(y|x)$ for all $i \in [0 : n)$.

To compress X^n, the rate-R (infinite-precision) DAC encoder iteratively maps source symbols 0 and 1 onto partially overlapped intervals $[0, 2^{-R})$ and $[(1 - 2^{-R}), 1)$, respectively, where $R \in (0, 1)$ [35, 36]. According to the SW theorem, the prerequisite to the lossless recovery of X^n is $R \geq H(X|Y) = H_b(p)$ bits per symbol.

Let us explain in more detail how the DAC bitstream of X^n is formed by the encoder. Let $[L_i, H_i)$ be the interval after coding $X^i \triangleq (X_0, \cdots, X_{i-1})$. Initially, $[L_0, H_0) \equiv [0, 1)$. Notice that symbols 0 and 1 are mapped onto intervals of the same length 2^{-R}, so the interval length after coding X^i must always be 2^{-iR}, that is, $H_i \equiv L_i + 2^{-iR}$. Hence, only L_i is traced below. If $X_i = 0$, the lower bound remains the same after coding X_i, that is, $L_{i+1} = L_i$; otherwise, the lower bound will be updated by $L_{i+1} = L_i + (H_i - L_i)(1 - 2^{-R}) = L_i + (1 - 2^{-R})2^{-iR}$. The relation between L_{i+1} and L_i can then be expressed briefly by

$$L_{i+1} = L_i + X_i(1 - 2^{-R})2^{-iR}$$

$$= (1 - 2^{-R}) \sum_{i'=0}^{i} X_{i'} 2^{-i'R}. \tag{6.1}$$

After coding n symbols, the final interval is $[L_n, H_n)$, where

$$L_n = (1 - 2^{-R}) \sum_{i=0}^{n-1} X_i 2^{-iR} \tag{6.2}$$

and $H_n = L_n + 2^{-nR}$. As n goes to infinity, 2^{-nR} will tend to 0. Thus the final interval $[L_n, H_n)$ will converge to a real number

$$U_0 \triangleq (1 - 2^{-R}) \sum_{i=0}^{\infty} X_i 2^{-iR}. \tag{6.3}$$

In other words, each infinite-length binary sequence X^∞ will be mapped onto a real number $U_0 \in [0, 1)$. Therefore, we call $f(u)$, the *probability density function* (pdf) of U_0, the **initial DAC spectrum** [41].

Now we multiply both L_n and H_n by 2^{nR}, then $2^{nR}H_n \equiv 2^{nR}L_n + 1$, implying that $\lceil 2^{nR}L_n \rceil = \lfloor 2^{nR}H_n \rfloor$ holds almost always. Let $M(X^n) \triangleq \lceil 2^{nR}L_n \rceil$, then $M(X^n) \in [0 : 2^{nR})$, where nR is assumed to be an integer for simplicity. Finally, $M(X^n)$ is binarized into nR bits that are output as the DAC bitstream of X^n. Notice that $M(X^n)$ is often reduced to M below for simplicity.

6.3.3 Depth-*i* DAC Spectrum

At the receiver, the rate-R (infinite-precision) DAC decoder tries to recover X^n from its DAC bitstream $M(X^n)$ with the help of side information Y^n, which is equivalent to finding a binary sequence $\tilde{X}^n \in \mathbb{B}^n$ satisfying

$$\left\lceil (1 - 2^{-R}) \sum_{i=0}^{n-1} \tilde{X}_i 2^{(n-i)R} \right\rceil = M(X^n). \tag{6.4}$$

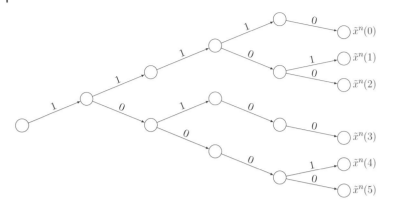

Figure 6.5 An example of DAC decoding tree, where $n = 5$.

For $R \in (0, 1)$, there are usually multiple sequences satisfying (6.4). From those candidate sequences, the decoder selects the sequence best matching Y^n as the estimate of X^n. The solutions to (6.4) actually form a depth-n incomplete binary tree, as shown by Figure 6.5, where each path from the root node to one leaf node corresponds to a binary sequence satisfying (6.4). A practical DAC decoder can then be implemented by searching the decoding tree either in the breadth-first order or in the depth-first order.

The breadth-first implementation of the DAC decoder is as follows [35, 36]. The decoding is decomposed into n stages and each stage corresponds to one level of the decoding tree. Since each path from the root node (depth-0 node) to a depth-i node corresponds to a length-i binary sequence, we use $\tilde{X}^i \triangleq (\tilde{X}_0, \cdots, \tilde{X}_{i-1})$ to stand for a depth-i node. Let $[\tilde{L}_i, \tilde{H}_i)$ be the decoder interval at node $\tilde{X}^i$. It is easy to find $[\tilde{L}_0, \tilde{H}_0) \equiv [0, 1)$ and $(\tilde{H}_i - \tilde{L}_i) \equiv 2^{-iR}$. By recursion, we have

$$\tilde{L}_i = (1 - 2^{-R}) \sum_{i'=0}^{i-1} \tilde{X}_{i'} 2^{-i'R}. \tag{6.5}$$

The decoding starts from the root node of the decoding tree. At node $\tilde{X}^i$, the decoder computes

$$\tilde{U}_i = \frac{M(X^n)2^{-nR} - \tilde{L}_i}{\tilde{H}_i - \tilde{L}_i}$$
$$= (M(X^n)2^{-nR} - \tilde{L}_i)2^{iR}. \tag{6.6}$$

Apparently, as n goes to infinity, $M(X^n)2^{-nR}$ will tend to U_0. Let

$$U_i \triangleq (U_0 - \tilde{L}_i)2^{iR}. \tag{6.7}$$

We call $g_i(u)$, the pdf of U_i, the **depth-i DAC spectrum** [42]. Obviously, $g_0(u)$ is just the initial DAC spectrum, that is, $g_0(u) = f(u)$.

Then, depending on $\tilde{U}_i$, the decoder creates a 0-branch and/or a 1-branch:

- $\tilde{U}_i \in [0, (1 - 2^{-R}))$: decide $\tilde{X}_i$ to 0 and create the 0-branch;
- $\tilde{U}_i \in [(1 - 2^{-R}), 2^{-R})$: create both 0-branch and 1-branch corresponding to possible source symbols 0 and 1;
- $\tilde{U}_i \in [2^{-R}, 1)$: decide $\tilde{X}_i$ to 1 and create the 1-branch.

Correspondingly, the decoder interval is updated to $[\tilde{L}_{i+1}, \tilde{H}_{i+1})$, where $\tilde{L}_{i+1} = \tilde{L}_i + \tilde{X}_i(1 - 2^{-R})2^{-iR}$ and $\tilde{H}_{i+1} = \tilde{L}_{i+1} + 2^{-(i+1)R}$. After branching, the decoder calculates the *a posteriori* probability of each path and sorts all paths according to their *a posteriori* probabilities. No more than K paths with the maximum *a posteriori* probabilities are retained, while other inferior paths are pruned, where K is a preset parameter to control decoder complexity. Such operations are iterated. Finally, after n stages, from all surviving full paths, the full path with the maximum *a posteriori* probability is output as the estimate of X^n.

6.3.4 Some Simple Properties of the DAC Spectrum

According to the properties of pdf, it is easy to know

$$\int_0^1 g_i(u)du = 1 \qquad (6.8)$$

and

$$\begin{cases} g_i(u) \geq 0, \ 0 \leq u < 1 \\ g_i(u) = 0, \ u < 0 \text{ or } u \geq 1 \end{cases} \qquad (6.9)$$

Since the symmetry holds for equiprobable binary sources, we have

$$g_i(u) = g_i(1 - u), \quad 0 < u < 1. \qquad (6.10)$$

Up to now, all what we know about $g_i(u)$ is just (6.8), (6.9), and (6.10). Below we will make use of a recursive formulation to deduce the explicit form of $g_i(u)$. Our first goal is to obtain the explicit form of $f(u)$, the initial DAC spectrum at root nodes. Then, we aim at expressing $g_{i+1}(u)$ as a function of $g_i(u)$ in a recursive way.

6.4 Formulation of the Initial DAC Spectrum

For brevity, we will use q to stand for 2^{-R} in the following.

Theorem 6.1 *(Formulation of the initial DAC spectrum).* The initial DAC spectrum of equiprobable binary sources satisfies [41]

$$2qf(u) = f\left(\frac{u}{q}\right) + f\left(\frac{u - (1 - q)}{q}\right). \qquad (6.11)$$

Proof. We divide $\mathbb{B}^\infty$ into two subsets $\mathbb{B}_0^\infty$ and $\mathbb{B}_1^\infty$, where $\mathbb{B}_0^\infty$ ($\mathbb{B}_1^\infty$, resp.) contains all infinite-length binary sequences beginning with symbol 0 (symbol 1, resp.) in $\mathbb{B}^\infty$. Denote the initial DAC spectrum of $X^\infty \in \mathbb{B}_0^\infty$ ($\mathbb{B}_1^\infty$, resp.) by $f_0(u)$ ($f_1(u)$, resp.). It is obvious that

$$f(u) = Pr\{X_0 = 0\}f_0(u) + Pr\{X_0 = 1\}f_1(u)$$
$$= \frac{f_0(u) + f_1(u)}{2}. \tag{6.12}$$

Now the problem is how to link $f_0(u)$ and $f_1(u)$ with $f(u)$. Let us begin with $f_0(u)$. We first discard the leading symbol 0 of each binary sequence in $\mathbb{B}_0^\infty$ to obtain a new subset $\hat{\mathbb{B}}_0^\infty$ and then map each binary sequence in $\hat{\mathbb{B}}_0^\infty$ onto interval $[0, q)$. Thus the initial DAC spectrum of $\hat{\mathbb{B}}_0^\infty$ is just $f_0(u)$. Because both $\hat{\mathbb{B}}_0^\infty$ and $\mathbb{B}^\infty$ are the sets of infinite-length binary sequences, $\hat{\mathbb{B}}_0^\infty$ is actually equivalent to $\mathbb{B}^\infty$. Therefore, the initial DAC spectrum of $\hat{\mathbb{B}}_0^\infty$ must have the same shape as that of $\mathbb{B}^\infty$, and the only difference between them is that $f(u)$ is defined over $0 \le u < 1$, while $f_0(u)$ is defined over $0 \le u < q$. Based on this analysis, we can obtain

$$f_0(u) = \frac{f(u/q)}{\int_0^q f(u/q)du} = \frac{f(u/q)}{q}, \quad 0 \le u < q. \tag{6.13}$$

Similarly, we can get

$$f_1(u) = \frac{f((u-(1-q))/q)}{q}, \quad (1-q) \le u < 1. \tag{6.14}$$

Finally, substituting (6.13) and (6.14) into (6.12) gives (6.11). □

Example 6.4 *(Formulation of the initial DAC spectrum).* For better understanding of the relations between $f(u)$, $f_0(u)$, and $f_1(u)$, Figure 6.6 gives an example when $q = 1/\sqrt{2}$. It can be observed that both $f_0(u)$ and $f_1(u)$ are scaled versions of $f(u)$, and $f_1(u)$ is a right-shifted version of $f_0(u)$. More precisely, it can be deduced from (6.13) and (6.14) that $f_1(u) = f_0(u-(1-q))$. In addition, it can also be observed that $f(u)$, $f_0(u)$, and $f_1(u)$ intersect at $u = \frac{1}{2}$. This point can be proved briefly as below. From (6.10) and (6.11), we can deduce $qf(\frac{1}{2}) = f(\frac{1}{2q}) = f(1 - \frac{1}{2q})$. Further, from (6.13) and (6.14), we get $f(\frac{1}{2}) = f_0(\frac{1}{2}) = f_1(\frac{1}{2})$.

According to (6.9) and (6.11), we can obtain

$$2qf(u) = f(u/q), \quad 0 \le u < (1-q). \tag{6.15}$$
$$2qf(u) = f((u-(1-q))/q), \quad q \le u < 1. \tag{6.16}$$

From (6.15), we can get $f(0) = 2qf(0)$, which is followed by $f(0) = 0$ if $\frac{1}{2} < q < 1$ [41]. Therefore, for $\frac{1}{2} < q < 1$, $f(u)$ is strictly symmetric around $u = \frac{1}{2}$ and (6.15)

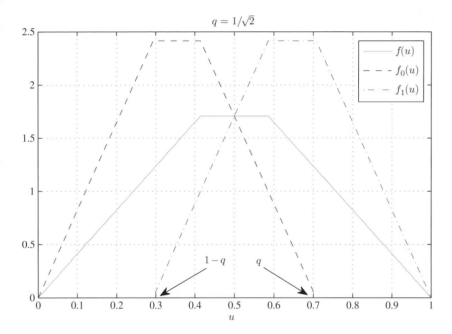

Figure 6.6 Illustrations of the relations between $f(u)$, $f_0(u)$, and $f_1(u)$. Both $f_0(u)$ and $f_1(u)$ have the same shape as $f(u)$. $f_0(u)$ is obtained by scaling $f(u)$ by q times along the u-axis. $f_1(u)$ is obtained by shifting $f_0(u)$ right by $(1 - q)$.

can be rewritten as

$$f(u) = 2qf(qu), \quad 0 \le u \le (1 - q)/q. \tag{6.17}$$

The general solution to (6.17) is

$$f(u) = \Theta(u)u^{(1-R)/R}, \quad 0 \le u \le (1 - q)/q, \tag{6.18}$$

where $\Theta(u)$ is some function satisfying $\Theta(u) = \Theta(uq^k)$, $\forall k \in \mathbb{Z}$ (see [43] for the details). From (6.17) and (6.18), we can get the following corollary.

Corollary 6.1 *(**Zero points of the initial DAC spectrum**).* For $q \in (\frac{1}{2}, \frac{\sqrt{5}-1}{2}]$,

$$f\left(\frac{q^k}{q+1}\right) = f\left(1 - \frac{q^k}{q+1}\right) = 0, \quad \forall k \in \mathbb{N}. \tag{6.19}$$

Proof. We first prove that if $\frac{1}{2} < q \le \frac{\sqrt{5}-1}{2}$, then $f(\frac{1}{q+1}) = 0$. Let $v = \frac{1}{q+1}$, then $v = 1 - qv$. According to (6.17), if $0 \le v \le \frac{1-q}{q}$, then

$$f(v) = 2qf(qv) = 2qf(1 - qv) = 2qf(v).$$

Hence, $f(v) = 0$ if $q > \frac{1}{2}$. Further, solving $v = \frac{1}{q+1} \leq \frac{1-q}{q}$ gives $q \leq \frac{\sqrt{5}-1}{2}$.

According to (6.18), $f(v) = \Theta(v)v^{(1-R)/R} = 0$ and hence $\Theta(v) = \Theta(vq^k) = 0$. Thus,

$$f(1 - vq^k) = f(vq^k) = \Theta(vq^k)(vq^k)^{(1-R)/R} = 0.$$

To assure $vq^k \leq v \leq \frac{1-q}{q}$, k must be not less than 0, that is, $k \in \mathbb{N}$. $\square$

6.5 Explicit Form of the Initial DAC Spectrum

Although (6.11) gives the implicit form of $f(u)$, its explicit form (especially the closed form) is more convenient for better understanding its properties. To get the explicit form of $f(u)$, we need to solve functional equation (6.11). Below we will first deduce a special closed form of $f(u)$ when $q = \frac{1}{2}$ (the classic AC) by directly solving (6.11) and then deduce the general explicit form of $f(u)$ with the help of Fourier transform.

Corollary 6.2 *(Spectrum of classic AC).* The spectrum of the classic AC is a shifted unit gate function, that is, if $q = \frac{1}{2}$, then $f(u) = G_1(u - \frac{1}{2})$ [41], where gate function $G_T(u)$ is defined as

$$G_T(u) = \begin{cases} 1/T, -\frac{T}{2} \leq u < \frac{T}{2} \\ 0, u < -\frac{T}{2} \text{ or } u \geq \frac{T}{2}. \end{cases} \tag{6.20}$$

Proof. From (6.17), if $q = \frac{1}{2}$, then $f(u) = f(u/2)$, $0 \leq u < 1$. Recursively, we will obtain $f(u) = \lim_{k\to\infty} f(u/2^k) = f(0)$, $0 \leq u < 1$. On substituting $f(u) = f(0)$ into (6.8), we obtain $f(u) \equiv f(0) = 1$, $0 \leq u < 1$. $\square$

This is in fact the only explicit form that we can obtain by directly solving (6.11), while to obtain the general explicit form of $f(u)$ needs more mathematical skills. Fortunately, this problem can be solved by Fourier transform. To facilitate the deduction, we shift $f(u)$ left by $\frac{1}{2}$ to get $\hat{f}(u) = f(u + \frac{1}{2})$. The benefit of such a shift lies in that it makes $\hat{f}(u)$ symmetric around $u = 0$ without changing its shape. Denote the inverse Fourier transform by $\mathcal{F}^{-1}\{\cdot\}$. The following theorem states the general explicit form of $\hat{f}(u)$.

Theorem 6.2 *(General explicit form of the initial DAC spectrum).*

$$\hat{f}(u) = \mathcal{F}^{-1}\left\{ \prod_{k=0}^{\infty} \cos\left(\frac{\omega(1-q)q^k}{2}\right) \right\}. \tag{6.21}$$

Proof. It is easy to obtain from (6.11) that

$$2q\hat{f}(u) = \hat{f}\left(\frac{u + \frac{1-q}{2}}{q}\right) + \hat{f}\left(\frac{u - \frac{1-q}{2}}{q}\right).$$

We perform a Fourier transform on both sides of the above equation, then

$$2q\mathcal{F}\{\hat{f}(u)\} = \mathcal{F}\left\{\hat{f}\left(\frac{u + \frac{1-q}{2}}{q}\right)\right\} + \mathcal{F}\left\{\hat{f}\left(\frac{u - \frac{1-q}{2}}{q}\right)\right\}.$$

Let $\hat{F}(\omega) \triangleq \mathcal{F}\{\hat{f}(u)\}$, then

$$2q\hat{F}(\omega) = q\hat{F}(q\omega)e^{i\omega(1-q)/2} + q\hat{F}(q\omega)e^{-i\omega(1-q)/2},$$

where i denotes the imaginary unit. Thus

$$\hat{F}(\omega) = \hat{F}(q\omega)\cos\left(\frac{\omega(1-q)}{2}\right)$$

$$= \hat{F}(\omega q^k)\prod_{k'=0}^{k-1}\cos\left(\frac{\omega(1-q)q^{k'}}{2}\right)$$

$$= \hat{F}(0)\prod_{k=0}^{\infty}\cos\left(\frac{\omega(1-q)q^k}{2}\right).$$

Remembering that $\hat{F}(0) = \int_{-\infty}^{\infty}\hat{f}(u)du = 1$, hence

$$\hat{F}(\omega) = \prod_{k=0}^{\infty}\cos\left(\frac{\omega(1-q)q^k}{2}\right).$$

The inverse Fourier transform of $\hat{F}(\omega)$ is just $\hat{f}(u)$. □

From (6.21), we can get the following interesting corollaries.

Corollary 6.3 *(Limit of the initial DAC spectrum).* As q approaches 1, $\hat{f}(u)$ will converge to the unit Dirac delta function $\delta(u)$.

Proof. If $q = 1$, then $\cos(\frac{\omega(1-q)q^k}{2}) \equiv 1$ and $\hat{F}(\omega) \equiv 1$. Thus $\hat{f}(u) = \mathcal{F}^{-1}\{1\} = \delta(u)$. □

Corollary 6.4 *(Expansion of sinc function).*

$$\text{sinc}(\omega) = \frac{\sin(\omega)}{\omega} = \prod_{k=1}^{\infty}\cos(\omega/2^k). \tag{6.22}$$

Proof. If $q = \frac{1}{2}$, then $\hat{f}(u)$ is the unit gate function (Corollary 6.2), whose Fourier transform is $\text{sinc}(\omega/2)$. From (6.21), we can get

$$\hat{F}(\omega) = \prod_{k=2}^{\infty} \cos(\omega/2^k) = \text{sinc}(\omega/2),$$

which is just equivalent to (6.22). □

However, we must point out that this is not the first time that (6.22) has been found. In fact, Euler was the first to discover (6.22) [44]. From (6.22), we can get the special closed form of $f(u)$ for rate-$\frac{1}{m}$ DAC as stated by the following corollary.

Corollary 6.5 *(Special closed form of the initial DAC spectrum).* Denote the convolution operation by $\otimes$ and let $T_{m'} = (1 - q)q^{-m'}$. Then the closed form of $\hat{f}(u)$ for rate-$\frac{1}{m}$ DAC, where $m \in \mathbb{Z}^+$, is

$$\hat{f}(u) = \bigotimes_{m'=1}^{m} G_{T_{m'}}(u). \tag{6.23}$$

Proof. For rate-$\frac{1}{m}$ DAC, $R = \frac{1}{m}$ and $q = 2^{-1/m}$, hence

$$\hat{F}(\omega) = \prod_{k=0}^{\infty} \cos\left(\frac{\omega(1 - q)}{2^{k/m+1}}\right).$$

Let $\frac{k}{m} = k' + \frac{m'}{m}$, where both k' and m' are integers and $0 \le m' < m$. Then

$$\hat{F}(\omega) = \prod_{m'=0}^{m-1}\left\{ \prod_{k=1}^{\infty} \cos\left(\frac{\omega(1 - q)}{2^{k+m'/m}}\right) \right\}.$$

From (6.22) and $q^{m'} = 2^{-m'/m}$, we can obtain

$$\hat{F}(\omega) = \prod_{m'=0}^{m-1} \text{sinc}(\omega(1 - q)q^{m'})$$

$$= \prod_{m'=1}^{m} \text{sinc}\left(\frac{\omega(1 - q)q^{-m'}}{2}\right).$$

Because $\mathcal{F}\{G_T(u)\} = \text{sinc}(\omega T/2)$, we have

$$\mathcal{F}^{-1}\left\{ \text{sinc}\left(\frac{\omega(1 - q)q^{-m'}}{2}\right) \right\} = G_{T_{m'}}(u),$$

where $T_{m'} = (1 - q)q^{-m'}$. Finally, (6.23) holds due to the duality between convolution and multiplication in Fourier transform. □

From corollaries 6.2, 6.3, and 6.5, it can be found that as q increases from $\frac{1}{2}$ to 1, $\hat{f}(u)$ will be gradually concentrated around $u = 0$ (changing gradually from the uniform distribution to a spike at $u = 0$). This point is also validated by a large number of simulations.

Example 6.5 *(Closed form of the initial DAC spectrum).* For rate-$\frac{1}{2}$ DAC, we have $R = \frac{1}{2}$ and $q = \frac{1}{\sqrt{2}}$. It can be deduced from (6.23) that $\hat{f}(u) = G_{T_1}(u) \otimes G_{T_2}(u)$, where $T_1 = \sqrt{2} - 1$ and $T_2 = 2 - \sqrt{2}$. Hence [41]

$$
f(u) = \begin{cases}
\dfrac{u}{3\sqrt{2} - 4}, & 0 \leq u < \sqrt{2} - 1 \\[2mm]
\dfrac{1}{2 - \sqrt{2}}, & \sqrt{2} - 1 \leq u < 2 - \sqrt{2} \\[2mm]
\dfrac{1 - u}{3\sqrt{2} - 4}, & 2 - \sqrt{2} \leq u < 1
\end{cases}
\tag{6.24}
$$

6.6 Evolution of the DAC Spectrum

Having obtained the initial DAC spectrum $f(u)$, now the remaining problem is how to deduce the stage-$(i + 1)$ DAC spectrum $g_{i+1}(u)$ from the stage-i DAC spectrum $g_i(u)$. With the help of Figure 6.7, we show below how $g_i(u)$ evolutes as the ideal DAC decoding proceeds.

Theorem 6.3 *(Evolution of the DAC spectrum).* If we have obtained the stage-i DAC spectrum $g_i(u)$, then the stage-$(i + 1)$ DAC spectrum $g_{i+1}(u)$ can be deduced from $g_i(u)$ by

$$
g_{i+1}(u) = \frac{q(g_i(qu) + g_i(qu + (1 - q)))}{1 + \int_{1-q}^{q} g_i(v)dv}.
\tag{6.25}
$$

Proof. We classify all stage-$(i + 1)$ nodes into 0-branch nodes (coming from their parents through 0-branches) and 1-branch nodes (coming from their parents through 1-branches). Let $g_{i+1}^0(u)$ be the spectrum of 0-branch nodes and $g_{i+1}^1(u)$ the spectrum of 1-branch nodes. Due to the symmetry, for equiprobable binary sources,

$$
g_{i+1}(u) = \frac{g_{i+1}^0(u) + g_{i+1}^1(u)}{2}.
\tag{6.26}
$$

Let us consider $g_{i+1}^0(u)$ first. Obviously, the random variables U_i at the parents of 0-branch nodes must fall into interval $[0, q)$, and after renormalization, interval $[0, q)$ at stage-i nodes will be mapped onto interval $[0, 1)$ at stage-$(i + 1)$

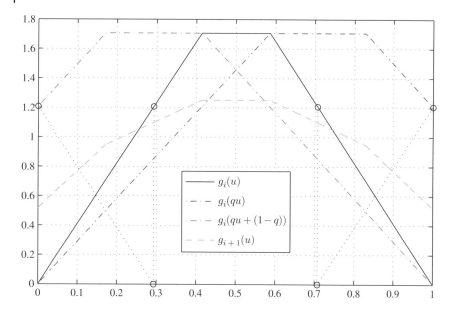

Figure 6.7 Illustration of DAC spectrum evolution. At stage-i nodes, if $0 \leq u < q$, a 0-branch will be created and then $g_i(u)$, $0 \leq u < q$ will be mapped onto $g_i(qu)$, $0 \leq u < 1$ at stage-$(i + 1)$ nodes; if $(1 - q) \leq u < 1$, a 1-branch will be created and then $g_i(u)$, $(1 - q) \leq u < 1$ will be mapped onto $g_i(qu + (1 - q))$, $0 \leq u < 1$ at stage-$(i + 1)$ nodes. Thus, $g_{i+1}(u)$ will be the normalized sum of $g_i(qu)$ and $g_i(qu + (1 - q))$.

nodes. Correspondingly, $g_i(u)$, $0 \leq u < q$ at stage-i nodes will be mapped onto $g_i(qu)$, $0 \leq u < 1$ at stage-$(i + 1)$ nodes. This mapping can be well illustrated by Figure 6.7. Hence

$$g_{i+1}^0(u) = \frac{g_i(qu)}{\int_0^1 g_i(qv)dv} = \frac{qg_i(qu)}{\int_0^q g_i(v)dv}. \tag{6.27}$$

Similarly,

$$g_{i+1}^1(u) = \frac{qg_i(qu + (1 - q))}{\int_{1-q}^1 g_i(v)dv}. \tag{6.28}$$

Due to the symmetry,

$$\int_0^q g_i(u)du = \int_{1-q}^1 g_i(u)du = \frac{1 + \int_{1-q}^q g_i(u)du}{2}. \tag{6.29}$$

□

It is important to know the limit property of $g_i(u)$ when i approaches infinity. For simplicity, we abbreviate $g_\infty(u)$ to $h(u)$, then

$$h(u) = \frac{q(h(qu) + h(qu + (1-q)))}{1 + \int_{1-q}^{q} h(v)dv}. \tag{6.30}$$

This is a functional equation with respect to $h(u)$. Before solving (6.30), we need to prove the following lemma.

Lemma 6.2 If $\frac{1}{2} < q < 1$, then $\int_{1-q}^{q} h(u)du < 1$.

Proof. Let us prove $\int_{1-q}^{q} h(u)du = 1$ to absurdity. If $\int_{1-q}^{q} h(u)du = 1$, then $\int_{0}^{1-q} h(u)du = 0$, which is followed by

$$h(u) \equiv 0, \quad 0 \le u < (1-q), \tag{6.31}$$

or equivalently $h(qu) \equiv 0, 0 \le u < \frac{1-q}{q}$. Since $(1-q) < \frac{1-q}{q}$, we can deduce from (6.30) that $h(qu + (1-q)) \equiv 0, 0 \le u < (1-q)$, or equivalently

$$h(u) \equiv 0, \quad (1-q) \le u < (1-q^2). \tag{6.32}$$

Combining (6.31) with (6.32), we get

$$h(u) \equiv 0, \quad 0 \le u < (1-q^2). \tag{6.33}$$

Recursively, $h(u) \equiv 0, 0 \le u < (1-q^k)$. As k approaches infinity, it becomes $h(u) \equiv 0, 0 \le u < 1$, which contradicts (6.8). Hence, we draw the conclusion that $\int_{1-q}^{q} h(u)du < 1$. □

Corollary 6.6 *(Final DAC spectrum).* As i approaches infinity, $g_i(u)$ will converge to the shifted unit gate function $G_1(u - \frac{1}{2})$.

Proof. For simplicity, we let $\hat{h}(u) = h(u + \frac{1}{2})$, then (6.30) becomes

$$\hat{h}(u) = \frac{q(\hat{h}(qu - \frac{1-q}{2}) + \hat{h}(qu + \frac{1-q}{2}))}{1 + \int_{1-q}^{q} h(u)du}. \tag{6.34}$$

We first prove that $\hat{h}(u)$ holds constant, where the dual problem is that the Fourier transform of $\hat{h}(u)$ is the Dirac delta function. Let us perform a Fourier transform on both sides of (6.34), then

$$\hat{H}(\omega) = \frac{2\hat{H}(\omega/q)\cos\frac{\omega(1-q)}{2q}}{1 + \int_{1-q}^{q} h(u)du}, \tag{6.35}$$

where $\hat{H}(\omega) \triangleq \mathcal{F}\{\hat{h}(u)\}$. Hence

$$\hat{H}(0) = \frac{2\hat{H}(0/q)}{1 + \int_{1-q}^{q} h(u)du}. \tag{6.36}$$

To make it hold, $\hat{H}(\omega)$ must have a Dirac delta spike at $\omega = 0$, that is,

$$\hat{H}(0) = c\delta(0), \tag{6.37}$$

where c is a constant and $\delta(\omega)$ is the Dirac delta function. Otherwise, either $\hat{H}(0) = \int_{-\infty}^{\infty} \hat{h}(u)du = 0$, which contradicts (6.8), or $\int_{1-q}^{q} h(u)du = 1$, which contradicts Lemma 6.2.

Further, because $\delta(0/q) = q\delta(0)$, we can get $1 + \int_{1-q}^{q} h(u)du = 2q$ from (6.36). Thus

$$\hat{H}(\omega) = \frac{\hat{H}(\frac{\omega}{q}) \cos\frac{\omega(1-q)}{2q}}{q}$$

$$= \lim_{k \to \infty} \hat{H}(\omega/q^k) \frac{\prod_{k'=1}^{k} \cos\frac{\omega(1-q)}{2q^{k'}}}{q^k}$$

$$= \hat{H}(\infty) \lim_{k \to \infty} \frac{\prod_{k'=1}^{k} \cos\frac{\omega(1-q)}{2q^{k'}}}{q^k}, \quad \omega \neq 0. \tag{6.38}$$

Since $\cos\frac{\omega(1-q)}{2q^k}$ does not converge as k approaches infinity, (6.38) holds only if

$$\hat{H}(\omega) = \hat{H}(\infty) = 0, \omega \neq 0. \tag{6.39}$$

Based on (6.37) and (6.39), we get $\hat{H}(\omega) = c\delta(\omega)$, which implies that $\hat{h}(u)$ is a constant.

Finally, combining with constraints (6.8) and (6.9), we can draw the conclusion $\hat{h}(u) = G_1(u)$, that is, $h(u) = G_1(u - \frac{1}{2})$. □

6.7 Numerical Calculation of the DAC Spectrum

Although (6.21) gives the explicit form of the initial DAC spectrum $f(u)$, it is usually very complex to calculate $f(u)$ directly by (6.21) as it contains the product of infinite terms. Hence, we propose below a numerical method to calculate $f(u)$, whose convergency is proved. Then we propose a numerical method to mimic the evolution of the DAC spectrum. Before introducing the algorithms, we define two notations: $\text{clip}_a^b(x) \triangleq \min(\max(a, x), b)$ and $\text{round}(x) \triangleq \lfloor x + 0.5 \rfloor$.

6.7.1 Numerical Calculation of the Initial DAC Spectrum

1) **Discretization.** The interval $[0, 1)$ is divided into N equal segments. For N sufficiently large, $f(u)$ can be approximated by $f(n/N)$, where $n \in [0 : N)$. For simplicity, $f(n/N)$ is abbreviated to $f(n)$ below.

2) **Initialization.** Let $f^{(t)}(n)$ be the approximate of $f(n)$ after t iterations. Initially, $f^{(0)}(n)$ can be arbitrarily set, if only $\sum_{n=0}^{N-1} f^{(0)}(n) = N$.

3) **Iteration.** Let $L = \text{round}(N(1-q))$ and $H = \text{round}(Nq)$.

- $0 \le n < L$: This corresponds to interval $[0, (1-q))$, thus

$$f^{(t)}(n) = \frac{f^{(t-1)}(\text{clip}_0^{N-1}(\text{round}(\frac{n}{q})))}{2q}.$$

- $L \le n < H$: This corresponds to interval $[(1-q), q)$, thus

$$f^{(t)}(n) = \frac{f^{(t-1)}(\text{clip}_0^{N-1}(\text{round}(\frac{n}{q}))) + f^{(t-1)}(\text{clip}_0^{N-1}(\text{round}(\frac{n-L}{q})))}{2q}.$$

- $H \le n < N$: This corresponds to interval $[q, 1)$, thus

$$f^{(t)}(n) = \frac{f^{(t-1)}(\text{clip}_0^{N-1}(\text{round}(\frac{n-L}{q})))}{2q}.$$

4) **Normalization.** Recall $\int_0^1 f(u)du = 1$. Hence, $f^{(t)}(n)$ should be normalized by

$$f^{(t)}(n) = \frac{Nf^{(t)}(n)}{\sum_{n=0}^{N-1} f^{(t)}(n)}.$$

5) **Termination.** The *mean squared error* (MSE) between two successive iterations is used to terminate the iteration. Let Δ be a small quantity. The iteration is terminated if

$$\text{MSE}^{(t)} = \frac{1}{N} \sum_{n=0}^{N-1} (f^{(t)}(n) - f^{(t-1)}(n))^2 < \Delta.$$

The following corollary justifies the above algorithm.

Corollary 6.7 (*Convergency of the numerical calculation of the initial DAC spectrum*). If only $\sum_{n=0}^{N-1} f^{(0)}(n) = N$, $f^{(t)}(n)$ will finally converge to $f(u)$ as t and N go to infinity.

Proof. As N goes to infinity, $f^{(t)}(n)$ will converge to a continuous function $f^{(t)}(u)$. For simplicity, we let $\hat{f}^{(t)}(u) = f^{(t)}(u + \frac{1}{2})$. Then

$$\hat{F}^{(t)}(\omega) = \hat{F}^{(t-1)}(q\omega) \cos\left(\frac{\omega(1-q)}{2}\right)$$

$$= \hat{F}^{(0)}(\omega q^t) \prod_{t'=0}^{t-1} \cos\left(\frac{\omega(1-q)q^{t'}}{2}\right),$$

where $\hat{F}^{(t)}(\omega) \triangleq \mathcal{F}\{\hat{f}^{(t)}(u)\}$. As t goes to infinity,

$$\hat{F}^{(\infty)}(\omega) = \hat{F}^{(0)}(0) \prod_{t=0}^{\infty} \cos\left(\frac{\omega(1-q)q^t}{2}\right).$$

If $\hat{F}^{(0)}(0) = \int_0^1 f^{(0)}(u)du = 1$, then

$$\hat{F}^{(\infty)}(\omega) = \prod_{t=0}^{\infty} \cos\left(\frac{\omega(1-q)q^t}{2}\right) = \hat{F}(\omega).$$

□

6.7.2 Numerical Estimation of DAC Spectrum Evolution

1) **Discretization**. The interval $[0, 1)$ is divided into N equal segments. For N sufficiently large, $g_i(u)$ can be approximated by $g_i(n/N)$, where $n \in [0 : N)$. For simplicity, $g_i(n/N)$ is abbreviated to $g_i(n)$ below.
2) **Initialization**. Initially, $g_0(n) = f^{(t)}(n)$, where $f^{(t)}(n)$ is the numerical result of $f(u)$ obtained by the above method.
3) **Iteration**. Recursively, $g_i(n)$ can be deduced from $g_{i-1}(n)$ by

$$g_i(n) = g_{i-1}(\text{clip}_0^{N-1}(\text{round}(nq))) + g_{i-1}(\text{clip}_0^{N-1}(\text{round}(nq + N(1-q)))).$$

4) **Normalization**. Since $\int_0^1 g_i(u)du = 1$, $g_i(n)$ must be normalized by

$$g_i(n) = \frac{Ng_i(n)}{\sum_{n=0}^{N-1} g_i(n)}.$$

Example 6.6 Figure 6.8(a)–(d) gives some examples of $f(u)$ that are obtained by the numerical algorithm in Section 6.7.1 with $N = 10^5$ and $f^{(0)}(n) \equiv 1$, for all $n \in [0 : N)$.

Figure 6.8(a) verifies the convergency of the numerical algorithm in Section 6.7.1 through the shape change of $f^{(t)}(n)$ with respect to t when $q = 1/\sqrt{2}$. It can be observed that as t grows, $f^{(t)}(n)$ does gradually converge to $f(u)$ given by (6.24). After 38 iterations, there is almost no perceptible difference between $f^{(t)}(n)$ and $f(u)$ (More precisely, the successive MSE has been less than 10^{-10} after 38 iterations).

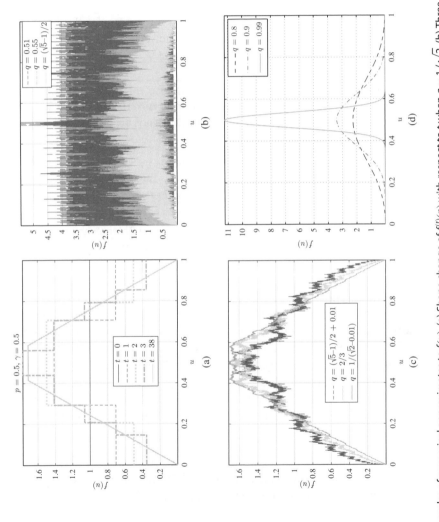

Figure 6.8 Examples of numerical approximates to $f(u)$. (a) Shape change of $f^{(t)}(n)$ with respect to t when $q = 1/\sqrt{2}$. (b) Three examples for $q \in (1/2, (\sqrt{5} - 1)/2]$. (c) Three examples for $q \in ((\sqrt{5} - 1)/2, 1/\sqrt{2})$. (d) Three examples for $q \in [1/\sqrt{2}, 1)$.

Figure 6.8(b)–(d) shows how the shape of $f(u)$ changes with respect to q. The details of simulations are as follows:

- Figure 6.8(b) includes three examples of $f(u)$ for $q \in (1/2, (\sqrt{5} - 1)/2]$. The threshold of successive MSE to cease the iteration is $\Delta = 10^{-4}$. The numbers of iterations are 586, 70, and 51 for $q = 0.51, 0.55$, and $(\sqrt{5} - 1)/2$, respectively.
- Figure 6.8(c) includes three examples of $f(u)$ for $q \in ((\sqrt{5} - 1)/2, 1/\sqrt{2})$. The threshold of successive MSE to cease the iteration is $\Delta = 10^{-9}$. The numbers of iterations are 85, 63, and 52 for $q = (\sqrt{5} - 1)/2 + 0.01, 2/3$, and $1/\sqrt{2} - 0.01$, respectively.
- Figure 6.8(d) includes three examples of $f(u)$ for $q \in [1/\sqrt{2}, 1)$. The threshold of successive MSE to cease the iteration is $\Delta = 10^{-10}$ (for $q = 0.8$ and 0.9) or 10^{-9} (for $q = 0.99$). The numbers of iterations are 39, 54, and 540 for $q = 0.8, 0.9$, and 0.99, respectively.

It can be observed from Figure 6.8(b)–(d) that $q = (\sqrt{5} - 1)/2$ and $q = 1/\sqrt{2}$ are two watersheds of $f(u)$ because $f(u)$ shows very different shapes when q falls into $(1/2, (\sqrt{5} - 1)/2]$, $((\sqrt{5} - 1)/2, 1/\sqrt{2})$, and $[1/\sqrt{2}, 1)$. Figure 6.8(b) confirms the zero points of $f(u)$ for $q \in (1/2, (\sqrt{5} - 1)/2]$, as predicted by Corollary 6.1. It can be seen that though q differs by only 0.01, the shape of $f(u)$ is very different when $q = (\sqrt{5} - 1)/2$ and $q = (\sqrt{5} - 1)/2 + 0.01$. Figure 6.8(d) shows that $f(u)$ becomes smooth for $q \in [1/\sqrt{2}, 1)$. It can also be found from Figure 6.8(d) that $f(u)$ does converge to the Dirac delta spike centered at $u = 0.5$ as q goes to 1, as predicted by Corollary 6.3.

Another finding from Figure 6.8(b)–(d) is that to achieve the same precision, the numerical algorithm in Section 6.7.1 shows different computational complexity for different q. In general, to achieve the same successive MSE, the fewest iterations are needed when $q = 1/\sqrt{2}$, while more iterations will be needed as q increases or decreases. For example, for $q \in (1/2, (\sqrt{5} - 1)/2]$, tens of iterations are needed to make the successive MSE less than 10^{-4}, while for $q \in ((\sqrt{5} - 1)/2, 1/\sqrt{2})$, tens of iterations have made the successive MSE less than 10^{-9}. In addition, it can be found that the computational complexity increases sharply when q approaches 0.5 or 1. For example, when $q = 0.9$, 54 iterations have made the successive MSE less than 10^{-10}, while when $q = 0.99$, even to make the successive MSE less than 10^{-9}, 540 iterations are needed. Similar phenomenon is also observed when q approaches 0.5.

6.8 Analyses on DAC Codes with Spectrum

It is easy to know that for a certain decoding tree of a given DAC code, the coding efficiency is subject to the following two factors:

- *tree density*: the sparser the decoding tree, that is, the fewer the paths, the better the coding efficiency, and vice versa;
- *tree structure*: the larger the Hamming distances between paths in the decoding tree, the better the coding efficiency, and vice versa.

Hence, a thorough analysis on decoding trees is the key to understanding the properties of DAC codes. This section will make use of the spectrum as a tool to analyze some properties of DAC codes. As above, we restrict our analyses to the equiprobable binary sources below. First, we reveal how DAC codes partition source space into codebooks (trees) and link codebook cardinalities with the initial spectrum. Using the final spectrum as a tool, which is a uniform distribution, we prove that due to the existence of "near" (in the sense of Hamming distance) codewords in each codebook (tree), the error probability of the ideal DAC decoder will not tend to zero as code length goes to infinity, even at rates greater than the SW limit.

6.8.1 Definition of DAC Codes

With the notations given in [21], we define the $(2^{nR}, n)$ DAC code of rate $R \in (0, 1)$ for equiprobable binary sources as

- an encoder $m : \mathbb{B}^n \to [0 : 2^{nR})$ that assigns index $m \in [0 : 2^{nR})$ (represented by a string of nR bits) to each source sequence $x^n \in \mathbb{B}^n$, and
- a decoder $\hat{x}^n : [0 : 2^{nR}) \to \mathbb{B}^n \cup \{e\}$ that assigns an estimate $\hat{x}^n \in \mathbb{B}^n$ or an error message e to each index $m \in [0 : 2^{nR})$.

The $(2^{nR}, n)$ DAC code partitions the source space $\mathbb{B}^n$ into 2^{nR} *codebooks* that are indexed from 0 to $(2^{nR} - 1)$. We call $[0 : 2^{nR})$ the *codebook index set* and the elements of each codebook the *codewords*. Denote the mth codebook by

$$C_m \triangleq \{\tilde{x}^n(0), \cdots, \tilde{x}^n(|C_m| - 1)\}, \tag{6.40}$$

where $\tilde{x}^n(j)$ is the jth codeword in C_m and $|\cdot|$ denotes set cardinality. Then

$$\mathbb{B}^n = \bigcup_{m=0}^{2^{nR}-1} C_m. \tag{6.41}$$

If x^n falls into C_m, then the binarized codebook index m is output as the DAC bitstream of x^n. Hence, each codeword in C_m just corresponds to a path from the root node to one leaf node in the decoding tree of m. Due to the one-to-one equivalence between each codebook and its decoding tree, the alternative factors deciding the achievable coding efficiency of DAC codes become *codebook cardinality* (the number of codewords in each codebook, equivalent to the density of its decoding tree) and *codeword distribution* (how codewords in each codebook are drawn from source space, equivalent to the structure of its decoding tree). According to the SW theorem, the SW limit is achievable if

- source spaces $\mathbb{B}^n$ are equally partitioned into 2^{nR} codebooks, that is, all codebooks are of the same cardinality $2^{n(1-R)}$, and
- codewords in each codebook are drawn from source space $\mathbb{B}^n$ randomly according to $p_{X^n}(x^n) = \prod_{i=0}^{n-1} p_X(x_i)$.

Unfortunately, the following deduction will show that DAC codes do not satisfy either of the above requirements.

6.8.2 Codebook Cardinality

As shown in Section 6.3.2, the $(2^{nR}, n)$ DAC encoder maps each source sequence (codeword) x^n onto interval $[l_n, l_n + \Delta u)$, where $\Delta u \triangleq 2^{-nR}$, and outputs $m = \lceil l_n/\Delta u \rceil \in [0 : 2^{nR})$ as the DAC bitstream of x^n, meaning that x^n belongs to C_m iff $l_n \in ((m-1)\Delta u, m\Delta u]$. Therefore, another physical meaning of $|C_m|$ is the number of codewords whose centers of final intervals fall into $((m-0.5)\Delta u, (m+0.5)\Delta u]$. From (6.41), we obtain $\sum_{m=0}^{2^{nR}-1} |C_m| = |\mathbb{B}^n| = 2^n$, which can be rewritten as

$$\sum_{m=0}^{2^{nR}-1} \frac{|C_m|}{2^{n(1-R)}} 2^{-nR} = 1. \tag{6.42}$$

Since there are in total 2^n codewords in $\mathbb{B}^n$, if we partition $\mathbb{B}^n$ into 2^{nR} codebooks, the average number of codewords in each codebook must be $2^n/2^{nR} = 2^{n(1-R)}$. As n increases, $2^{n(1-R)}$ will go to infinity so that $|C_m|$ does not converge in general cases. Let us define

$$\hat{f}(m\Delta u) \triangleq \frac{|C_m|}{2^{n(1-R)}}. \tag{6.43}$$

In other words, $\hat{f}(m\Delta u)$ is the scaled number of codewords whose centers of final intervals fall into $((m-0.5)\Delta u, (m+0.5)\Delta u]$. Then (6.42) will become

$$\sum_{m=0}^{(1/\Delta u)-1} \hat{f}(m\Delta u)\Delta u = 1. \tag{6.44}$$

As n goes to infinity, the interval $((m-0.5)\Delta u, (m+0.5)\Delta u]$ will converge to a real number $u \in [0, 1)$ and Δu will tend to du, the differential of u. Further, (6.44) will become

$$\int_0^1 \tilde{f}(u)du = 1, \tag{6.45}$$

where

$$\tilde{f}(u) \triangleq \lim_{n \to \infty} \hat{f}(m\Delta u). \tag{6.46}$$

It is obvious that $\tilde{f}(u)$ is just the scaled number of infinite-length source sequences (codewords) that are mapped onto u.

Let us recall the definitions of random variable U_0 and the initial spectrum $f(u)$ in Section 6.3.2. As n goes to infinity, the $(2^{nR}, n)$ DAC encoder maps each source sequence (codeword) onto a real number $u \in [0, 1)$. Thus the initial spectrum $f(u)$ actually reflects how many (in a relative sense) infinite-length source sequences (codewords) are mapped onto u. Now it is clear that both $\tilde{f}(u)$ and $f(u)$ have the same physical meaning and they are equivalent to each other, that is, $\tilde{f}(u) \equiv f(u)$. Therefore, as n goes to infinity,

$$|C_m| = \hat{f}(m\Delta u)2^{n(1-R)} \to f(m\Delta u)2^{n(1-R)}. \tag{6.47}$$

6.8.3 Codebook Index Distribution

With codebook cardinalities, we can obtain the pmf of codebook indices. Let random variable $M \sim p_M(m)$ be the codebook index output by the $(2^{nR}, n)$ DAC encoder. Since codebook index m will be output if $x^n \in C_m$,

$$p_M(m) = Pr\{x^n \in C_m\} = \sum_{x^n \in C_m} p_{X^n}(x^n)$$

$$\overset{(a)}{=} |C_m|2^{-n}$$

$$\overset{(b)}{=} \hat{f}(m\Delta u)2^{-nR} \to f(m\Delta u)2^{-nR}, \tag{6.48}$$

where (a) comes from $p_{X^n}(x^n) \equiv 2^{-n}$ for equiprobable binary sources and (b) comes from (6.47). Therefore, though the accurate value of $p_M(m)$ changes with n and thus is hard to compute, it can be well approximated by $f(m\Delta u)2^{-nR}$ for n sufficiently large. As shown in [41], the initial spectrum $f(u)$ is not a uniform function for $R \in (0, 1)$, hence the output of the $(2^{nR}, n)$ DAC encoder will not be uniformly distributed over codebook index set $[0 : 2^{nR})$, which implies a possible rate loss.

6.8.4 Rate Loss

To show how much rate loss may be caused by the nonuniformly distributed codebook index, we compute the entropy of encoder output M by

$$H(M) = -\sum_{m=0}^{2^{nR}-1} p_M(m)\log_2 p_M(m)$$

$$= -\sum_{m=0}^{2^{nR}-1} \hat{f}(m\Delta u)2^{-nR}\log_2[\hat{f}(m\Delta u)2^{-nR}]$$

$$= nR - \sum_{m=0}^{(1/\Delta u)-1} [\hat{f}(m\Delta u)\log_2\hat{f}(m\Delta u)]\Delta u. \tag{6.49}$$

Hence, the whole rate loss of the $(2^{nR}, n)$ DAC code will tend to a constant

$$\lim_{n\to\infty}(nR - H(M)) = \int_0^1 f(u)\log_2 f(u)du. \tag{6.50}$$

The right-hand side of (6.50) is actually the opposite of the differential entropy of U_0, so it will reach the minimum 0 only if $f_0(u)$ is a uniform function over $[0, 1)$. The rate loss per source symbol is $\frac{nR - H(M)}{n}$, tending to zero as n goes to infinity. Hence, though the encoder output of the $(2^{nR}, n)$ DAC code is not uniformly distributed, the rate loss per source symbol will vanish as code length increases.

6.8.5 Decoder Complexity

The decoder complexity of the $(2^{nR}, n)$ DAC code may be measured by the average number of searched codewords (or paths) during the decoding. Given $x^n \in C_m$, the optimal coding efficiency is achieved if all codewords in C_m are searched exhaustively. Hence, the decoder complexity of the $(2^{nR}, n)$ DAC code is upper bounded by

$$
E_M(|C_M|) = \sum_{m=0}^{2^{nR}-1} p_M(m)|C_m|
$$

$$
= 2^{n(1-R)} \left(\sum_{m=0}^{(1/\Delta u)-1} \hat{f}^2(m\Delta u)\Delta u \right)
$$

$$
\to 2^{n(1-R)} \int_0^1 f^2(u)du. \tag{6.51}
$$

To measure how fast the branches are created in the decoding trees formed by the ideal DAC decoder, we define the *expansion factor* as below.

Definition 6.1 *(Expansion factor).* The *depth-i expansion factor* is defined as the ratio of the number of depth-$(i + 1)$ nodes to that of depth-i nodes in the decoding trees formed by the ideal DAC decoder.

Denote the depth-i expansion factor by γ_i. It is easy to link γ_i with $g_i(u)$.

Corollary 6.8 *(Relation of the expansion factor to the DAC spectrum).* The depth-i expansion factor γ_i is related to the depth-i DAC spectrum $g_i(u)$ by $\gamma_i = 1 + \int_{1-q}^q g_i(u)du$.

Proof. At depth-i nodes, if U_i falls into $[(1 - q), q)$, then two branches are created, otherwise one branch is created. Thus

$$
\gamma_i = \int_0^{1-q} g_i(u)du + 2\int_{1-q}^q g_i(u)du + \int_q^1 g_i(u)du
$$

$$
= 1 + \int_{1-q}^q g_i(u)du.
$$

□

It is important to know the limit property of γ_i as i goes to infinity.

Corollary 6.9 *(Final expansion factor).* As i goes to infinity, the expansion factor γ_i will converge to 2^{1-R}.

Proof. $\lim_{i \to \infty} \gamma_i = 1 + \int_{1-q}^{q} h(u)du = 2q = 2^{1-R}$. □

This means that if a length-n binary sequence is compressed by the rate-R DAC encoder, the complexity of the corresponding ideal DAC decoder is approximately $O(2^{n(1-R)})$ for n sufficiently large.

According to the definition of expansion factor, $E_M(|C_M|) = \prod_{i=0}^{n-1} \gamma_i$, hence

$$\lim_{n \to \infty} \prod_{i=0}^{n-1} \frac{\gamma_i}{2^{1-R}} = \int_0^1 f^2(u)du. \tag{6.52}$$

This point can be explained in an intuitive way as below. The codebooks of larger cardinalities are more likely to appear and further, more codewords must be searched for those codebooks of larger cardinalities, so the upper bound of DAC decoding complexity is roughly $\int_0^1 f^2(u)du$ times $2^{n(1-R)}$ in the asymptotic sense.

Example 6.7 *(Expansion factor).* Figure 6.9 gives some theoretical and experimental results for the expansion factor. To obtain theoretical results, we first calculate $f(u)$ through the numerical algorithm in Section 6.7.1, with $N = 10^5$ and $f^{(0)}(n) \equiv 1$, for $n \in [0, N)$. Then seeded with $f(u)$, the numerical algorithm in Section 6.7.2 is run to obtain $g_i(u)$. Finally, the depth-i expansion factor γ_i is computed by

$$\gamma_i = 1 + \frac{\sum_{n=L}^{H-1} g_i(n)}{N}, \tag{6.53}$$

where $L = \text{round}(N(1 - q))$ and $H = \text{round}(Nq)$.

As for experimental results, we first generate 10^4 length-1024 equiprobable binary sequences and then encode each of them into a DAC code by a 31-bit DAC encoder. Notice that the practical DAC codec is of finite precision, thus the *renormalization* and *underflow* problems must be addressed as shown in [40]. These DAC codes are then parsed by the ideal DAC decoder (without branch pruning) to form 10^4 incomplete binary trees. In practice, the ideal DAC decoder is implemented by a depth-first recursive programming. Due to the exponentially increasing complexity, the maximum decoding depth is restricted to 20. We count all depth-i ($1 \leq i \leq 20$) nodes in all decoding trees (it is unnecessary to count root nodes as the number is just 10^4). Then the depth-i expansion factor γ_i ($0 \leq i \leq 19$) equals the ratio of the number of depth-$(i + 1)$ nodes to that of depth-i nodes.

From Figure 6.9 we find that theoretical results coincide with experimental results perfectly, and both theoretical and experimental curves converge to 2^{1-R}

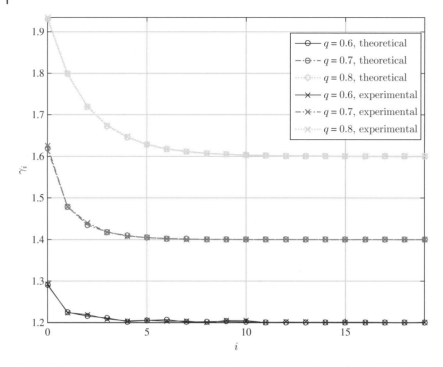

Figure 6.9 Theoretical and experimental results of the expansion factor for $q = 0.6, 0.7$, and 0.8.

rapidly as the decoding proceeds. These findings confirm our theoretical analyses very well.

6.8.6 Decoding Error Probability

Before analyzing the decoding error probability of the $(2^{nR}, n)$ DAC code, we need some knowledge about its decoding trees.

Definition 6.2 Given $x^n \in C_m$, we define $k_d(x^n)$ as the number of codewords in C_m that are d-away (in Hamming distance) from x^n, that is,

$$k_d(x^n) \triangleq |\{ j : \langle \tilde{x}^n(j), x^n \rangle = d, \forall \tilde{x}^n(j) \in C_m \}|, \tag{6.54}$$

where $\tilde{x}^n(j)$ is the jth codeword in C_m, $|\cdot|$ denotes set cardinality, and $\langle \cdot, \cdot \rangle$ denotes the Hamming distance between two binary sequences.

Obviously, $d \in [0 : n] \triangleq \{0, \cdots, n\}$ and $k_0(x^n) \equiv 1$, $\sum_{d=1}^{n} k_d(x^n) \equiv |C_m| - 1$. Actually, $\{k_0(x^n), \cdots, k_n(x^n)\}$ just forms the so-called *distance spectrum* of C_m with respect to x^n.

Definition 6.3   We classify all leaf nodes (codewords) in C_m into two subsets: D_m, the subset of twin leaf nodes, and S_m, the subset of single leaf nodes.

For example, in Figure 6.5, $D_m = \{\tilde{x}^n(1), \tilde{x}^n(2), \tilde{x}^n(4), \tilde{x}^n(5)\}$, and $S_m = \{\tilde{x}^n(0), \tilde{x}^n(3)\}$. For $|D_m|$ and $|S_m|$, it is easy to prove the following lemma.

Lemma 6.3   As n goes to infinity, $\frac{|D_m|}{|C_m|}$ will tend to $(2 - 2^R)$ and $\frac{|S_m|}{|C_m|}$ will tend to $(2^R - 1)$.

Proof. For each leaf node in D_m, the random variable U_{n-1} at its parent node must fall into $[(1 - 2^{-R}), 2^{-R})$, while for each leaf node in S_m, the random variable U_{n-1} at its parent node must fall into $[0, (1 - 2^{-R}))$ or $[2^{-R}, 1)$. Hence,

$$\frac{|D_m|}{|S_m|} = \frac{2\int_{1-2^{-R}}^{2^{-R}} g_{n-1}(u)du}{\int_0^{1-2^{-R}} g_{n-1}(u)du + \int_{2^{-R}}^1 g_{n-1}(u)du}. \tag{6.55}$$

Since the final spectrum $g_\infty(u)$ is a uniform distribution,

$$\lim_{n\to\infty} \frac{|D_m|}{|S_m|} = \frac{2^{1-R} - 1}{1 - 2^{-R}}. \tag{6.56}$$

Since $C_m = D_m \cup S_m$, $|C_m| = |D_m| + |S_m|$. Therefore, as n goes to infinity, $\frac{|D_m|}{|C_m|}$ will tend to $(2 - 2^R)$ and $\frac{|S_m|}{|C_m|}$ will tend to $(2^R - 1)$.   □

We are particularly interested in D_m as its elements have the following important property.

Lemma 6.4   Given $x^n \in D_m$, then $k_1(x^n) \geq 1$.

Proof. Because each pair of twin leaf nodes (codewords) in binary trees share the same parent node, their exclusive or (XOR) sequence must be $(0^{n-1}, 1)$, that is, their Hamming distance is always 1. Hence if $x^n \in D_m$, there is at least one codeword (its sibling) that is 1-away (in Hamming distance) from it.   □

Based on the above lemmas, we can now reach the following theorem.

Theorem 6.4   (Lower bound of decoding error probability).   Let $X \sim p_X(x)$ be the source to be compressed at the encoder and Y the side information available only at the decoder. If $p_X(x) = \frac{1}{2}$ for $x \in \mathbb{B}$ and $Pr\{X \neq Y\} = p$, the decoding error probability of the $(2^{nR}, n)$ DAC code is lower bounded by $p(2 - 2^R)$ as n goes to infinity.

Proof. Let $Pr\{\mathcal{E}\}$ be the decoding error probability of a $(2^{nR}, n)$ BDAC code, then

$$Pr\{\mathcal{E}\} = E_M(Pr\{\mathcal{E}|m\}), \tag{6.57}$$

where $Pr\{\mathcal{E}|m\}$ denotes the conditional decoding error probability given m received at the decoder. Let $\tilde{x}^n(j) \in \mathbb{B}^n$ for $j \in [0 : |C_m|)$ be the jth codeword of C_m, then

$$
\begin{aligned}
Pr\{\mathcal{E}|m\} &= \sum_{j=0}^{|C_m|-1} Pr\{x^n = \tilde{x}^n(j)|x^n \in C_m\}Pr\{\mathcal{E}|x^n = \tilde{x}^n(j)\} \\
&\overset{(a)}{=} \frac{1}{|C_m|} \sum_{j=0}^{|C_m|-1} Pr\{\mathcal{E}|x^n = \tilde{x}^n(j)\},
\end{aligned} \tag{6.58}
$$

where (a) comes from $Pr\{x^n = \tilde{x}^n(j)|x^n \in C_m\} \equiv \frac{1}{|C_m|}$. Obviously,

$$
\sum_{j=0}^{|C_m|-1} Pr\{\mathcal{E}|x^n = \tilde{x}^n(j)\} = \sum_{\tilde{x}^n(j) \in D_m} Pr\{\mathcal{E}|x^n = \tilde{x}^n(j)\} \\
+ \sum_{\tilde{x}^n(j) \in S_m} Pr\{\mathcal{E}|x^n = \tilde{x}^n(j)\}. \tag{6.59}
$$

Thus,

$$Pr\{\mathcal{E}|m\} \geq \frac{1}{|C_m|} \sum_{\tilde{x}^n(j) \in D_m} Pr\{\mathcal{E}|x^n = \tilde{x}^n(j)\}. \tag{6.60}$$

Without loss of generality, let us consider $Pr\{\mathcal{E}|x^n = \tilde{x}^n(0)\}$. Given $\{x^n = \tilde{x}^n(0)\}$, only if $\langle x^n, y^n \rangle = \langle \tilde{x}^n(0), y^n \rangle \leq \langle \tilde{x}^n(j), y^n \rangle$ for all $j \in [1 : |C_m|)$, the decoding *may but does not necessarily* succeed. Hence, the decoding success probability is upper bounded by

$$1 - Pr\{\mathcal{E}|x^n = \tilde{x}^n(0)\} \leq \prod_{j=1}^{|C_m|-1} Pr\{\langle x^n, Y^n \rangle \leq \langle \tilde{x}^n(j), Y^n \rangle\}. \tag{6.61}$$

Let $\mathcal{E}(j) \triangleq \{\langle x^n, Y^n \rangle > \langle \tilde{x}^n(j), Y^n \rangle\}$. The decoding error probability is lower bounded by

$$Pr\{\mathcal{E}|x^n = \tilde{x}^n(0)\} \geq 1 - \prod_{j=1}^{|C_m|-1} (1 - Pr\{\mathcal{E}(j)\}). \tag{6.62}$$

To compute $Pr\{\mathcal{E}(j)\}$, we need the aid of $\langle x^n, \tilde{x}^n(j) \rangle$. Given $\langle x^n, \tilde{x}^n(j) \rangle = d$, there are d pairs of symbols flipped between x^n and $\tilde{x}^n(j)$. We draw these d pairs of flipped symbols from x^n and $\tilde{x}^n(j)$ to form x^d and $\tilde{x}^d(j)$, then $x^d \oplus \tilde{x}^d(j) \equiv 1^d$. Let Y^d be the tuple of d random variables drawn from Y^n that are colocated with the elements of x^d and $\tilde{x}^d(j)$, then $\langle x^d, Y^d \rangle + \langle \tilde{x}^d(j), Y^d \rangle \equiv d$. Let $Pr\{\mathcal{E}(j)|d\} \triangleq Pr\{\mathcal{E}(j)|\langle x^n, \tilde{x}^n(j)\rangle = d\}$, then

$$
\begin{aligned}
Pr\{\mathcal{E}(j)|d\} &= Pr\{\langle x^d, Y^d \rangle > \langle \tilde{x}^d(j), Y^d \rangle\} \\
&= Pr\{\langle x^d, Y^d \rangle > d - \langle x^d, Y^d \rangle\} \\
&= Pr\{\langle x^d, Y^d \rangle \geq \lceil (d+1)/2 \rceil\}.
\end{aligned}
\tag{6.63}
$$

Since $Pr\{X \neq Y\} = p$, we have

$$
Pr\{\langle x^d, Y^d \rangle = d'\} = \binom{d}{d'} p^{d'}(1-p)^{d-d'},
\tag{6.64}
$$

where $0 \leq d' \leq d$. Hence,

$$
\begin{aligned}
Pr\{\mathcal{E}(j)|d\} &= \sum_{d'=\lceil (d+1)/2 \rceil}^{d} Pr\{\langle x^d, Y^d \rangle = d'\} \\
&= \sum_{d'=0}^{\lfloor (d-1)/2 \rfloor} \binom{d}{d'} p^{d-d'}(1-p)^{d'}.
\end{aligned}
\tag{6.65}
$$

Since (6.65) has nothing to do with j, we reduce $Pr\{\mathcal{E}(j)|d\}$ to $Pr\{\mathcal{E}|d\}$ below. Hence,

$$
\begin{aligned}
Pr\{\mathcal{E}|x^n = \tilde{x}^n(0)\} &\geq 1 - \prod_{d=1}^{n}(1 - Pr\{\mathcal{E}|d\})^{k_d(\tilde{x}^n(0))} \\
&\geq 1 - (1 - Pr\{\mathcal{E}|1\})^{k_1(\tilde{x}^n(0))} \\
&\overset{(a)}{=} 1 - (1-p)^{k_1(\tilde{x}^n(0))},
\end{aligned}
\tag{6.66}
$$

where (a) comes from $Pr\{\mathcal{E}|1\} = p$. Similarly, we can obtain

$$
Pr\{\mathcal{E}|x^n = \tilde{x}^n(j)\} \geq 1 - (1-p)^{k_1(\tilde{x}^n(j))}.
\tag{6.67}
$$

By Lemma 6.4, $k_1(\tilde{x}^n(j)) \geq 1$ for $\tilde{x}^n(j) \in D_m$, thus

$$
Pr\{\mathcal{E}|x^n = \tilde{x}^n(j)\} \geq p, \forall \tilde{x}^n(j) \in D_m.
\tag{6.68}
$$

Therefore,

$$
\begin{aligned}
Pr\{\mathcal{E}|m\} &\geq \frac{1}{|C_m|} \sum_{\tilde{x}^n(j) \in D_m} Pr\{\mathcal{E}|x^n = \tilde{x}^n(j)\} \\
&\geq \frac{p|D_m|}{|C_m|}.
\end{aligned}
\tag{6.69}
$$

By Lemma 6.3, $\frac{|D_m|}{|C_m|}$ will tend to $(2 - 2^R)$ as code length goes to infinity, thus

$$
\lim_{n \to \infty} Pr\{\mathcal{E}|m\} \geq p(2 - 2^R).
\tag{6.70}
$$

Finally, we obtain

$$\lim_{n\to\infty} Pr\{\mathcal{E}\} = \lim_{n\to\infty} E_M(Pr\{\mathcal{E}|m\})$$
$$\geq p(2 - 2^R). \tag{6.71}$$

This theorem states clearly that the decoding error probability of the $(2^{nR}, n)$ DAC code will not tend to zero as code length n goes to infinity, even at rates greater than the SW limit.

6.9 Improved Binary DAC Codec

On the basis of theoretical analyses, this section will present two techniques to improve the coding efficiency of binary DAC (BDAC) codec, that is, the permutation technique, which removes "near" codewords in each codebook, and the weighted branching (WB) technique, which reduces the mis-pruning risk of proper paths during the decoding.

6.9.1 Permutated BDAC Codec

As shown by the proof in the appendix, the reason why the decoding errors of DAC codes do not vanish as code length goes to infinity for rates greater than the SW limit is that there always exist "near" codewords in each codebook. Hence, the key to improving the coding efficiency of DAC codes is to remove "near" codewords in each codebook, which can be achieved by permutating the codewords in each codebook. Below, we will first explain the principle of *permutated BDAC* (PBDAC) codec and then prove its SW limit is achievable.

Principle

The PBDAC encoder replaces each original source sequence $x^n \in \mathbb{B}^n$ with an intermediate sequence $z^n \in \mathbb{B}^n$ by a permutation $x^n \leftrightarrow z^n$ and then compresses z^n with the BDAC encoder in Section 6.3.2 to get $m(z^n)$.

The PBDAC decoder parses $m(z^n)$ to obtain its decoding tree, which corresponds to the intermediate codebook $C'_m \triangleq \{\tilde{z}^n(0), \cdots, \tilde{z}^n(|C'_m| - 1)\}$. By the inverse permutation, C'_m can be converted back into the original codebook $C_m \triangleq \{\tilde{x}^n(0), \cdots, \tilde{x}^n(|C_m| - 1)\}$, where $\tilde{z}^n(j) \leftrightarrow \tilde{x}^n(j)$. Finally, from C_m, the codeword best matching y^n is selected as the estimate of x^n.

Through permutation, the codewords in each codebook are randomized so that those "near" codewords will vanish as code length n goes to infinity due to the *law of large numbers* (LLN). Thus given $R \geq H(X|Y)$, the decoding error probability of a $(2^{nR}, n)$ PBDAC code will tend to zero as n goes to infinity, just as proved below.

Proof of SW Limit Achievability
Basically, the following proof is similar to that of SW theorem [45] but with two differences: *unequal codebook partitioning* and *codeword permutation*. The following proof makes use of the properties of typical sets. The reader may refer to [21] for the related background knowledge.

Encoding. To send message $x^n \in \mathbb{B}^n$, transmit $m(z^n) \in [0 : 2^{nR})$, where $x^n \leftrightarrow z^n$.

Decoding. The decoder parses $m(z^n)$ to obtain the intermediate codebook C'_m and then inversely permutates each codeword in C'_m to obtain the original codebook C_m. The decoder declares that x^n is sent if it is the unique codeword in C_m such that $(x^n, y^n) \in \mathcal{T}_\epsilon^{(n)}(X, Y)$, where $\mathcal{T}_\epsilon^{(n)}(X, Y)$ denotes the jointly ϵ-typical set of X and Y, otherwise if there is none or more than one such codewords in C_m, it declares an error.

Decoding Error Probability. Denote the decoding error probability by $Pr\{\mathcal{E}\}$, then

$$Pr\{\mathcal{E}\} = E_M(Pr\{\mathcal{E}|m\}), \tag{6.72}$$

where $Pr\{\mathcal{E}|m\}$ denotes the decoding error probability if m is received. The decoder makes an error if $(x^n, y^n) \notin \mathcal{T}_\epsilon^{(n)}(X, Y)$ or there exists another codeword $\tilde{x}^n \in C_m$ such that $(\tilde{x}^n, y^n) \in \mathcal{T}_\epsilon^{(n)}(X, Y)$. Hence

$$Pr\{\mathcal{E}|m\} = Pr\{\mathcal{E}_1 \cup \mathcal{E}_2|m\} \leq Pr\{\mathcal{E}_1|m\} + Pr\{\mathcal{E}_2|m\}, \tag{6.73}$$

where

$$\begin{cases} \mathcal{E}_1 \triangleq \{(x^n, y^n) \notin \mathcal{T}_\epsilon^{(n)}(X, Y)\} \\ \mathcal{E}_2 \triangleq \bigcup_{\tilde{x}^n \neq x^n} \{(\tilde{x}^n, y^n) \in \mathcal{T}_\epsilon^{(n)}(X, Y)\}. \end{cases} \tag{6.74}$$

By the LLN, $Pr\{\mathcal{E}_1|m\}$ will tend to zero as n goes to infinity. As for $Pr\{\mathcal{E}_2|m\}$, since the codewords of C_m are drawn from $\mathbb{B}^n$ randomly,

$$Pr\{(\tilde{x}^n, y^n) \in \mathcal{T}_\epsilon^{(n)}(X, Y)\} \overset{(a)}{\leq} 2^{-n(I(X;Y)-\delta(\epsilon))}$$

$$\overset{(b)}{=} 2^{-n(1-H(X|Y)-\delta(\epsilon))}, \tag{6.75}$$

where (a) comes from the properties of jointly typical set, (b) comes from $I(X; Y) = H(X) - H(X|Y)$, and $\delta(\epsilon)$ is a function tending to zero as $\epsilon \to 0$ (see [21] for the details). By the union of events bound,

$$Pr\{\mathcal{E}_2|m\} \leq (|C_m| - 1)2^{-n(1-H(X|Y)-\delta(\epsilon))}. \tag{6.76}$$

Therefore,

$$Pr\{\mathcal{E}|m\} < |C_m|2^{-n(1-H(X|Y)-\delta(\epsilon))}$$

$$= \hat{f}(m\Delta u)2^{-n(R-H(X|Y)-\delta(\epsilon))}. \tag{6.77}$$

Finally,

$$Pr\{\mathcal{E}\} = \sum_{m=0}^{2^{nR}-1} p_M(m)Pr\{\mathcal{E}|m\}$$

$$< \left(\sum_{m=0}^{(1/\Delta u)-1} \hat{f}^2(m\Delta u)\Delta u \right) 2^{-n(R-H(X|Y)-\delta(\epsilon))}. \tag{6.78}$$

As n goes to infinity, $\delta(\epsilon)$ will tend to zero [21] and

$$\sum_{m=0}^{(1/\Delta u)-1} \hat{f}^2(m\Delta u)\Delta u \rightarrow \int_0^1 f^2(u)du < \infty. \tag{6.79}$$

Hence, $Pr\{\mathcal{E}\}$ will tend to zero if $R \geq H(X|Y)$. This completes the proof.

Remark 6.1 The complexity of codeword permutation grows exponentially as code length increases. For a large n, the codec complexity and delay will be unacceptable. Hence in practice, codeword permutation must be implemented in a segmented way: each source sequence x^n is divided into n/b (assumed to be an integer for simplicity) length-b segments and then each segment is permutated to obtain its intermediate segment. According to the LLN, as segment size b increases, the probability that "near" codewords belong to the same codebook will be reduced. Hence, the coding efficiency of PBDAC codec is subject to segment size b: the larger b, the better the coding efficiency, and vice versa.

6.9.2 BDAC Decoder with Weighted Branching

To maintain linear complexity, the BDAC decoder in [35, 36] prunes decoding trees after each stage to retain no more than K paths with the best metrics, where K is the preset maximum number of surviving paths. However, this pruning depends only on side information y^n, without considering the redundancy of the DAC bitstream. Actually, due to the nonuniformity of the initial spectrum, better coding efficiency can be achieved if one takes the intrinsic property of DAC decoding trees into consideration during the pruning.

Without loss of generality, we take the ith decoding stage as an example. Assume that source sequence x^n falls into the jth path $\tilde{x}^i(j)$ before the ith stage, that is, $x^i = \tilde{x}^i(j)$. Denote the decoder interval of $\tilde{x}^i(j)$, which corresponds to a depth-i node, by $[\tilde{l}_i(j), \tilde{l}_i(j) + 2^{-iR})$. The decoder computes $u_i(j) = (m2^{(i-n)R} - \tilde{l}_i(j)2^{iR})$ and then creates branches depending on $u_i(j)$: if $u_i(j)$ falls into $[0, (1 - 2^{-R}))$ or $[2^{-R}, 1)$, either an 0-branch or a 1-branch will be created, otherwise, both an 0-branch and a 1-branch will be created. Notice that if two branches are created, they will be treated equally regardless of $u_i(j)$. However, given $u_i(j)$, the conditional probabilities that x^n falls into an 0-branch and a 1-branch at node $\tilde{x}^i(j)$ are different. Let $p_i(x|u) \triangleq Pr\{X_i = x|U_i = u\}$ and let $g_i(u|x)$ be the conditional pdf of U_i given $\{X_i = x\}$. By Bayes' theorem,

$$\begin{cases} p_i(0|u) = \frac{g_i(u|0)}{g_i(u|0)+g_i(u|1)} \\ p_i(1|u) = \frac{g_i(u|1)}{g_i(u|0)+g_i(u|1)} \end{cases}. \tag{6.80}$$

To obtain $g_i(u|x)$, we should exploit the equivalence between the initial spectrum and the *spectrum along proper decoding paths* [42]. Due to the recursive equivalence of the *spectrum along proper decoding paths* (see [41] for the details), as $(n-i)$ goes to infinity, if $X_i = 0$, the initial spectrum of $X_{i+1}^n \triangleq (X_{i+1}, \cdots, X_{n-1})$ will tend to $f(u2^R)$, otherwise the initial spectrum of X_{i+1}^n will tend to $f((1-u)2^R)$, that is,

$$\begin{cases} g_i(u|0) \to f(u2^R) \\ g_i(u|1) \to f((1-u)2^R) \end{cases}. \tag{6.81}$$

Further, as $(n-i)$ goes to infinity,

$$\begin{cases} p_i(0|u) \to \frac{f(u2^R)}{f(u2^R)+f((1-u)2^R)} \\ p_i(1|u) \to \frac{f((1-u)2^R)}{f(u2^R)+f((1-u)2^R)} \end{cases}. \tag{6.82}$$

Notice that $p_i(x|u)$ has nothing to do with i for $(n-i)$ sufficiently large, so we will below replace it with $p_{X|U}(x|u)$. Recursively,

$$p_{X^i|U^i}(\tilde{x}^i(j)|u^i(j)) = \prod_{i'=0}^{i-1} p_{X|U}(\tilde{x}_{i'}(j)|u_{i'}(j)). \tag{6.83}$$

Finally, the metric calculation equation for each path is updated to

$$p_{X^i|Y^i,U^i}(\tilde{x}^i(j)|y^i, u^i(j)) \propto p_{X^i|Y^i}(\tilde{x}^i(j)|y^i)p_{X^i|U^i}(\tilde{x}^i(j)|u^i(j)), \tag{6.84}$$

where

$$p_{X^i|Y^i}(\tilde{x}^i(j)|y^i) = \prod_{i'=0}^{i-1} p_{X|Y}(\tilde{x}_{i'}(j)|y_{i'}). \tag{6.85}$$

For $Pr\{X \neq Y\} = p$,

$$p_{X|Y}(\tilde{x}_i(j)|y_i) = p^{(\tilde{x}_i(j)\oplus y_i)}(1-p)^{1-(\tilde{x}_i(j)\oplus y_i)}. \tag{6.86}$$

We call $p_{X^i|Y^i}(\tilde{x}^i(j)|y^i)$ the *extrinsic metric* as it is the *a posteriori* probability of $\{X^i = \tilde{x}^i(j)\}$ given $\{Y^i = y^i\}$, and $p_{X^i|U^i}(\tilde{x}^i(j)|u^i(j))$ the *intrinsic metric* as it is the *a posteriori* probability of $\{X^i = \tilde{x}^i(j)\}$ provided with $\{U^i = u^i(j)\}$. This metric updating method is named the *weighted branching* (WB) technique. Obviously, the original BDAC decoder in [35, 36] makes use of only the extrinsic metric of each path.

Remark 6.2 The WB technique makes use of the initial spectrum to compute the intrinsic metric of each path, which is based on the assumption that $(n - i)$ is sufficiently large. Hence, the WB technique should be bypassed for symbols too close to the ends of source sequences.

Remark 6.3 For equiprobable binary sources, all codewords are of the same probability 2^{-n}, so the WB technique actually does not work for the ideal decoder. However, for small K, the WB technique will benefit the proper path so that the proper path is more likely to be retained after pruning. Further, the mis-pruning risk of the proper path during the decoding will be reduced. Hence, the WB technique is more useful for a low-complexity (smaller K) decoder.

6.10   Implementation of the Improved BDAC Codec

The inventors of DAC codes noticed the fact that the residual errors of DAC codes cannot be removed by increasing the rate [35, 36], so they proposed a special ending strategy of mapping the last several symbols of each source sequence onto non-overlapped intervals, just as the classical AC. Although this ending strategy will unavoidably result in a longer bitstream, it does reduce the residual errors significantly. Hence, we refer to this BDAC codec [35, 36] as the *standard BDAC codec* and use it as the platform to implement our algorithms. Below, we integrate the WB technique into the PBDAC decoder to achieve better coding efficiency and put this codec into effect. Before going into details, we define the following parameters:

- b: segment size for the permutation technique;
- T: the number of symbols at the end of each source sequence that are mapped onto non-overlapped intervals;
- K: the maximum number of surviving paths after pruning;
- W: the number of symbols at the end of each source sequence for which the WB technique is bypassed.

6.10.1   Encoder

Principle

The sender divides each source sequence x^n into n/b (assumed to be an integer for simplicity below) length-b segments $(\mathbf{x}_0, \cdots, \mathbf{x}_{n/b-1})$, where $\mathbf{x}_{i'} \triangleq (x_{i'b}, \cdots, x_{i'b+b-1})$. Then each segment is permuted to obtain its intermediate segment $\mathbf{z}_{i'}$. Finally, $z^n \triangleq (\mathbf{z}_0, \cdots, \mathbf{z}_{n/b-1})$ is compressed by the standard BDAC encoder. Thus, the PBDAC encoder can be parameterized by the tuple of $\{n, R, b, T\}$.

```
void encoder(byte *bitstr, char *x, uint *map, int n, float R, int b, int T){
    /* definitions of local variables */
    ...
    char *z = new char[n];   // intermediate sequence
    /* segmented codeword permutation */
    for(int i=0; i<n; i+=b)  // for each segment
        permutate(z+i, x+i, map, b);
    /* standard BDAC encoding of z */
    for(int i=0; i<(n-T); i++){
    /* map symbols 0 and 1 onto [0,2^{-R}) and [1-2^{-R},1), respectively */
        ...
    }
    for(int i=(n-T); i<n; i++){
    /* map symbols 0 and 1 onto [0,2^{-1}) and [2^{-1},1), respectively */
        ...
    }
    done_encoding(bitstr, ...);
    delete[] z;
}

void permutate(char *z, char *x, uint *map, int b){
    uint temp = 0;
    for(int k=0; k<b; k++)
        temp |= (x[k]<<k);
    for(int k=0; k<b; k++)
        z[k] = (map[temp]>>k)&1;
}
```

Figure 6.10 C++ implementation of the PBDAC encoder for equiprobable binary sources.

Implementation

The C++implementation of the $\{n, R, b, T\}$ PBDAC encoder is sketched in Figure 6.10. Among the arguments of the encoder function, bitstr is the output DAC bitstream, x is the input source sequence, and map is the input permutation array. In the body of the encoder function, the permutate function is called to permutate each segment of x. As for the details of range renormalization, underflow handling, and the done_encoding function, the reader may refer to the classical paper [40].

6.10.2 Decoder

Principle

The decoding contains n stages that are grouped into n/b *super-stages*. During each super-stage, the decoder runs full search for each surviving path of the previous super-stage. After each super-stage, the decoding tree is pruned to retain no more than K paths. Therefore, at the end of the current super-stage, each surviving path of the previous super-stage will have at most 2^b descendants and there will be at most $K2^b$ paths in total, each corresponding to an intermediate sequence. The practical decoder complexity is hence subject to b and K. Further, the PBDAC decoder with WB can be parameterized by the tuple of $\{n, R, b, T, K, W\}$.

After each stage, the intrinsic metric of each path is updated based on its intermediate sequence. Let $\tilde{z}^i(j)$ be the intermediate sequence of the jth path, then

$$p_{X^i|U^i}(\tilde{z}^i(j)|u^i(j)) = \prod_{i'=0}^{i-1} p_{X|U}(\tilde{z}_{i'}(j)|u_{i'}(j)). \qquad (6.87)$$

After each super-stage, the intermediate sequence of each path is inversely permutated to get its original sequence and then the extrinsic metric is updated based on the original sequence and side information. For the jth path,

$$p_{X^i|Y^i}(\tilde{x}^i(j)|y^i) = \prod_{i'=0}^{i-1} p_{X|Y}(\tilde{x}_{i'}(j)|y_{i'}), \qquad (6.88)$$

where $\tilde{x}^i(j) \leftrightarrow \tilde{z}^i(j)$. The overall metric of the jth path is then

$$p_{X^i|Y^i,U^i}(\tilde{x}^i(j)|y^i, u^i(j)) \propto p_{X^i|Y^i}(\tilde{x}^i(j)|y^i) p_{X^i|U^i}(\tilde{z}^i(j)|u^i(j)). \qquad (6.89)$$

Further, if there are more than K paths, the decoding tree will be pruned to retain K best paths with the maximum overall metrics.

The above process is iterated. Finally, after n stages, the original sequence of the best path with the maximum overall metric is output as the estimate of x^n.

Implementation

Although the DAC decoding forms a tree, it is found that tree data structure will complicate the programming. Hence, we use array data structure to implement the decoder. To begin with, we define a structure named tagPath (see Figure 6.11) for decoding paths. Then with structure tagPath, we sketch the C++ implementation of the decoder in Figure 6.11.

Among the arguments of the decoder function, xr is the output optimal estimate of x, bitstr is the input DAC bitstream, y is side information, demap is the inverse permutation array, that is, demap[map[a]] = a, and p is the crossover probability between x and y.

Before iteration, the decoder allocates the memory for BUDGET paths and sets n_paths to 1 as the decoding always starts from the root node. At each stage, the decoder calls the branch function, which is detailed in Figure 6.12, to extend each active path along the 0-branch and/or the 1-branch, and then updates the intrinsic metric of each path for symbols not too close to the end of the sequence (i<(n-W)). The cal_metric_in function is based on (6.87) and (6.82), so we omit its detailed C++ implementation in this paper. After each super-stage ((i+1)%b==0), the intermediate segments are inversely permutated to obtain the original segments. Because demap[map[a]] = a, forward permutation and inverse permutation can be realized with the same permutate function (see Figure 6.10). The extrinsic metrics are then updated according to the original symbols and side information. If there are more than K paths, these paths will be sorted in descending order of overall metric and then only the first K paths are retained (n_paths=K). Finally, after n stages, the original sequence of the 0th path is output as the optimal estimate of x. As for the

```
typedef struct tagPath{
    uint low, high;    // lower and upper bounds of the interval
    uint code;         // bits in the finite-precision decoder buffer
    uint ptr;          // bit pointer to the input bitstream
    float metric;      // overall path metric
    float u;           // u = (code-low)/(high-low+1)
    char z[b];         // intermediate segment of one super-stage
    char x[n];         // original sequence
};

void decoder(char *xr, byte *bitstr, char *y, uint *demap, float p,
             int n, float R, int b, int T, int K, int W){
    /* definitions of local variables */
    ...
    tagPath *paths = new tagPath[BUDGET];
    int n_paths = 1;   // number of active paths
    /* initialize paths[0] */
    ...
    for(int i=0; i<n; i++){ // for each stage
        n_paths = branch(paths, bitstr, n_paths, i%b, (i<n-T)?R:1);
        if(i<(n-W)){ // for symbols not too close to the end of sequence
            /* update intrinsic metric */
            for(int j=0; j<n_paths; j++)   // for each active path
                paths[j].metric *= cal_metric_in(paths[j].z[i%b], paths[j].u);
        }
        if((i+1)%b==0){ // after each super-stage
            for(int j=0; j<n_paths; j++){ // for each active path
                /* inverse permutation */
                permutate(paths[j].x+i-b+1, paths[j].z, demap, b);
                /* update extrinsic metric */
                for(int k=i; k>(i-b); k--)
                    paths[j].metric *= (paths[j].x[k]==y[k]) ? (1-p):p;
            }
            if(n_paths>K){ // there are more than K paths
                /* sort paths in descending order of overall metric */
                ...
                /* retain K best paths and prune other inferior paths */
                n_paths = K;
            }
        }
    }
    memcpy(xr, paths[0].x, n);
    delete[] paths;
}
```

Figure 6.11 C++ implementation of the PBDAC+WB decoder for equiprobable binary sources.

details of range renormalization and underflow handling, the reader may refer to the classical paper [40].

Decoder Memory Budget Theoretically, at the end of the current super-stage, there will be at most $K2^b$ paths in total, which requires a huge memory budget for large b and K. However, in practice, a much smaller memory budget will be enough. As we know, the spectrum converges to the uniform distribution rapidly, so at the end of the current super-stage each surviving path of the previous super-stage will have on average $2^{b(1-R)}$ descendants and there will be on average $K2^{b(1-R)}$ paths in total. Therefore, this paper recommends allocating the memory of BUDGET = $\min(K2^b, 2K2^{b(1-R)})$ paths at the decoder, which is affordable for K, b, and $(1-R)$ not too large. Such a setting will not cause memory overflow in practice.

```
int n_sons = branch(tagPath paths[], byte *bitstr, int n_parents, int k,
                    float alpha){
    /* definitions of local variables */
    ...
    int n_sons = n_parents;
    for(int j=0; j<n_parents; j++){  // for each parent node
        tagPath *p = paths + j;
        p->u = (p->code - p->low)/(p->high - p->low + 1);
        if(p->u < 1-pow(2,-alpha))
            p->z[k] = 0;   // extend the j-th path along 0-branch
        else if(p->u >= pow(2,-alpha))
            p->z[k] = 1;   // extend the j-th path along 1-branch
        else{
            tagPath *q = paths + n_sons;
            memcpy(q, p, sizeof(tagPath));   // duplicate the j-th path
            p->z[k] = 0;   // extend the j-th path along 0-branch
            q->z[k] = 1;   // extend the n_sons-th path along 1-branch
            n_sons++;
        }
        /* range renormalization and underflow handling */
        ...
    }
    return n_sons;
}
```

Figure 6.12 C++ implementation of the branch function.

6.11 Experimental Results

This section gives several experiments to verify the effectiveness of the permutation and WB techniques. The following four DAC codecs will be assessed:

- BDAC codec: the standard BDAC codec, without using the WB technique at the decoder;
- PBDAC codec: the permutated BDAC codec, without using the WB technique at the decoder;
- BDAC+WB codec: the encoder is the same as the BDAC encoder, but the WB technique is used at the decoder;
- PBDAC+WB codec: the encoder is the same as the PBDAC encoder, but the WB technique is used at the decoder.

We compare the above four DAC codecs from four aspects. First, we show how the permutation technique is subject to segment size by comparing the PBDAC codec with the BDAC codec. Second, we show how the WB technique is subject to the number of surviving paths after each super-stage by comparing the BDAC+WB codec with the BDAC codec. Third, we compare these four DAC codecs together with the LDPC codec. Finally, we attempt the application of the PBDAC codec to nonuniform binary sources. In the following experiments, the precision of DAC codecs is fixed to 16 bits and the decoder allocates the memory for BUDGET= 2^{17} paths. No memory overflow is found in any experiment. Only the results for rate-1/2 code are presented below, while the results for other rates are very similar. When $R = 1/2$, the closed form of the initial spectrum is given by (6.24). Hence, it is easy to obtain

$$p_{X|U}(1|u) = \begin{cases} 0, & 0 \le u < (1 - \frac{1}{\sqrt{2}}) \\ 1 - \frac{\sqrt{2}-1}{\sqrt{2}u}, & (1 - \frac{1}{\sqrt{2}}) \le u < (\sqrt{2} - 1) \\ \frac{\sqrt{2}u}{2-\sqrt{2}} - \frac{1}{\sqrt{2}}, & (\sqrt{2} - 1) \le u < (2 - \sqrt{2}). \\ \frac{\sqrt{2}-1}{\sqrt{2}(1-u)}, & (2 - \sqrt{2}) \le u < \frac{1}{\sqrt{2}} \\ 1, & \frac{1}{\sqrt{2}} \le u < 1 \end{cases} \tag{6.90}$$

Further, $p_{X|U}(0|u) = 1 - p_{X|U}(1|u)$. The results are presented in terms of *frame-error rate* (FER) below, while the results of *symbol-error-rate* (SER) are very similar.

Effect of Segment Size on Permutation Technique

As the first experiment, the effect of segment size b on the permutation technique is assessed. To achieve this goal, we compare the PBDAC codec with the BDAC codec in Figure 6.13(a). Other codec parameters are fixed as $n = 192$, $R = \frac{1}{2}$, $T = 12$, and $K = 64$. Three segment sizes (8, 12, and 16) are tested. Correspondingly, each source sequence is divided into $192/8 = 24, 192/12 = 16$, or $192/16 = 12$ segments, respectively. It can be seen from Figure 6.13(a) that as b increases, the gain of PBDAC over BDAC becomes more and more significant. This phenomenon verifies the results given at the end of Section 6.9.1.

Effect of Surviving-Path Number on WB Technique

The second experiment studies the effect of K, the number of surviving paths after each super-stage, on the WB technique. Other codec parameters are fixed as $n = 192$, $R = 1/2$, $b = 12$, $T = 12$, and $W = 24$. Three surviving-path numbers (32, 128, and 512) are tested. The BDAC codec and the BDAC+WB codec are compared and the results are presented in Figure 6.13(b). It can be seen that both the BDAC codec and the BDAC+WB codec show lower FER as K increases. In addition, Figure 6.13(b) shows that the gain of the BDAC+WB codec over the BDAC codec is more significant for small K and becomes trivial for large K. This phenomenon verifies the results given at the end of Section 6.9.2.

Comparison with LDPC Codes

The third experiment compares various codec schemes. Besides DAC codes, we also consider the rate-adaptive LDPC codes proposed in [46], whose software implementation is publicly available on [47]. We use the two length-396 LDPC codes (one is regular and the other is irregular). The DAC codec parameters are fixed as $n = 396$, $R = 1/2$, $b = 12$, $T = 12$, $K = 512$, and $W = 24$. Notice that for DAC codes, the length of the resulting bitstream is actually $(n - T)R + T =$

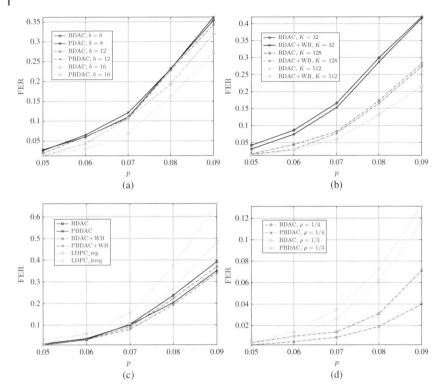

Figure 6.13 The *y*-axis is the residual FER and the *x*-axis is the crossover probability between source and side information, varying from 0.05 to 0.09 with increment 0.01. Each data point is the average over 1000 trials. (a) Effect of *b* on the permutation technique. Other codec parameters are fixed as $(n, R, T, K) = (192, 1/2, 12, 64)$. (b) Effect of *K* on the WB technique. Other codec parameters are fixed as $(n, R, b, T, W) = (192, 1/2, 12, 12, 24)$. (c) Comparison of four DAC codecs with two LDPC codecs. The DAC codec parameters are fixed as $(n, R, b, T, K, W) = (396, 1/2, 12, 12, 512, 24)$. (d) Application of PBDAC to nonuniform sources. The codec parameters are fixed as $(n, R, b, T, K) = (192, 1/2, 12, 12, 512)$.

204 bits. Hence, the rates of LDPC codes are fixed to 34/66 [47] in the experiment to get length-204 syndromes. It can be seen from Figure 6.13(c) that DAC codes significantly outperform LDPC codes. Among those DAC codecs, the BDAC codec performs worst and the PBDAC+WB performs best. In addition, both the PBDAC codec and the BDAC+WB codec show significant gains over the BDAC codec. These results convincingly confirm the effectiveness of the permutation technique and the WB technique.

Application of PBDAC to Nonuniform Sources

Finally, we attempt to apply the permutation technique to nonuniform sources. In Figure 6.13(d), the PBDAC codec is compared with the BDAC codec for sources with various bias probabilities. Notice that direct application of the permutation technique to nonuniform sources will cause a serious problem: even though the original sequence is not uniformly distributed, the intermediate sequence after permutation will become equiprobable, which will cause the change in bitstream length. For a fair comparison, we must guarantee an equi-length bitstream for both the BDAC codec and the PBDAC codec. To solve this problem, we implement the BDAC codec and the PBDAC codec in a special way, as described below. Let us begin with the BDAC codec. We first generate X with bias probability ρ and then generate Y according to X with crossover probability p. The encoder maps symbols 0 and 1 onto intervals $[0, 2^{-R})$ and $[(1 - 2^{-R}), 1)$, respectively. At the decoder, the metric of each path is computed by

$$p_{X^i|Y^i}(\tilde{x}^i(j)|y^i) = \prod_{i'=0}^{i-1} p_{X|Y}(\tilde{x}_{i'}(j)|y_{i'}).$$ (6.91)

It is easy to know $p_{X|Y}(0|0) \propto (1 - \rho)(1 - p)$, $p_{X|Y}(1|0) \propto \rho p$, $p_{X|Y}(0|1) \propto (1 - \rho)p$, and $p_{X|Y}(1|1) \propto \rho(1 - p)$. The implementation of the PBDAC codec is similar, except that a permutation process is added at the encoder and an inverse permutation process at the decoder. In this way, both the BDAC codec and the PBDAC codec will produce an equi-length bitstream so that a fair comparison can be made. The codec parameters are fixed as $n = 192$, $R = 1/2$, $b = 12$, $T = 12$, and $K = 512$. Two source bias probabilities, $1/3$ and $1/4$, are tested. The residual FER with respect to p is plotted in Figure 6.13(d). It can be seen that the permutation technique does involve a significant gain in coding efficiency, even for nonuniform sources.

6.12 Conclusion

This chapter introduces the DAC, a source-coding implementation of SW coding. To study the properties of DAC codes, a tool named the DAC spectrum is developed, which can be obtained by a recursive formulation. With the help of Fourier transform, we get the closed form of the initial DAC spectrum. To calculate the DAC spectrum, a numerical method is proposed, whose convergency is proved. We also show how DAC codes partition source space into codebooks and link codebook cardinalities with the initial spectrum. To measure the complexity of the full-search DAC decoder, we define the expansion factor and link it with the DAC spectrum. Further, by exploiting the final spectrum, we prove that the decoding errors of DAC codes do not vanish as code length goes to infinity. Based on theoretical analyses, two techniques are proposed to improve the coding efficiency of BDAC codes, that is, the permutation technique that randomizes the codewords in each codebook to remove "near" codewords, and the WB technique that exploits the nonuniform distribution of codebook cardinalities to reduce the mis-pruning risk of proper paths during the decoding.

There are still lots of open problems regarding the properties of DAC. First, this chapter answers only the complexity problem of the full-search DAC decoder, while the coding efficiency of the full-search/partial-search DAC decoder still remains a very difficult issue. Second, this chapter treats only a special kind of source, that is, equiprobable binary sources, while the concept of DAC spectrum can be generalized in parallel to other sources, for example binary sources with unequally-likely symbols and even nonbinary sources. These problems will be tackled in the future.

The reader may refer to [48–51] for more advances in DAC.

7

Wyner–Ziv Code Design

In Chapter 3, we introduced the main theoretical results of WZ coding. In this chapter, we will consider the practical design of WZ codes. Figure 7.1 depicts the block diagram of a generic WZ codec, where X is first source coded, that is, quantized, as in a conventional lossy source coding system. However, there is usually correlation remaining between the quantized version of X and the side information Y, so SW coding should be employed to exploit this correlation to reduce the rate of X. Since SW coding is essentially equivalent to channel coding, WZ coding is actually a source-channel coding problem. Therefore, both quantization loss due to source coding and binning loss due to channel coding coexist in WZ coding. In order to reach the WZ limit, one needs to employ both source codes (i.e., quantization techniques) that can achieve the granular gain and channel codes (e.g., turbo and LDPC codes) that can approach the SW limit. In the following we will introduce two important vector quantization methods, that is, lattice quantization (LQ) and trellis coded quantization (TCQ), and outline two WZ code designs based on LQ/TCQ and LDPC codes.

7.1 Vector Quantization

Consider the problem that we have to transmit a discrete-time continuous source $X_1^n = X_1, \cdots, X_n$ over a digital channel. Since the source is continuous, we have to first digitize or *quantize* it. The simplest quantizer to achieve this will be a uniform scalar quantizer, that is, the input is quantized independently one part at a time and the quantization step size is uniform. For example, consider that X_1^n is generated from a memoryless Gaussian source with zero mean and unit variance. Since we know that the chance of X to be three times of the variance (i.e., 3) away from the mean will be very small, for a 3-bit quantizer, we may set $(2^3 = 8)$ *quantization points* as $\{-3.5, -2.5, -1.5, -0.5, 0.5, 1.5, 2.5, 3.5\}$. An input X_i will then be quantized to the quantization point closest to it, for example if $X_i = 0.3$, it should be quantized to 0.5.

Distributed Source Coding: Theory and Practice, First Edition. Shuang Wang, Yong Fang, and Samuel Cheng.
© 2017 John Wiley & Sons Ltd. Published 2017 by John Wiley & Sons Ltd.
Companion Website: www.wiley.com/go/cheng/dsctheoryandpractice

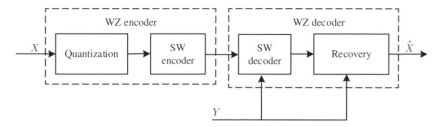

Figure 7.1 Systematic framework of WZ codec.

More concretely, let us define eight quantization *regions*, $(-\infty, -3]$, $(-3, -2]$, $(-2, -1]$, $(-1, 0]$, $(0, 1]$, $(1, 2]$, $(2, 3]$, $(3, \infty)$, and assign indices one to eight, respectively, to the regions. During encoding, the quantizer identifies which region the input lies in and outputs the corresponding indices. During decoding, the dequantizer simply returns the quantization point of the region of the received index. The first and last regions, $(-\infty, -3]$ and $(3, \infty)$, are called the overloading regions and the rest are called the granular regions. Note that quantization errors at the overloading regions can be significantly larger than those at the granular regions. However, the probability that an input lies in the overloading regions can be much smaller if we design our quantizer well. The quantization errors at the granular regions can also be very similar if the regions have the same size and the rate is sufficiently high (the regions are small in size in the sense that input distribution does not change significantly inside a region).

While uniform scalar quantization is very easy to understand and use, it may not be very efficient if it is only used alone. This is especially easy to see when we consider the statistics of pairs of X. For example, in Figure 7.2 we simulate pairs of X with the statistics described previously and we can see that very few pairs will be quantized to the "corner" points such as $(3.5, 3.5)$. Note that if we store and transmit the quantization indices without further processing, we will need 6 bits for each pair. However, since the statistics is so skewed, we can simply reduced the number of bits needed by applying entropy coding (e.g., Huffman coding) on the quantization index to reduce code rate.

Entropy coding helps to reduce the bit rate significantly. However, more gain can still be achieved by improving the quantization strategy. To explain this, let us consider a moderate to high coding rate case where the quantization region is quite small with respect to the change in the density of the distribution. Therefore, given that an input is quantized to a particular region, we can assume that the source is equally likely to appear anywhere in the region. Thus, to minimize the quantization error (in terms of MSE), one should pick the center of mass as the quantization point for a region (that is exactly what we did!). Then for two dimensions, assuming the region has unit area (see Figure 7.2), the MSE will be $\int_{1/2}^{1/2} \int_{1/2}^{1/2} (x^2 + y^2) dxdy = \left(\int_{-1/2}^{1/2} x^2 dx \right) + \left(\int_{-1/2}^{1/2} y^2 dy \right) = 1/6$.

Figure 7.2 Simulated pairs out of a memoryless Gaussian source with zero mean and unit variance. One can see that the distribution of a pair of quantization indices will be very skewed. For example, a pair will be unlikely to quantize to the index pair corresponding to the quantization point $(3.5, 3.5)$.

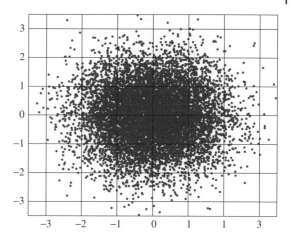

Per each dimension, the quantization error is $1/12$. In general, we see that the quantization error excluding the overloading regions is $N/12$ if the granular regions are N-dimensional hypercube. Note that the error per sample (dimension) is still $1/12$ and we don't have any gain since the quantizer treats each dimension independent of the rest in the hypercube case.

Instead, if we return to the two-dimensional case and use a hexagonal granular region (see the next section on lattice theory), we can easily see that the quantization error will decrease for the same granular region size. This is because given that the input source is uniformly distributed inside a granular region, for two small areas of the same size in the region the contribution to the MSE will only depend on the distance from the quantization point (i.e., the center of mass). Therefore, the MSE error will always decrease if an area in the region can move closer to the quantization point. Essentially, this is equivalent to making the region more spherical. Since a hexagonal granular region is more spherical than a square region, we expect that the MSE will be smaller. As a limit, if we can really pack the space in two dimensions using unit circles, the quantization error will be $\int_0^{1/\sqrt{\pi}} 2\pi r(r^2)\,dr = \frac{1}{2\pi}$. So per dimension, the MSE is $\frac{1}{4\pi} \approx 12.57^{-1} \leq 1/12$ as expected. We can further decrease the quantization error if we consider spherical granular regions in higher dimensional space. For example, in three dimensions the same unit size granular region will lead to quantization error $\int_0^{\left(\frac{3}{4\pi}\right)^{1/3}} 4\pi r^2(r^2)\,dr = \frac{3}{5}\left(\frac{3}{4\pi}\right)^{2/3}$, or $\frac{1}{5}\left(\frac{3}{4\pi}\right)^{2/3} \approx 12.99^{-1}$ per dimension.

The example above is the simplest kind of *vector quantizer*. Unlike a scalar quantizer, a component cannot be quantized independently for vector quantization. For instance, in the hexagonal case, to find the region that encloses a particular input pair (X_1, X_2) we have to process (X_1, X_2) at the same time. The potential gain (decrease in quantization error) by introducing a more spherical granular region in contrast to cubic lattice is known as granular gain. For low dimensions, vector quantization can be realized with lattices, while for high dimensions, TCQ is a good technique for realizing vector quantization.

7.2 Lattice Theory

7.2.1 What is a Lattice?

Let $\mathbf{V} \triangleq (\mathbf{v}_1, \cdots, \mathbf{v}_n)$ be an $n \times n$ matrix formed by n linearly *independent* (not necessary *orthogonal*) column vectors $\mathbf{v}_i \in \mathbb{R}^{n \times 1}$. That means all n column vectors $\mathbf{v}_i$ of $\mathbf{V}$ spans $\mathbb{R}^n$, and $\mathbf{V}$ forms a basis of $\mathbb{R}^n$. Therefore, each point in space $\mathbb{R}^n$ can be represented by a linear combination of $\mathbf{v}_i$ for $1 \leq i \leq n$. We consider the following special linear combination:

$$\Lambda \triangleq \left\{ \lambda \, : \, \lambda = \sum_{i=1}^{n} a_i \mathbf{v}_i, \forall a_i \in \mathbb{Z} \right\}. \tag{7.1}$$

In plain words, λ is a special point in $\mathbb{R}^n$ that is the linear combination of $\mathbf{v}_i$ with integral weights, and all such points form the set $\Lambda \in \mathbb{R}^n$. We call Λ the n-dimensional (n-D) lattice generated by $\mathbf{V}$. Correspondingly, $\mathbf{V}$ is called the *generator* matrix of Λ. We call $\mathbf{G} = \mathbf{V}^T \mathbf{V}$ the Gram matrix of Λ.

An n-D lattice Λ is a discrete subgroup of $\mathbb{R}^n$, and its elements are discrete points regularly distributed in $\mathbb{R}^n$. One can also view the n-D lattice Λ as a regular tiling of space $\mathbb{R}^n$ by a primitive cell (an n-D polyhedron). Thus the n-D lattice Λ divides the whole of $\mathbb{R}^n$ into equal polyhedra (copies of the primitive cell). The n-D volume of the primitive cell is called the covolume of Λ, which is equal to the absolute value of $|\mathbf{V}|$, the determinant of generator matrix $\mathbf{V}$. However, note that different bases may generate the same lattice Λ, but the absolute values of the determinants of all these generator matrices must be the same. Hence, the covolume of Λ is usually denoted by $d(\Lambda)$. If $d(\Lambda) = 1$, Λ is called *unimodular*.

Examples
Let us give two examples to illustrate the concepts of lattice.

Square 2D Lattice The simplest lattice in $\mathbb{R}^n$ is $\Lambda = \mathbb{Z}^n$, whose generator matrix $\mathbf{V}$ is the $n \times n$ unit matrix I_n. Let $\mathbf{v}_1 = (1,0)^T$ and $\mathbf{v}_2 = (0,1)^T$. Then $\lambda = (a_1, a_2)^T$ for all $a_i \in \mathbb{Z}$ and $\Lambda = \mathbb{Z}^2$. Since $|(\mathbf{v}_1, \mathbf{v}_2)| = 1$, Λ is a unimodular lattice. As shown in Figure 7.3(a), the primitive cell of Λ is a square of area 1 (the shaded region), which gives the name of square 2D lattice.

Hexagonal 2D Lattice Let $\mathbf{v}_1 = (1,0)^T$ and $\mathbf{v}_2 = (1/2, \sqrt{3}/2)^T$. Then

$$\lambda = a_1 \mathbf{v}_1 + a_2 \mathbf{v}_2 = (a_1 + a_2/2, \sqrt{3}a_2/2)^T \tag{7.2}$$

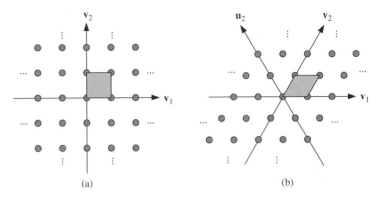

Figure 7.3 Two lattices in $\mathbb{R}^2$. (a) Square 2D lattice $\Lambda = \mathbb{Z}^2$, where $\mathbf{v}_1 = (1,0)^\mathsf{T}$ and $\mathbf{v}_2 = (0,1)^\mathsf{T}$. (b) Hexagonal 2D lattice, where $\mathbf{v}_1 = (1,0)^\mathsf{T}$ and $\mathbf{v}_2 = (1/2, \sqrt{3}/2)$. In this case, $\Lambda = a_1\mathbf{v}_1 + a_2\mathbf{v}_2$ for all $a_i \in \mathbb{Z}$. An alternative basis is $(\mathbf{v}_1, \mathbf{u}_2)$, where $\mathbf{u}_2 = (-1/2, \sqrt{3}/2)$.

for all $a_i \in \mathbb{Z}$. Since $|(\mathbf{v}_1, \mathbf{v}_2)| = \sqrt{3}/2 \neq 1$, Λ is not a unimodular lattice. As shown in Figure 7.3(b), the primitive cell of Λ is a parallelogram of area $\sqrt{3}/2$ (the shaded region). We call Λ the hexagonal 2D lattice (the reason will be explained below). As shown in Figure 7.3(b), an alternative basis for the hexagonal 2D lattice is $(\mathbf{v}_1, \mathbf{u}_2)$, where $\mathbf{u}_2 = (-1/2, \sqrt{3}/2)$. It can be seen that both bases produce exactly the same lattice.

Dual Lattice

The dual lattice (or reciprocal lattice sometimes) of Λ is a lattice whose elements all have integral inner products with all elements in Λ. Let Λ^* denote the dual lattice of Λ. Then

$$\Lambda^* \triangleq \{\lambda^* : \lambda^* \in \mathbb{R}^n \text{ and } \langle \lambda^*, \lambda \rangle \in \mathbb{Z}, \forall \lambda \in \Lambda\}. \tag{7.3}$$

As suggested by its name, the dual lattice of Λ is unique. Let $\mathbf{v}_i^*$ be the ith basis vector of Λ^*. Although Λ^* can be generated from a different basis, there always exists a basis satisfying $\langle \mathbf{v}_i, \mathbf{v}_j^* \rangle = \delta_{ij}$, where $\delta_{ij} = 1$ for $i = j$ and $\delta_{ij} = 0$ for $i \neq j$. Let $\mathbf{V}^* \triangleq (\mathbf{v}_1^*, \cdots, \mathbf{v}_n^*)$. Then $\mathbf{V}^\mathsf{T}\mathbf{V}^* = I_n$ and $\mathbf{V}^* = (\mathbf{V}^\mathsf{T})^{-1} = (\mathbf{V}^{-1})^\mathsf{T}$. The Gram matrix of Λ^* is $\mathbf{G}^* = (\mathbf{V}^*)^\mathsf{T}\mathbf{V}^* = \mathbf{G}^{-1}$.

Example of dual lattice The dual lattice of a hexagonal 2D lattice can be generated from the basis vectors $\mathbf{v}_1^* = (1, -1/\sqrt{3})^\mathsf{T}$ and $\mathbf{v}_2^* = (0, 2/\sqrt{3})^\mathsf{T}$. It is easy to check $\mathbf{G}^* = \mathbf{G}^{-1}$.

Integral Lattice

The lattice Λ is called integral when the inner products between each two lattice vectors are all integral, that is, $\langle \lambda_1, \lambda_2 \rangle \in \mathbb{Z}$ for all $\lambda_1, \lambda_2 \in \Lambda$. The sufficient and

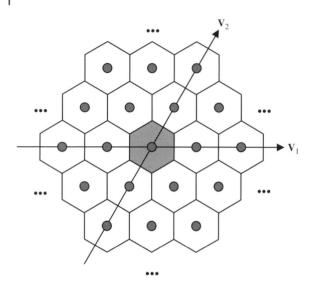

Figure 7.4 2D space division with the hexagonal 2D lattice, where $\mathbf{v}_1 = (1,0)^{\mathsf{T}}$ and $\mathbf{v}_2 = (1/2, \sqrt{3}/2)^{\mathsf{T}}$. Each point is an element of the hexagonal 2D lattice. Each hexagon around a point λ is a Voronoi cell $\mathcal{V}_\lambda$ and the shaded hexagon is the basic Voronoi cell $\mathcal{V}_0$.

necessary condition for an integral lattice is that all entries of $\mathbf{G}$ are integral. If Λ is an integral lattice, $\Lambda \subseteq \Lambda^*$. For an integral lattice Λ, if $\|\lambda\|^2$ is an even for all $\lambda \in \Lambda$, we say that Λ is an even lattice, otherwise we say that Λ is an odd lattice (note that an odd lattice is still an integral lattice). Integral lattices are important in practice because there exist fast algorithms to find the nearest lattice point of each vector for integral lattices.

Lattice Quantization

The *nearest neighbor quantizer* (NNQ) $Q(\cdot)$ associated with Λ is defined by

$$Q(\mathbf{x}) \triangleq \lambda \in \Lambda, \text{if } \|\mathbf{x} - \lambda\| \leq \|\mathbf{x} - \lambda'\| \ \forall \lambda' \in \Lambda. \tag{7.4}$$

In plain words, $Q(\mathbf{x})$ is a point in Λ that is the nearest to $\mathbf{x}$.

As shown in Figure 7.4, the *Voronoi cell* associated with each $\lambda \in \Lambda$, which is denoted by $\mathcal{V}_\lambda$, is the set of points $\mathbf{x}$ such that $Q(\mathbf{x}) = \lambda$, that is,

$$\mathcal{V}_\lambda \triangleq \{\mathbf{x} : Q(\mathbf{x}) = \lambda\}. \tag{7.5}$$

That means, for any $\mathbf{x} \in \mathcal{V}_\lambda$, $\|\mathbf{x} - \lambda\| \leq \|\mathbf{x} - \lambda'\|$ for all $\lambda' \in \Lambda$. In particular, the *basic* Voronoi cell of Λ is the set of points in $\mathbb{R}^n$ closest to $\mathbf{0}$:

$$\mathcal{V}_0 \triangleq \{\mathbf{x} : Q(\mathbf{x}) = \mathbf{0}\}. \tag{7.6}$$

See Figure 7.4 for an example of a basic Voronoi cell. It is easy to see that $\mathcal{V}_\lambda$ is the shifted version of $\mathcal{V}_0$ by λ. The modulo-Λ operation is defined as

$$\mathbf{x} \bmod \Lambda \triangleq \mathbf{x} - Q(\mathbf{x}), \tag{7.7}$$

which is actually the quantization error of $\mathbf{x}$ w.r.t. Λ.

7.2.2 What is a Good Lattice?

In space $\mathbb{R}^n$, there are infinite-number distinct lattices. Among these lattices, a natural problem is: which one is the best? There are mainly four criteria for assessing a lattice, as described below.

Packing Efficiency

Let B be the closed unit sphere centered at $\mathbf{0}$, that is, for any $\mathbf{x} \in B$, $\|\mathbf{x}\| \leq 1$. Similarly, rB is a closed radius-r sphere centered at $\mathbf{0}$, that is, for any $\mathbf{x} \in rB$, $\|\mathbf{x}\| \leq r$. Then for $\lambda \in \Lambda$, $\lambda + rB$ is a closed radius-r sphere centered at λ. Further, we use $\Lambda + rB$ to denote the set of all closed radius-r spheres centered at $\lambda \in \Lambda$. We say that the set $\Lambda + rB$ is a packing in $\mathbb{R}^n$ if for all lattice points $\lambda, \lambda' \in \Lambda$, where $\lambda \neq \lambda'$, we have

$$(\lambda + rB) \cap (\lambda' + rB) = \emptyset, \tag{7.8}$$

that is, all spheres in $\Lambda + rB$ do not intersect.

We define the packing radius of Λ as the radius of the largest n-D sphere contained in each Voronoi cell, that is,

$$r_\Lambda^{\text{pack}} \triangleq \sup\{r : \Lambda + rB \text{ is a packing}\}. \tag{7.9}$$

An example of r_Λ^{pack} for hexagonal 2D lattice can be found in Figure 7.5.

Let us consider such a sphere that has the same volume as $\mathcal{V}_0$. We call its radius the "effective radius" of Λ and denote it by r_Λ^{effec}, that is,

$$\text{Vol}(r_\Lambda^{\text{effec}} B) = \text{Vol}(\mathcal{V}_\lambda) = \text{Vol}(\mathcal{V}_0), \tag{7.10}$$

where $\text{Vol}(\cdot)$ denotes the volume of a space. An example of r_Λ^{effec} for a hexagonal 2D lattice can be found in Figure 7.5.

Figure 7.5 Examples of packing radius r_Λ^{pack}, covering radius r_Λ^{cov}, and effective radius r_Λ^{effec} for hexagonal 2D lattice.

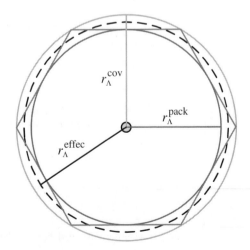

To evaluate Λ, we define its packing efficiency as

$$\rho_{\text{pack}}(\Lambda) \triangleq \frac{r_\Lambda^{\text{pack}}}{r_\Lambda^{\text{effec}}}. \tag{7.11}$$

Obviously, $0 \leq \rho_{\text{pack}} \leq 1$. Intuitively, the larger $\rho_{\text{pack}}(\Lambda)$, the better Λ.

Covering Efficiency

The covering problem is the counterpart of the packing problem. The set $\Lambda + r\mathcal{B}$, which is composed of all radius-r spheres centered at the lattice points in Λ, is a covering of $\mathbb{R}^n$ if

$$\mathbb{R}^n \subseteq \Lambda + r\mathcal{B}, \tag{7.12}$$

that is, each point in space $\mathbb{R}^n$ is covered by at least one sphere.

We define the covering radius of Λ as the radius of the minimum sphere containing $\mathcal{V}_\lambda$, that is,

$$r_\Lambda^{\text{cov}} \triangleq \min\{r : \Lambda + r\mathcal{B} \text{ is a covering}\}. \tag{7.13}$$

An example of r_Λ^{cov} for hexagonal 2D lattice can be found in Figure 7.5.

To evaluate Λ, we define the covering efficiency of Λ as

$$\rho_{\text{cov}}(\Lambda) \triangleq \frac{r_\Lambda^{\text{cov}}}{r_\Lambda^{\text{effec}}}. \tag{7.14}$$

Obviously, $\rho_{\text{cov}}(\Lambda) \geq 1$. Intuitively, the smaller $\rho_{\text{cov}}(\Lambda)$, the better Λ.

Normalized Second Moment

The *normalized second moment* (NSM) of a lattice is defined as the normalized second moment per dimension of a uniform distribution over $\mathcal{V}_\lambda$, that is,

$$G(\Lambda) \triangleq \frac{1}{n\text{Vol}(\mathcal{V}_\lambda)^{1+2/n}} \int_{\mathcal{V}_\lambda} \|\mathbf{x} - \lambda\|^2 d\mathbf{x}$$

$$= \frac{1}{n\text{Vol}(\mathcal{V}_0)^{1+2/n}} \int_{\mathcal{V}_0} \|\mathbf{x}\|^2 d\mathbf{x}. \tag{7.15}$$

The minimum possible value of $G(\Lambda)$ over all lattices in $\mathbb{R}^n$ is denoted by G_n. The NSM of an n-D sphere, denoted by G_n^*, approaches $1/2\pi e$ as the dimension $n \to \infty$. Therefore, $G_n > G_n^* > 1/2\pi e$ for all n. As we know, $G_n \leq G_1 = 1/12$. It means, by using vector quantization, one can achieve up to $10\log_{10}(G_1/G_\infty^*) = 1.533$ dB gain over scalar quantization.

Kissing Number

Kissing number is originally a geometric concept, which is defined as the maximum number of unit rigid spheres that can be arranged such that each of them touches another given unit rigid sphere. The kissing number may vary

Table 7.1 Some of the best lattices.

Dimension	1	2	3	4	5	6	7	8	12	16	24
Densest packing	$\mathbb{Z}$	A_2	A_3	D_4	D_5	E_6	E_7	E_8	K_{12}	Λ_{16}	Λ_{24}
Highest kissing	$\mathbb{Z}$	A_2	A_3	D_4	D_5	E_6	E_7	E_8	P_{12a}	Λ_{16}	Λ_{24}
number	2	6	12	24	40	72	126	240	840	4320	196560
Thinnest covering	$\mathbb{Z}$	A_2	A_3^*	A_4^*	A_5^*	A_6^*	A_7^*	A_8^*	A_{12}^*	A_{16}^*	Λ_{24}
Best quantizer	$\mathbb{Z}$	A_2	A_3^*	D_4	D_5^*	E_6^*	E_7^*	E_8	K_{12}	Λ_{16}	Λ_{24}

from one sphere to another, but for a lattice sphere packing (the center of each sphere is located at a lattice point) the kissing number is the same for every sphere. Intuitively, the larger the kissing number, the better Λ.

Some Good Lattices

In Table 7.1 we list some of the best lattices. It can be seen that for each dimension n, the best lattice may differ in different criteria. For the definitions of these lattices and their properties, the reader may refer to [52].

7.3 Nested Lattice Quantization

The first WZ code design is nested lattice quantization (NLQ), which was first introduced by Zamir et al. [36, 37]. We say that a coarse lattice Λ_2 is nested in a fine lattice Λ_1 if each point of the coarse lattice Λ_2 is also a point of the fine lattice Λ_1 but not vice versa, that is, $\Lambda_2 \subset \Lambda_1$. Figure 7.6 shows a pair of 2D nested lattices [37] based on a hexagonal 2D lattice, which well illustrates the basic concept of NLQ. The fine lattice points corresponds to the centers of small hexagons (numbered by $0, 1, \cdots, 8$), while the coarse lattice points correspond to the centers of large hexagons (numbered by 8). Each small hexagon is a Voronoi cell of the fine lattice Λ_1, while each large hexagon is a Voronoi cell of the coarse lattice Λ_2. The nesting ratio is defined as the nth root of the ratio of the covolume of large Voronoi cells to that of small Voronoi cells. Denote the basic Voronoi cell of Λ_1 by $\mathcal{V}_{0,1}$ and the basic Voronoi cell of Λ_2 by $\mathcal{V}_{0,2}$. Then

$$\gamma \triangleq \sqrt[n]{\mathrm{Vol}(\mathcal{V}_{0,2})/\mathrm{Vol}(\mathcal{V}_{0,1})}. \tag{7.16}$$

For example, in Figure 7.6, the area of each large Voronoi cell is nine times that of each small Voronoi cell, so the nesting ratio is 3. The resulting code rate is the base-2 logarithm of nesting ratio:

$$R = \log_2 \gamma = \frac{1}{n} \log_2 \frac{\mathrm{Vol}(\mathcal{V}_{0,2})}{\mathrm{Vol}(\mathcal{V}_{0,1})}. \tag{7.17}$$

For example, the code rate is $\log_2 3$ in Figure 7.6.

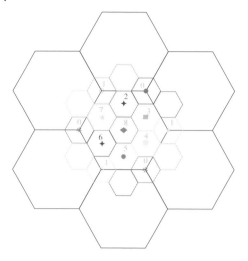

Figure 7.6 Example of 2D nested lattices based on a hexagonal 2D lattice.

Encoding/decoding

At the encoder, $\mathbf{x} \in \mathbb{R}^n$ is first quantized with respect to the fine lattice Λ_1 to get $\lambda_1(\mathbf{x})$, resulting in quantization loss $(\mathbf{x} - \lambda_1(\mathbf{x}))$. Then $\lambda_1(\mathbf{x})$ is quantized with respect to the coarse lattice Λ_2 to get $\lambda_2(\lambda_1(\mathbf{x}))$. Finally, the residue $\mathbf{s} = (\lambda_1(\mathbf{x}) - \lambda_2(\lambda_1(\mathbf{x})))$ is transmitted to the decoder.

At the decoder, $\mathbf{s}$ is subtracted from side information $\mathbf{y} \in \mathbb{R}^n$ to get $(\mathbf{y} - \mathbf{s})$, which is then quantized with respect to the coarse lattice Λ_2 to get $\lambda_2(\mathbf{y} - \mathbf{s})$. Finally, the sum of $\mathbf{s}$ and $\lambda_2(\mathbf{y} - \mathbf{s})$ is output as the estimate of $\mathbf{x}$.

The following deduction explains why the above scheme works

$$\hat{\mathbf{x}} = \mathbf{s} + \lambda_2(\mathbf{y} - \mathbf{s})$$
$$= \mathbf{s} + \lambda_2(\mathbf{y} - \lambda_1(\mathbf{x}) + \lambda_2(\lambda_1(\mathbf{x})))$$
$$= \mathbf{s} + \lambda_2(\lambda_1(\mathbf{x}))) + \lambda_2(\mathbf{y} - \lambda_1(\mathbf{x}))$$
$$= \lambda_1(\mathbf{x}) + \lambda_2(\mathbf{y} - \lambda_1(\mathbf{x})). \tag{7.18}$$

If only $\lambda_2(\mathbf{y} - \lambda_1(\mathbf{x})) = \mathbf{0}$, we can obtain $\hat{\mathbf{x}} = \lambda_1(\mathbf{x})$, which is just the quantized version of $\mathbf{x}$ w.r.t. the fine lattice Λ_1. We call $(\mathbf{x} - \lambda_1(\mathbf{x}))$ the *quantization loss*. Further, to make $\lambda_2(\mathbf{y} - \lambda_1(\mathbf{x})) = \mathbf{0}$, side information $\mathbf{y}$ must fall within the large Voronoi cell (w.r.t. the coarse lattice Λ_2) centered at $\lambda_1(\mathbf{x})$. Otherwise, $\hat{\mathbf{x}} \neq \lambda_1(\mathbf{x})$. In the latter case, $(\mathbf{x} - \hat{\mathbf{x}})$ is called *binning loss*.

Coset Binning

Consider the example in Figure 7.6 again. Point 8 belongs to both the fine lattice Λ_1 and the coarse lattice Λ_2. There are seven small Voronoi cells (centered at solid points 2 to 8) falling fully within the large Voronoi cell centered at point 8. There are also six small Voronoi cells (centered at points 0 and 1) falling partially within the large Voronoi cell centered at point 8. For fairness (each large Voronoi cell should contain nine small Voronoi cells), mapping of border fine lattice points (e.g., points 0 and 1 of Λ_1 falling on the envelope of the large Voronoi cell centered at point 8) to coarse lattice points is done in a systematic fashion, so only solid fine lattice points 0 and 1 are quantized to the coarse lattice point 8 in Figure 7.6. All nine solid fine lattice points 0 to 8 in Figure 7.6 are quantized to the same coarse lattice point 8.

Note that only $\mathbf{s} = \lambda_1(\mathbf{x}) - \lambda_2(\lambda_1(\mathbf{x}))$, which is indexed from 0 to 8 in Figure 7.6, is coded to save rate. In general,

$$\mathbf{s} \in (\Lambda_1 \bmod \Lambda_2) = \Lambda_1 \cap \mathcal{V}_{0,2}. \tag{7.19}$$

Therefore, a pair of nested lattices in fact divides the fine lattice points into $|\Lambda_1 \bmod \Lambda_2|$ cosets (bins), where $|\cdot|$ denotes the cardinality of a set. Each coset (bin) is actually a shifted version of Λ_2. Thus the vth coset (bin), where $v \in (\Lambda_1 \bmod \Lambda_2)$, can be denoted by $v + \Lambda_2$. In particular, the points in $(\Lambda_1 \bmod \Lambda_2)$ are called the coset (bin) leaders of Λ_2 relative to Λ_1. The index $\mathbf{s}$ identifies the coset (bin) that contains the quantized version of $\mathbf{x}$ (w.r.t. Λ_1). Using this coded index, the decoder finds in the coset (bin) the point closest to the side information $\mathbf{y}$ as the best estimate of $\mathbf{x}$.

Quantization Loss and Binning Loss

Let us use the 1D nested lattice pair as an example to explain quantization loss and binning loss in more detail. In Figure 7.7, the fine lattice Λ_1 includes all points along the horizontal line, while the coarse lattice Λ_2 includes all points numbered by 0. A small Voronoi cell is the length-q segment around a point, while a large Voronoi cell is the length-$4q$ segment around a point numbered by 0. The covolumes of small and large Voronoi cells are the lengths of short and long segments, respectively, which are also called quantization stepsizes. Thus, the nesting ratio is 4 and the code rate is 2 in Figure 7.7. The source x is first quantized to the nearest diamond with index 1 and then only the coset index 1 is transmitted. At the decoder, if side information y is not far from x (the top of Figure 7.7), $\hat{x} = \lambda_1(x)$, that is, there is only quantization loss. However, if y is far from x (the bottom of Figure 7.7), $\hat{x} \neq \lambda_1(x)$, which causes binning loss.

From the viewpoint of reducing quantization loss, the covolume (quantization stepsize in the 1D case) of small Voronoi cells should be as small as possible. From the viewpoint of reducing binning loss, the covolume of large Voronoi cells should be as large as possible (so that the points in the same coset will be placed as far as possible). When the nesting ratio is fixed (meaning that code rate is also fixed), the covolume of large Voronoi cells is proportional to (γ^n times of) the covolume of small Voronoi cells. Thus, at each code rate, the

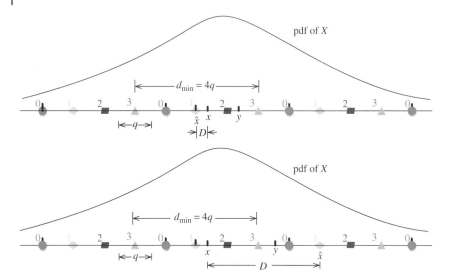

Figure 7.7 Example of 1D nested lattice pair to illustrate quantization loss and binning loss.

recovery quality of the source totally depends on the covolume of small Voronoi cells. The optimization of NLQ for WZ coding is a trade-off between quantization loss and binning loss by selecting the best covolume of small Voronoi cells.

SW Coded NLQ

It is proven in [37] that as dimension $n \to \infty$, NLQ will achieve the WZ limit asymptotically. In [38], the performance of finite-dimensional nested lattice quantizers is analyzed for continuous sources under the high-rate assumption, where the high-rate assumption is consistent with the one in classic quantization theory [18], that is, the source is uniformly distributed inside the small Voronoi cell of the quantizer. The distortion-rate (D-R) performance is analyzed for both the general and the quadratic Gaussian cases. For the quadratic Gaussian case, a tight lower bound of the D-R function is given, showing an increasing gap from the WZ limit as the rate increases. This increasing rate gap is attributed to the decreasing boundary gain as the rate increases. Thus a practical approach to boosting the overall performance is to increase the boundary gain with a second stage of binning, that is, SW coding. It is proved that at high rate, ideal SW coded 1D/2D NLQ performs 1.53/1.36 dB away from the theoretical limit.

7.3.1 Trellis Coded Quantization

While we can reduce quantization error by designing more spherical granular regions in high dimensions, we have two hurdles to overcome. First, pure spherical regions cannot pack a space. Even though we have approximately spherical regions such as hexagonal regions in two dimensions and leech lattices in 24 dimensions, we will have some loss compared to the ideal case. Second and more importantly, it is very difficult to encode an input into sphere-like granular regions. Unlike the hypercube regions where quantization can be done independently along each dimension, one has to consider all n samples at the same time for an n-dimensional vector quantizer. The encoding problem is essentially the same as asking which quantization point is closest to the input. But as the number of quantization points increases exponentially with n, the problem can be very computationally intensive if it is handled by brute force. Trellis coded quantization (TCQ) is essentially a very clever way to materialize the granular gain without exponential blow-up of computational complexity.

TCQ is closely related to convolutional codes (see also Section B.2) and the respective trellis coded modulation (TCM). The relationship to channel coding problem should not be too surprising. The channel coding problem and the lossy source coding (quantization problem) can be treated as dual problems to each other in the following sense. During source encoding (quantization), one needs to find a codeword (quantization point) closest to the input. The index corresponding to the quantization point is send to the decoder. Decoding is simply a dummy step of mapping the index back to the quantization point. On the other hand, in channel coding, the encoder simply maps an input message to a codeword, which is then sent through a noisy channel. On receiving a noisy code vector, the decoder will try to find the codeword closest to the received code vectors (if we assume independent Gaussian noise). One can see that source encoding step is essentially identical to the channel decoding step. Therefore, it should be intuitive that decoder used in the channel coding problem can be applied to the quantization problem.

7.3.2 Principle of TCQ

The best way to understand the principle of TCQ is to use an example. Let us consider a special application of TCQ to phase quantization, which is useful in MIMO systems. The reason why we choose phase quantization as an example lies in the periodicity of phases: $e^{j\phi} = e^{j(\phi+2k\pi)}$, where j is the imaginary unit and $k \in \mathbb{Z}$. It will be seen that TCQ is based on uniform scalar quantization (USQ), so we first explain the USQ of phases. For an n-bit USQ quantizer, there are $N = 2^n$ quantization levels, corresponding to N **codewords**. These quantization levels are separated by $(N - 1)$ **thresholds**. Let L and U be the lower and upper bounds of source symbols. Due to the periodicity of phases, we only need to consider the region $[-\pi, \pi)$, giving $L = -\pi$ and $U = \pi$. Hence, the quantization step is $\Delta = (U - L)/N = 2\pi/N$. The lth threshold, where $1 \leq l \leq N - 1$, is $t_l = L + l\Delta$ and the lth codeword, where $0 \leq l \leq N - 1$, is $c_l = L + (l + 0.5)\Delta$. The set of codewords is called a **codebook** and is denoted by $C = \{c_l\}_{l=0}^{N-1}$. After quantization, source symbol $x \in [L, U)$ is represented by an n-bit integer $l' = \lfloor (x - L)/\Delta \rfloor$, where l' is called the codeword index. At the dequantizer, x is reconstructed by $c_{l'}$ and the squared-error (SE) is $\xi = |c_{l'} - x|^2$. Obviously, $0 \leq \xi \leq (\Delta/2)^2$. As an example, let us consider the USQ of $\vartheta = (-\pi, -\pi/2, 0, \pi/2)$. For a 2-bit USQ quantizer, $N = 4$ and $\Delta = \pi/2$, so $t_l = -\pi + l\pi/2$ and $c_l = -\pi + (l + 0.5)\pi/2$. After quantization, 8 bits "00 01 10 11" will be output for ϑ, corresponding to codeword indexes $\{0, 1, 2, 3\}$. At the dequantizer, ϑ is reconstructed by codewords $\{c_0, c_1, c_2, c_3\}$, giving mean SE (MSE) $(\pi/4)^2$. After understanding the USQ of phases, we are ready to study the TCQ of phases.

Generation of Codebooks

Now we consider an n-bit TCQ quantizer, whose codebook includes $2^{n+1} = 2N$ codewords, where $N = 2^n$. For simplicity, we assume that the codebook is divided into four sub-codebooks, each with $N/2$ codewords. We denote the ith ($0 \leq i \leq 3$) sub-codebook by $C_i = \{c_{i,l}\}_{l=0}^{N/2-1}$, where $c_{i,l}$ is the lth codeword in C_i. These sub-codebooks are generated in the following way. Let L and U be the lower and upper bounds of source symbols. Let $\Delta = \frac{U-L}{N/2}$ and $\delta = \Delta/4 = \frac{U-L}{2N}$. Then $c_{i,l} = L + l\Delta + (i + 0.5)\delta$.

An example of 2-bit TCQ codebook is given in Figure 7.8. There are four sub-codebooks, each with two codewords. Hence, $\Delta = \pi$, $\delta = \pi/4$, and therefore $c_{i,l} = -\pi + l\pi + (i + 0.5)\pi/4$. Since $e^{jc_{i,l}} = e^{jc_{i,l+2k}}$, where $k \in \mathbb{Z}$, we replace $c_{0,2}, c_{1,2}, c_{2,-1}$, and $c_{3,-1}$ by $c_{0,0}, c_{1,0}, c_{2,1}$, and $c_{3,1}$, respectively.

Generation of Trellis from Convolutional Codes

A memory-m convolutional encoder is a finite state machine with 2^m states. Figure 7.9 shows a diagram of a recursive, systematic convolutional encoder with memory 2 and rate 1/2, where x is the input bit, while $y_0 = x$ and y_1 are the output bits. Its corresponding trellis is depicted in Figure 7.10. This trellis includes four states: "00", "01", "10", and "11". Starting from each state, there are two branches, each labeled x/y_0y_1, where x corresponds to the input bit "0" or "1", while $y_0y_1 = xy_1$ are the output bits that depend on the input bit x and the preceding state.

Figure 7.8 Uniform codebook and partition for 2-bit TCQ. There are four sub-codebooks, each with two codewords. The source symbols to be quantized are phases. Due to the periodicity of phases, that is, $e^{j\phi} = e^{j(\phi+2k\pi)}$, $\forall k \in \mathbb{Z}$, we let $L = -\pi$ and $U = \pi$, thus $\Delta = \pi$, $\delta = \pi/4$, and $c_{i,l} = -\pi + l\pi + (i + 0.5)\pi/4$. It is easy to verify $e^{jc_{i,l}} = e^{jc_{i,l+2k}}$, $\forall k \in \mathbb{Z}$, so $c_{0,2}, c_{1,2}, c_{2,-1}$, and $c_{3,-1}$ are replaced by $c_{0,0}, c_{1,0}, c_{2,1}$, and $c_{3,1}$, respectively.

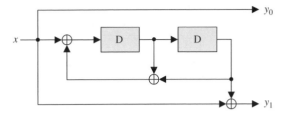

Figure 7.9 Diagram of a recursive, systematic convolutional encoder with memory 2 and rate 1/2, where each rectangle containing the letter D represents a delay component. For each input bit x, two bits are output: $y_0 y_1 = xy_1$.

Figure 7.10 Trellis generated from Figure 7.9, including four states. Starting from each state, there are two branches, each labeled by $x/y_0 y_1$, where x is the input bit, while $y_0 y_1 = xy_1$ are the output bits depending on the input bit x and the preceding state. Solid arrow lines are for 0-branches and dashed arrow lines for 1-branches. Branches with output bits "00", "01", "10", and "11" will use sub-codebooks C_0, C_1, C_3, and C_2, respectively.

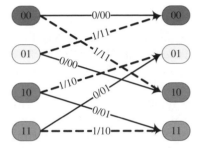

Mapping of Trellis Branches onto Sub-codebooks

A TCQ quantizer maps each branch of the trellis onto a sub-codebook indexed by the Gray code of the output bits of this branch. The Gray code of bit string $b_0 b_1$ is $b_0(b_0 \oplus b_1)$, so the Gray codes of bit strings "00", "01", "10", and "11" are "00" (0), "01" (1), "11" (3), and "10" (2), respectively. Therefore, in Figure 7.10, branches with output bits "00", "01", "10", and "11" will use sub-codebooks C_0, C_1, C_3, and C_2, respectively.

Quantization

The TCQ quantizer first calculates the nearest codewords of each source symbol in the used sub-codebooks. Let x be the source symbol and L the lower bound of source symbols. The index of the nearest codeword of x in sub-codebook C_i can be calculated by

$$l' = round\left(\frac{x - L - (i + 0.5)\delta}{\Delta}\right), \tag{7.20}$$

where $round(\cdot)$ is the round operation. On knowing the nearest codeword, the TCQ quantizer calculates the metric, that is, SE. Finally, the TCQ quantizer searches for the best path with the smallest total accumulated metric by the Viterbi algorithm.

Table 7.2 Nearest codewords and corresponding metrics.

Phase	C_0		C_1		C_2		C_3	
	NC	SE	NC	SE	NC	SE	NC	SE
$-\pi$	$c_{0,0}$	$(\pi/8)^2$	—	—	$c_{2,1}$	$(3\pi/8)^2$	—	—
$-\pi/2$	$c_{0,0}$	$(3\pi/8)^2$	$c_{1,0}$	$(\pi/8)^2$	$c_{2,0}$	$(\pi/8)^2$	$c_{3,0}$	$(3\pi/8)^2$
0	$c_{0,1}$	$(\pi/8)^2$	$c_{1,1}$	$(3\pi/8)^2$	$c_{2,0}$	$(3\pi/8)^2$	$c_{3,0}$	$(\pi/8)^2$
$\pi/2$	$c_{0,1}$	$(3\pi/8)^2$	$c_{1,1}$	$(\pi/8)^2$	$c_{2,1}$	$(\pi/8)^2$	$c_{3,1}$	$(3\pi/8)^2$

NC, nearest codeword; SE, squared error.

Example

Consider the phase vector $\vartheta = \{ -\pi, -\pi/2, 0, \pi/2 \}$. For each source symbol, the nearest codewords (NC) in each sub-codebook as shown in Figure 7.8 are listed in Table 7.2, as well as the corresponding metrics (i.e., SE). Note that the periodicity of phases is exploited in Table 7.2.

The TCQ quantizer searches for the best path through the Viterbi algorithm, as shown in Figure 7.11. This process starts from state "00". At this point, only two sub-codebooks C_0 and C_2 can be used, meaning that for the first symbol, $-\pi$, we only need to calculate the nearest codewords in C_0 and C_2 (Table 7.2). At the following stages, for each state, only the incoming branch with smaller partial accumulated metric (black arrow lines in Figure 7.11) is retained, while the other (gray arrow lines in Figure 7.11) is cut off. Finally, "0111", the hint of the best path with the smallest total accumulated metric (the thick black arrow lines in Figure 7.11), is output, accompanied by codeword indexes "0001".

As for the reconstruction, the TCQ dequantizer first recovers sub-codebook indexes {0, 2, 3, 2} from the best path hint "0111" by going through the trellis. Then, sub-codebook indexes {0, 2, 3, 2} are combined with codeword indexes

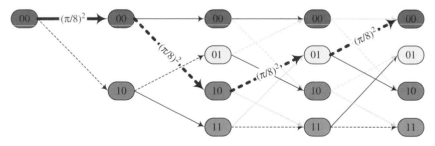

Figure 7.11 Searching for the best path (i.e., thick black arrow lines) with the smallest total accumulated metric by the Viterbi algorithm. At each stage, for each state, only the incoming branch with smaller partial accumulated metric (black lines) is retained, while the other (gray arrow lines) is cut off. Each branch along the optimal path is labeled by its metric (Table 7.2). Solid arrow lines are for 0-branches and dashed arrow lines for 1-branches.

$\{0, 0, 0, 1\}$ to obtain codewords $\{c_{0,0}, c_{2,0}, c_{3,0}, c_{2,1}\}$, which is just the recon-struction of ϑ. It is easy to find that the MSE is $(\pi/8)^2$. Compared to the same example of USQ-based phase quantization, the reader can easily find the sig-nificant advantage of TCQ over USQ.

7.4 WZ Coding Based on TCQ and LDPC Codes

A WZ code based on TCQ and SW coding is designed in [53], which defeats the WZ code based on NLQ and SW coding for quadratic Gaussian sources. The block diagram of SW coding–TCQ is depicted in Figure 7.12, which involves separate TCQ and SW encoding at the sender, and joint SW decoding and esti-mation at the receiver. Assume that source X is related to decoder side informa-tion Y by $X = Y + Z$, where $Y \sim N(0, \sigma_Y^2)$ and $Z \sim N(0, \sigma_Z^2)$ are independent. It is easy to determine $X \sim N(0, \sigma_Y^2 + \sigma_Z^2)$. The aim of this system is to minimize $E\{d(X, \hat{X})\}$ for a target bitrate R_X b/s, where $\hat{X}$ is the recovered version of X.

At the encoder, each L source sample $\mathbf{x} = \{x_0, \cdots, x_{L-1}\}$ is first quantitized by an R-bit TCQ, yielding a quantization index vector $Q(\mathbf{x}) = \mathbf{b} = \{b_0, \cdots, b_{L-1}\}$. Each R-bit index b_i, where $0 \le i < L$, can be represented as $b_i = b_i^{R-1} \cdots b_i^0$, where $b_i^r \in \mathbb{B}$ for $0 \le r < R$. In this way, $\mathbf{b}$ is decomposed into R bit-planes, with $\mathbf{b}^r = \{b_0^r, \cdots, b_{L-1}^r\}$, $0 \le r < R$. These bit-planes of $\mathbf{b}$ are then compressed via multilevel syndrome-based SW coding [54] (with side information Y at the decoder) to get $\mathbf{s} = \mathbf{s}^0|\mathbf{s}^1|\cdots|\mathbf{s}^{R-1}$, where $\mathbf{s}^r$ is the syndrome of the underlying channel code for the SW coding of $\mathbf{b}^r$.

At the decoder, side information sequence $\mathbf{y} = \{y_0, \cdots, y_{L-1}\}$ is used in con-junction with the received syndrome $\mathbf{s}$ to sequentially decode the bit-planes of $\mathbf{b}$, starting from trellis bit-plane $\mathbf{b}^0$. Hence, when decoding the rth bit-plane $\mathbf{b}^r$, the lower bit-planes $\mathbf{b}^0, \cdots, \mathbf{b}^{r-1}$ have already been recovered as $\hat{\mathbf{b}}^0, \cdots, \hat{\mathbf{b}}^{r-1}$ and the conditional probability $P(\mathbf{b}^r|\hat{\mathbf{b}}^{r-1}, \cdots, \hat{\mathbf{b}}^0, \mathbf{y})$ can be utilized to recover $\mathbf{b}^r$. Finally, the decoder produces an estimate $\hat{\mathbf{x}}$ of $\mathbf{x}$ based on $\hat{\mathbf{b}}$ and $\mathbf{y}$, namely $\hat{\mathbf{x}} = E\{\mathbf{X}|\hat{\mathbf{b}}, \mathbf{y}\}$.

Below, we will first introduce how to obtain the statistics of TCQ indices through the Monte-Carlo method. Based on the statistics of TCQ indices, the log-likelihood ratio (LLR) of each bit-plane can be calculated, which is needed for seeding the LDPC-based SW decoder. Then we introduce the minimum MSE estimation of the source and reveal how to determine the rate of each bit-plane. Finally, the experimental results are briefly discussed.

7.4.1 Statistics of TCQ Indices

According to the principle of TCQ, on knowing trellis structure, $\mathbf{b}^0$ uniquely determines $\mathbf{c} = \{c_0, \cdots, c_{L-1}\}$, the vector of sub-codebook indices, where $c_i \in \{0, 1, 2, 3\}$ is the corresponding TCQ sub-codebook index of x_i. Hence, we

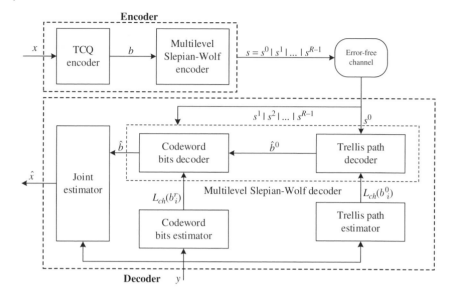

Figure 7.12 Block diagram of the SW coding–TCQ scheme.

call $\mathbf{b}^0$ the trellis bit-plane (Figure 7.13). Let $w_i = \sum_{r=1}^{R-1} b_i^r 2^{r-1}$. It is easy to determine that w_i is the codeword index of quantizing x_i in the c_ith sub-codebook. Hence, we call $\mathbf{b}^r$ for $1 \le r < R$ the rth codeword bit-plane (Figure 7.13). Since $P(c_i, w_i | y_i)$ fully determines $P(b_i^r = 1 | \hat{b}_i^{r-1}, \cdots, \hat{b}_i^1, \hat{c}_i, y_i), 1 \le r < R$, it is very important to find the probabilities $P(c_i, w_i | y_i)$. Because the Markov chain $\{C_i, W_i\} \to X_i \to Y_i$ holds for $0 \le i < L$, where C_i is the random variable associated with c_i (the meaning of W_i is similar), and the correlation between X_i and Y_i is already known, the key problem then becomes how to model the correlation between $\{C_i, W_i\}$ and X_i. The problem boils down to computing $P(c_i, w_i | x_i)$, which can be interpreted as the probability that x_i is quantized to $q_{c_i}^{w_i}$, where $q_{c_i}^{w_i}$ stands for the w_ith codeword in the c_i sub-codebook. Let

$$\mathcal{W}(x_i, c_i) \triangleq \arg \min_{0 \le w < 2^{R-1}} \|x_i - q_{c_i}^w\|^2. \tag{7.21}$$

Then

$$P(c_i, w_i | x_i) = P(c_i | x_i) I(w_i = \mathcal{W}(x_i, c_i)), \tag{7.22}$$

where $I(\cdot)$ is the indicator function, taking 1 if its argument is true, and 0 otherwise. Hence we only need to look into $P(c_i | x_i)$.

Note that $P(c_i | x_i)$ varies with i due to the state *start-up* problem of the Viterbi algorithm: $P(c_i | x_i)$ values for small i values (the trellis is partially spreaded) differ from those for large i values (the trellis is fully spread). The length of this start-up procedure is decided by TCQ memory size. Fortunately, practical TCQ

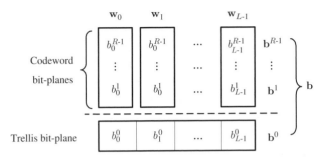

Figure 7.13 R bit-planes of TCQ index vector $\mathbf{b} = \{b_0, \cdots, b_{L-1}\}$.

usually has block length L much larger than memory size. Thus the subscript i in $P(c_i|x_i)$ can be ignored without much penalty by assuming $P(c_i|x_i) \equiv P(c|x)$ for all i. Hence we can use the empirical statistics between C and X to approximate $P(c|x)$. Similarly, we assume $P(c_i, w_i|x_i) \equiv P(c, w|x)$ for all i.

It is very hard to determine $P(c|x)$ analytically because TCQ is equivalent to a high-dimensional vector quantizer. Hence, the Monte-Carlo method can be used. Let us set a real number $A > 0$ such that $P(X \notin [-A, A]) < \epsilon$, and partition the range $[-A, A]$ into M length-δ mini-cells, $\Delta_1, \cdots, \Delta_M$, with $\delta = \frac{2A}{M}$. Define $\Delta_0 = (-\infty, -A)$ and $\Delta_{M+1} = (A, \infty)$, also denote t_m as the mid-point of Δ_m for $1 \leq m \leq M$, and $t_0 = -A$, $t_{M+1} = A$. This partition procedure is illustrated by Figure 7.14. The conditional probability $P(c|x)$ can then be approximated by $P(c|x \in \Delta_m)$ as $\epsilon \to 0$ and $M \to \infty$:

$$P(c|x) \approx P(c|x \in \Delta_m) \approx \lim_{L \to \infty} \frac{\sum_{i=0}^{L-1} I(c_i = c, x_i \in \Delta_m)}{\sum_{i=0}^{L-1} I(x_i \in \Delta_m)}. \qquad (7.23)$$

To compute the statistics in (7.23), Monte-Carlo simulations can be run on the training data drawn from $X \sim N(0, \sigma_X^2)$. Let us count the number of occurrences for each possible input–output pair $\{(m, c) : x_i \in \Delta_m, c_i = c\}$, which is denoted by count(m, c). Then the desired probability becomes

$$P(c|x \in \Delta_m) = \frac{P(c, x \in \Delta_m)}{P(x \in \Delta_m)} \approx \frac{\text{count}(m, c)}{\sum_{c'=0}^{3} \text{count}(m, c')}. \qquad (7.24)$$

The value of $P(c|x \in \Delta_m)$, where $0 \leq m \leq M + 1$ and $0 \leq c \leq 3$, can be shared by both encoder and decoder using a look-up table.

As $\delta \to 0$, $\mathcal{W}(x, c) \approx \mathcal{W}(t_m, c)$ for all $x \in \Delta_m$, then

$$P(w, c|x) \approx P(w, c|x \in \Delta_m) \approx P(c|x \in \Delta_m)I(w = \mathcal{W}(t_m, c)). \qquad (7.25)$$

We can also estimate the conditional pdf $f(x|w, c)$ based on count(m, c), where $0 \leq m \leq M + 1$, $0 \leq c \leq 3$, because this conditional pdf can be approximated

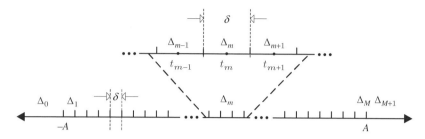

Figure 7.14 Partition of the real line into $M + 1$ mini-cells.

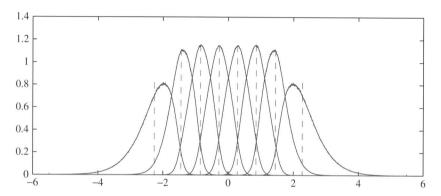

Figure 7.15 An example of $f(x|c, w)$ generated from a 2-bit TCQ with quantization stepsize 0.575 for $\sigma_X^2 = 1.28$. Dashed lines mark the centroids used in the quantizer decoder.

by $\frac{1}{\delta}P(x \in \Delta_m|w, c)$ when δ is very small. This means for $x \in \Delta_m$,

$$f(x|c, w) \approx \frac{1}{\delta}P(x \in \Delta_m|c, w)$$

$$\approx \frac{1}{\delta}I(w = \mathcal{W}(x, c))\frac{\text{count}(m, c)}{\sum_{m' : \mathcal{W}(t_{m'}, c) = w}\text{count}(m', c)}. \tag{7.26}$$

This PDF is actually the conditional distribution of the TCQ input, given that it is quantized to q_c^w. From Figure 7.15 we can clearly see the nonGaussian shape of $f(x|c, w)$ for the boundary cells of a 2-bit TCQ.

7.4.2 LLR of Trellis Bits

Because $\mathbf{b}^0$ has memory (whereas all the other bit-planes are sample-wise independent given $\mathbf{b}^0$), we have to treat it carefully. Specifically, we first use the *trellis path estimator* to compute the LLR of each trellis bit b_i^0 with side information $\mathbf{y}$, which is defined as

$$L_{ch}(b_i^0) \triangleq \log\frac{p(\mathbf{y}|b_i^0 = 1)}{p(\mathbf{y}|b_i^0 = 0)} = \log\frac{P(b_i^0 = 1|\mathbf{y})}{P(b_i^0 = 0|\mathbf{y})}, \tag{7.27}$$

where the second equation is due to $P(b_i^0 = 0) = P(b_i^0 = 1) = \frac{1}{2}$ (note that the PDFs of both source X and TCQ codebook are symmetric with respect to the origin). We use the probabilities of b_i^0 given the whole side information sequence $\mathbf{y}$ instead of only y_i because of the memory in $\mathbf{b}^0$. Hence computations of $L_{ch}(b_i^0)$ are carried out block by block. This is done by randomly generating realizations $\mathbf{z}' = \{z_0', \cdots, z_{L-1}'\}$ of $Z' \sim N(0, \sigma_Z^2)$, quantizing $\mathbf{x}' = \mathbf{y} + \mathbf{z}'$ with the same TCQ used in the encoder, and counting the occurrences of 0 and 1 in each trellis bit b_i^0 to obtain $P(b_i^0 = 1|\mathbf{y})$ and $P(b_i^0 = 0|\mathbf{y})$. For simplicity, we can assume $P(b_i^0 = 1|\mathbf{y}) \equiv P(b^0 = 1|\mathbf{y})$ and $P(b_i^0 = 0|\mathbf{y}) \equiv P(b^0 = 0|\mathbf{y})$.

7.4.3 LLR of Codeword Bits

With $\hat{\mathbf{b}}^0$ available at the decoder, the sub-codebook index vector $\mathbf{c}$ can be recovered as $\hat{\mathbf{c}}$. Since all the codeword bit-planes of $\mathbf{b}$ are sample-wise independent given $\mathbf{b}^0$, the *codeword bits estimator* can be sample based rather than block based. Let $\hat{c}_i$ be the recovered sub-codebook index of x_i. Then

$$
L_{ch}(b_i^r) \triangleq \log \frac{p(\hat{b}_i^{r-1}, \cdots, \hat{b}_i^1, \hat{c}_i, y_i | b_i^r = 1)}{p(\hat{b}_i^{r-1}, \cdots, \hat{b}_i^1, \hat{c}_i, y_i | b_i^r = 0)}
$$

$$
= \log \frac{P(b_i^r = 1 | \hat{b}_i^{r-1}, \cdots, \hat{b}_i^1, \hat{c}_i, y_i)}{P(b_i^r = 0 | \hat{b}_i^{r-1}, \cdots, \hat{b}_i^1, \hat{c}_i, y_i)}, \tag{7.28}
$$

for each codeword bit b_i^r, given $\{\hat{b}_i^{r-1}, \cdots, \hat{b}_i^1, \hat{c}_i, y_i\}$ as decoder side information, where $1 \le r < R$ and $0 \le i < L$. Again the second equation in (7.28) comes from $P(b_i^r = 1) = P(b_i^r = 0) = \frac{1}{2}$. Using this LLR, the *codeword bits decoder* sequentially recovers $\mathbf{b}^1, \cdots, \mathbf{b}^{R-1}$ (and hence $\mathbf{b}$) based on side information $\mathbf{y}$ and received syndrome vectors $\mathbf{s}^1, \cdots, \mathbf{s}^{R-1}$.

7.4.4 Minimum MSE Estimation

The *loint estimator* jointly recovers $\hat{\mathbf{x}} = E\{\mathbf{X}|\hat{\mathbf{b}}, \mathbf{y}\}$ from both $\hat{\mathbf{b}}$ and $\mathbf{y}$ at the decoder. The usual estimator is linear, which is good only when the quantization error $\hat{X} - X$ is a sequence of independent Gaussian random variables. However, we know from classic source coding that this is true only when the source code achieves the rate-distortion bound, which requires infinite-dimensional vector quantizers. Although TCQ is an efficient quantization technique, its quantization error is still not Gaussian, especially when the TCQ rate R is low. Hence, an optimal estimator nonlinear was proposed as

$$
\hat{x}_i \approx \sum_{m=0}^{M+1} t_m \frac{I(w_i = \mathcal{W}(t_m, c_i)) P(c_i | x_i \in \Delta_m) f_Z(t_m - y_i)}{\sum_{m: \mathcal{W}(t_m, c_i) = w_i} P(c_i | x_i \in \Delta_m) f_Z(t_m - y_i)}. \tag{7.29}
$$

The estimator in (7.29) is nonlinear in general and universal as it does not assume Gaussianity of the quantization error $\hat{X} - X$ or independence of X

and Z. It works well even if the noise Z is not Gaussian because the distribution of Z is involved in (7.29). It also outperforms the usual estimator which linearly combines $E\{\mathbf{X}|\hat{\mathbf{b}}\}$ and $\mathbf{y}$, especially at low rate.

7.4.5 Rate Allocation of Bit-planes

Because the WZ decoder has access to the side information $\mathbf{y} = \{y_0, \cdots, y_{L-1}\}$, which is correlated to $\mathbf{x}$, the SW compression limit of $\mathbf{b}$ is $H(Q(X)|Y) = \frac{1}{L}H(\mathbf{B}|\mathbf{Y})$ [55], where $\mathbf{B} = \{B_0, \cdots, B_{L-1}\}$ is the discrete random vector associated with $\mathbf{b}$, and $\mathbf{Y} = \{Y_0, \cdots, Y_{L-1}\}$ is the continuous random vector associated with $\mathbf{y}$. We then have $R_X \geq \frac{1}{L}H(\mathbf{B}|\mathbf{Y})$. The *multilevel SW encoder* is in fact an R-level LDPC encoder defined by R parity-check matrices $\mathbf{H}_0, \cdots, \mathbf{H}_{R-1}$, where $\mathbf{H}_r$, $0 \leq r < R$, corresponds to an (n, k_r) binary LDPC code of rate $R_r^{CC} = \frac{k_r}{n}$. For notational convenience, we assume $L = n$. The SW encoder compresses $\mathbf{b}^r$ to a length-$(n - k_r)$ syndrome vector $\mathbf{s}^r = \mathbf{b}^r \mathbf{H}_r^T$ and concatenates all R syndrome vectors $\mathbf{s} = \mathbf{s}^0 |\mathbf{s}^1| \cdots |\mathbf{s}^{R-1}$. The total length of $\mathbf{s}$ is $l = \sum_{r=0}^{R-1}(n - k_r)$, which results in the SW coding–TCQ rate of

$$R_X = \frac{l}{L} = \frac{l}{n} = \sum_{r=0}^{R-1} R_r^{SW} \text{ b/s,} \qquad (7.30)$$

where $R_r^{SW} = 1 - R_r^{CC}$ is the rate allocated with $\mathbf{b}^r$.

The goal of multilevel SW coding is to approach the conditional entropy $H(Q(X)|Y) = \frac{1}{L}H(\mathbf{B}|\mathbf{Y})$ b/s. Because $\mathbf{b}$ is a vector of L elements, each with R-bit resolution, we can use the chain rule on $H(\mathbf{B}|\mathbf{Y})$ to get

$$\frac{1}{L}H(\mathbf{B}|\mathbf{Y}) = \frac{1}{L}\left[H(\mathbf{B}^0|\mathbf{Y}) + \sum_{r=1}^{R-1} H(\mathbf{B}^r|\mathbf{B}^{r-1}, \cdots, \mathbf{B}^0, \mathbf{Y}) \right]$$

$$= \frac{1}{L}\left[H(\mathbf{B}^0|\mathbf{Y}) + \sum_{r=1}^{R-1}\sum_{i=0}^{L-1} H(B_i^r|B_i^{r-1}, \cdots, B_i^1, C_i, Y_i) \right],$$

$$\stackrel{(a)}{\approx} \frac{1}{L}H(\mathbf{B}^0|\mathbf{Y}) + \sum_{r=1}^{R-1} H(B^r|B^{r-1}, \cdots, B^1, C, Y), \qquad (7.31)$$

where $\mathbf{B}^r$ is the binary random vector associated with bit-plane $\mathbf{b}^r$ and (a) is true if we assume that the conditional entropies are invariant among samples, that is, $H(B_i^r|B_i^{r-1}, \cdots, B_i^1, C_i, Y_i) \equiv H(B^r|B^{r-1}, \cdots, B^1, C, Y)$ for all $0 \leq i < L$ and $1 \leq r < R$. Thus the zeroth and first level SW encoders are designed to approach the conditional entropies $\frac{1}{L}H(\mathbf{B}^0|\mathbf{Y})$ and $H(B^1|C, Y)$, respectively, while the rth level ($2 \leq r < R$) SW encoder targets at rate $H(B^r|B^{r-1}, \cdots, B^1, C, Y)$. This is illustrated in Figure 7.16.

Experiments show that the conditional entropy $\frac{1}{L}H(\mathbf{B}^0|\mathbf{Y})$ approaches 1b/s as the TCQ rate R increases, where $\mathbf{B}^0$ is the binary random vector associated

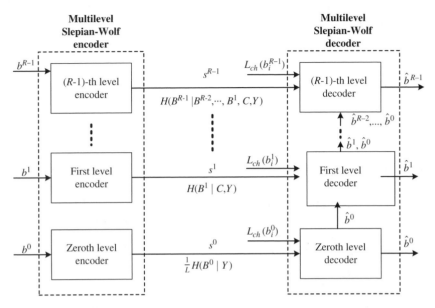

Figure 7.16 Multilevel SW coding.

with $\mathbf{b}^0$. With $R_0^{SW} \geq \frac{1}{L}H(\mathbf{B}^0|Y)$ approaching 1b/s, it is reasonable to assume that $\mathbf{b}^0$ can be recovered error free when decoding the $R - 1$ higher bit-planes $\mathbf{b}^1, \cdots, \mathbf{b}^{R-1}$. The remaining thing is to measure the conditional entropy $H(B^r|B^{r-1}, \cdots, B^1, C, Y)$. We first need to find the conditional probabilities $P(c_i, w_i|y_i), 0 \leq i \leq L - 1$. As mentioned in Section 7.4.1, using Markov chains $\{C_i, W_i\} \rightarrow X_i \rightarrow Y_i$, $P(c_i, w_i|y_i)$ can be computed via the conditional probabilities $P(c|x \in \Delta_m)$ in (7.24), then

$$P(c_i, w_i|y_i) = \sum_{m=0}^{M+1} P(c_i, w_i, x_i \in \Delta_m|y_i)$$

$$= \sum_{m=0}^{M+1} P(c_i, w_i|x_i \in \Delta_m)P(x_i \in \Delta_m|y_i)$$

$$= \sum_{m=0}^{M+1} P(c_i|x_i \in \Delta_m)I(w_i = \mathcal{W}(t_m, c_i)) \int_{\Delta_m} f_Z(x - y_i)dx$$

$$\approx \delta \sum_{m:\mathcal{W}(t_m,c_i)=w_i} P(c_i|x_i \in \Delta_m)f_Z(t_m - y_i). \tag{7.32}$$

In (7.32), we set x_i to t_m and the pdf $f_Z(x - y_i)$ to $f_Z(t_m - y_i)$ when $x \in \Delta_m$. This approximation is accurate for large M. Note that $P(c_i|x_i \in \Delta_m)$ in (7.32) comes from the look-up table indexed by (m, c_i). Another table for the exponential function in $f_Z(z)$ can also be used to speed up the computation.

Now we define

$$p_i^r(1) \triangleq \sum_{0 \leq w < 2^{R-1}} I(b^r = 1, b^{r-1} = \hat{b}_i^{r-1}, \cdots, b^1 = \hat{b}_i^1) P(c_i, w | y_i) \qquad (7.33)$$

and let $p_i^r(0) = 1 - p_i^r(1)$. Then

$$H(B^r | B^{r-1}, \cdots, B^1, C, Y) \approx \lim_{L \to \infty} \frac{1}{L} \sum_{i=0}^{L-1} \mathcal{H}(p_i^r(1)), \qquad (7.34)$$

where $\mathcal{H}(p) = p\log_2\frac{1}{p} + (1-p)\log_2\frac{1}{1-p}$. Since both the encoder and the decoder know the joint distribution of X and Y, the above conditional entropy $H(B^r | B^{r-1}, \cdots, B^1, C, Y)$ can be computed off-line at both encoder and decoder by randomly generating realizations of $Y' \sim N(0, \sigma_Y^2)$ and $Z' \sim N(0, \sigma_Z^2)$ before quantizing $X' = Y' + Z'$ with TCQ and invoking (7.34). Note that based on the probabilities $P(c_i, w_i | y_i)$, the *codeword bits estimator* can also compute $L_{ch}(b_i^r)$ defined in (7.28) by using $L_{ch}(b_i^r) = \log(p_i^r(1)/p_i^r(0))$.

7.4.6 Experimental Results

Assuming 256-state TCQ and ideal SW coding which can achieve the theoretical limit $H(Q(X)|Y)$, it was reported that SW coded TCQ performs 0.2 dB away from the WZ D-R function $D_{WZ}(R)$ at high rate, which mirrors that of entropy-constrained TCQ in classic source coding of Gaussian sources. Further, with 8,192-state TCQ and ideal SW coding, simulations show that SW-coded TCQ performs only 0.1 dB away from $D_{WZ}(R)$ at high rate. As for practical designs, with 8,192-state TCQ, irregular LDPC codes for SW coding, and optimal nonlinear estimation at the decoder, the performance gap to $D_{WZ}(R)$ is 0.20, 0.22, 0.30, and 0.93 dB at 3.83, 1.83, 1.53, and 1.05 b/s, respectively. Better results can be obtained by using trellis-coded vector quantization.

Part III

Applications

This part is dedicated to applications of DSC. We split all applications into those related to multimedia (audio/image/video) coding and processing, and those related to general communications.

We will start with multimedia-related applications and first review the basics of image and video coding. Then we will describe in detail one of the most promising DSC applications, WZ video coding or distributed video coding. We will then showcase several other multimedia applications: solar image coding, secure image compression, and biometrics.

8

Wyner–Ziv Video Coding

8.1 Basic Principle

WZ video coding (sometimes known as distributed video coding) utilizes WZ coding principles to achieve video compression. As frames in a video sequence typically are highly correlated, taking advantage of this correlation is essential for efficient compression. Conventional video coding typically achieves this through compressing a target frame with the help of neighboring frames. As shown in Figure 8.1, a frame in the input sequence is first classified into an intra-coded frame (sometimes also known as a key frame or I-frame) and an inter-coded frame. An inter-coded frame is encoded with the help of its neighboring frames, which can either be intra-coded or inter-coded themselves. Typically, motion search is conducted to the target frame relative to its neighboring (reference) frames, which are first stored in a frame buffer. The difference between the motion-compensated reference and the target frame will be coded along with the corresponding motion vectors. At the decoder, such motion-compensated reference can be reconstructed with the received motion vectors and the already decoded neighboring frames. Moreover, one can decode the frame difference from the received code stream. Finally, the target frame can be recreated by adding the reference by the decoded frame difference.

Note that video coding is typically lossy and so the reference frames stored in the frame buffer at the decoder are generally not the same as those stored in the encoder frame buffer. This difference will result in small decoding errors but these errors can propagate and accumulate as frames are repeatedly stored in the frame buffer. Therefore, the introduction of intra-coded frame is necessary to stop this error propagation indefinitely.

The buffered reference frame used during intra-coding is essentially a piece of side information for source coding. This side information is utilized at both the encoder and the decoder. Therefore, as shown in Figure 8.2, a natural

Distributed Source Coding: Theory and Practice, First Edition. Shuang Wang, Yong Fang, and Samuel Cheng.
© 2017 John Wiley & Sons Ltd. Published 2017 by John Wiley & Sons Ltd.
Companion Website: www.wiley.com/go/cheng/dsctheoryandpractice

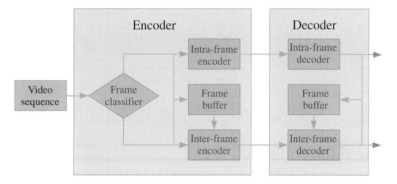

Figure 8.1 A simplified block diagram of a typical conventional video coding scheme.

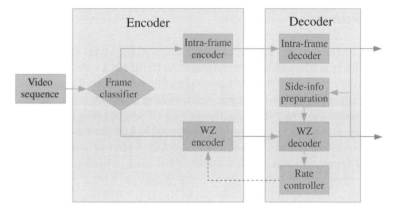

Figure 8.2 A typical WZ video coding scheme.

inspiration of utilizing WZ coding to video coding would be to ignore the reference frame at the encoder but only utilize the side information at the decoder. In fact the idea of taking advantage of WZ coding for video coding is not new. It was first described in a US patent [9] in 1980, but to the best of the authors' knowledge it remained without software implementation until early this century [11, 10]. The main drivers for rediscovering WZ video coding have been the practical DSC code designs described in Part II.

8.2 Benefits of WZ Video Coding

One may puzzle over the advantages of this artificial choice. Unlike a sensor network application, the side information (i.e., neighboring frames) is indeed *available* at the encoder. What would be the gain in ignoring it?

There are two main benefits of utilizing WZ coding principles for video coding. First, WZ video coding allows an option to trade-off the complexity of the encoder and decoder. More precisely, it pushes the computational complexity from the encoder to the decoder. This is beneficial to hand-held recorders such as smart phones where computing power is expensive, as one would like to extend battery life as much as possible. Moreover, this shift in complexity is better matched with the trend of the flood of capturing devices. In the past, video capturing capability was monopolized by the movie industry and broadcasting stations. Encoding was therefore typically performed infrequently and thus high encoding complexity was rarely a major concern. As capturing devices became ubiquitous and used everywhere from daily entertainment to medical endoscopy, the lack of balance in the encoding and decoding complexities had to be reconsidered.

Second, due to the lack of side information at the encoder, there is no frame-error propagation and thus the recovered videos tend to be more robust and less susceptible to undesirable artifacts. Of course, this benefit depends on the actual implementation and only tends to hold in general.

8.3 Key Components of WZ Video Decoding

In contrast with conventional video coding, WZ video coding simply replaces intra-coders with WZ coders. As mentioned earlier, this essentially pushes the complexity of the coder from the encoding side to the decoding side. As a result, a WZ video decoder is significantly more complex than that used in conventional video coding. In particular, the following components in the decoder are either completely unique or significantly different from those in conventional coders and thus deserved special attention:

- side-information preparation;
- correlation modeling;
- rate controller.

8.3.1 Side-information Preparation

In conventional video coding, neighboring frames can be treated as side information for intra-coding. As for WZ coding, in theory the decoder can also exploit all these neighboring frames as side information in a similar manner. For WZ coding, in theory the decoder can exploit all neighboring frames as side information at the same time, but in practice the resulting complexity will be too high to be manageable. Therefore, most WZ video-coding schemes will first generate a single reference frame as side information. The procedure is similar to the conventional case where motion search is applied to a neighboring frame and a motion-compensated reference frame is constructed.

Figure 8.3 Comparison of different motion compensation approaches of the Carphone sequence. Upper left: target image; upper right: encoder motion search; lower left: extrapolation (using motion vectors from previously decoded frames); lower right: bidirectional motion compensation.

The motion search can be implemented simply by searching for the minimum mean square difference of the target block from a reference block within a fixed search range. A typical block size is 4×4 or 8×8 and an often-used search range is 20×20. Despite the simplicity of the motion search operation, it contributes a significant amount of computational complexity to conventional video encoding. Thus, it defeats the purpose of reducing encoding complexity if we leave this operation at the encoder. The difficulty, however, is that the decoder cannot compute motion vector in a straightforward manner since the target frame is not available there. One simple way out is to assume motion does not change rapidly and use motion vectors recovered in the previous constructed frame. However, for frames with rapidly varying motion, there can be significant discrepancy in the motion vectors, which leads to inefficient compression (see Figure 8.3).

Bidirectional Motion Compensation

Rather than performing motion vector extrapolation, a better approach is to conduct bidirectional motion compensation [56]. Consider a simple example where the target WZ frame F_T is adjacent to two key frames, F_1 and F_2, one in front of and another behind F_T. Since the two key frames are recovered independently at the decoder, we can utilize them to estimate the motion vectors. Let $\mathbf{v}(\mathbf{x})$ be the motion vectors at location $\mathbf{x}$. In other words, $F_1(\mathbf{x} - \mathbf{v}(\mathbf{x})) \approx F_2(\mathbf{x})$. Assuming that the motion is reasonably smooth over the period from F_1 to F_2, we may then approximate $F_T(\mathbf{x}) \approx F_1(\mathbf{x} - \mathbf{v}(\mathbf{x})/2)$. Similarly, we may approximate $F_T(\mathbf{x}) \approx F_2(\mathbf{x} + \mathbf{v}(\mathbf{x})/2)$. Therefore, a bidirectional motion compensated frame is simply

$$\frac{1}{2}(F_1(\mathbf{x} - \mathbf{v}(\mathbf{x})/2) + F_2(\mathbf{x} + \mathbf{v}(\mathbf{x})/2)).$$

Figure 8.3 compares different motion compensation schemes with the Carphone sequence. We can see that bidirectional motion compensation performs significantly better than motion vector extrapolation. Of course, encoder motion search can perform even better since the target frame is known in that case.

The aforementioned scheme requires a key frame for every WZ frame. However, we can reduce the number of key frames by utilizing previously decoded WZ frames in a hierarchical manner [57]. For example, we can assign Frames 1 and 9 as key frames and Frames 2 to 8 as WZ frames. From Frames 1 and 9 we can recover Frame 5 using WZ decoding and the bidirectional motion compensation as mentioned above. Similarly, from Frames 1, 5, and 9 we can recover Frames 3 and 7. Finally we can recover all the even frames from the already reconstructed odd frames.

8.3.2 Correlation Modeling

The correlation between the target frame and the side information will depend on the methods of side-information preparation described in the previous section. However, this correlation can be rather accurately fitted using statistical models such as Laplace and Gaussian distributions. As shown in Figure 8.4, the prediction errors using encoder and extrapolation motion compensation are fitted with Laplace distributions

$$p(e) = \frac{\alpha}{2} \exp(-\alpha|e|). \tag{8.1}$$

where the prediction error e is simply computed as the difference between the intensity value of the target frame from the value of the respective pixel in the generated side information.

One can readily verify that the variance of the Laplace model variable is $2\alpha^2$. Therefore, we may approximate $\alpha \approx \sqrt{V_e/2}$, with V_e being the empirical variance of the prediction error. However, just as the motion vector cannot be computed directly at the decoder, the empirical variance also needs to estimated indirectly since the reconstructed target frame is not yet available. We may do an online estimation that performs parameter estimation and

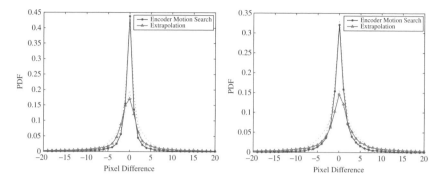

Figure 8.4 Prediction error of the Carphone sequence (left) and the Foreman sequence (right). The dotted lines are the corresponding fitted models with Laplace distributions.

WZ decoding at the same time, as described in the next chapter. We can also conduct an offline estimation that essentially uses estimations obtained from previously decoded frames. Alternatively, given the two adjacent key frames $F_1(\mathbf{x})$ and $F_2(\mathbf{x})$ before and after the target frame, we could approximate the error at pixel $\mathbf{x}$ as [58]

$$\hat{e}(\mathbf{x}) \triangleq \frac{F_1(\mathbf{x} - \mathbf{v}(\mathbf{x})/2) - F_2(\mathbf{x} + \mathbf{v}(\mathbf{x})/2)}{2},$$

where $\mathbf{v}(\mathbf{x})$ is the motion vector at pixel $\mathbf{x}$ estimated from the two key frames. From the estimated error, we can compute the empirical variance of the variable node as usual.

Since the correlation model depends on the dynamic of the scene and varies frame by frame, one may want to estimate α at different levels. For example, [58] considered estimation of α at the pixel, block, frame, and sequence levels. Moreover, α could change significantly for pixels in an occluded region. One possible solution is to adjust the estimate heuristically using $\hat{e}(\mathbf{x})$ [59]. For example, one could decrease the estimate of α (i.e., increase the variance of the model) when $\hat{e}(\mathbf{x})$ is large. The heuristic argument is that when $\hat{e}(\mathbf{x})$ is large, it is likely that the corresponding pixel is occluded and thus side information is unreliable and the actual prediction error should have a large variance.

Exploiting Spatial Redundancy
The discussion so far has only exploited the temporal redundancy and the redundancy between the target and the side-information frame. The spatial within the frame itself has not been considered. While one may consider spatial redundancy explicitly [60], it is more common to exploit this redundancy using block-based transforms such as discrete cosine transform (DCT) [61]. The resulting coder is very similar to the pixel-based coder and the same Laplace

distribution model could be employed. However, the coefficients of the same frequency band could be grouped and handled with the same α as the statistics of the same band would be similar.

8.3.3 Rate Controller

Unlike conventional video coding, where rate control is usually performed at the encoder, it is difficult to have the WZ video encoder to determine the rate unless the actual side information is available, but that will require motion compensation, which the encoder would like to avoid because of its high computational complexity.

Since an actual WZ coding is composed of the quantization step and SW coding, the encoder can indirectly affect the required code rate by adjusting the quantization step size. However, the SW rate limit depends on the correlation between the source and the side information. Having no access to the side information, the encoder cannot possibly determine the optimal code rate accurately. Preferably we want the code rate to be just slightly above the SW limit so that the SW decoding error is close to zero. Yet, the encoder does not want to have a rate too far from the limit so that it becomes wasteful.

When a feedback channel is available, the decoder may simply control the video coding rate according to the current correlation statistics, that is, when SW decoding fails, the decoder will send a signal to the encoder to request more SW coded bits. Note that rate-compatible codes are necessary in this case, for example rate-compatible punctured turbo (RCPT) codes are used for parity-based SW coding in [11]. The rate controller at the decoder ensures that only the minimum necessary rate is sent from the encoder. The encoder computes parities of the input blocks but only sends a fraction of them to the decoder and buffers the rest. If the decoder encounters a decoding failure with the received bits, it will request more bits from the encoder. The encoder will then send out a block of buffered parity bits to the decoder. The procedure will be repeated until the decoder recovers the source successfully.

Unfortunately, feedback is not always possible in all applications. Moreover, there has also been some debate about whether the presence of feedback goes against basic WZ coding principles where the encoder should not have any access to the side information. However, from a purely performance point of view, decoder rate control via feedback is beneficial, and should be used whenever possible.

8.4 Other Notable Features of Miscellaneous WZ Video Coders

Besides the bidirectional motion search described earlier, spatial smoothing can be applied to further improve the motion-compensated reference frame for the WZ decoder. For example, authors in [56] used weighted vector-median filtering [62] to improve the motion field spatial coherency.

While it may be considered *cheating* to conduct motion estimation at the encoder, the encoder may be able to ease the decoder's job by providing it with some hints [63]. One way is a "hash-based" approach. A hash of the target frame, which usually is just a coarsely quantized downsampled version of the target frame, may be sent to the decoder for this purpose. Such a hash may also be used to help estimate the original pixel after WZ decoding.

Another way to facilitate motion control at the encoder is to use a classifier to estimate the correlation between the quantized source and the respective side information [64]. Moreover, the encoder may purposefully select a key frame rather than doing this blindly. One way is to estimate both the variance of the current frame and the sum of absolute differences (SAD) between the current frame and previous frame [65]. If the variance is small and the SAD value is large, then the current frame will be selected as a key frame. The rationale is intuitive. If variance is small, then the frame can be compressed efficiently even by itself. If SAD is large, the previous frame cannot be used as a side-information frame efficiently. Thus, it would be better to consider the current frame as a key frame instead.

9

Correlation Estimation in DVC

9.1 Background to Correlation Parameter Estimation in DVC

9.1.1 Correlation Model in WZ Video Coding

As mentioned in the previous chapter for WZ coding, correlation between the source and side information (SI) is modeled as a virtual communication channel that can be expressed in the form $X = Y + N$, where X is the source to be recovered, Y is the SI, and N is the the virtual channel noise. Based on experimental observations, most DVC designs so far [66, 56, 58] (with few exceptions) model the correlation noise as the Laplace distribution as follows:

$$p[WZ(x,y) - SI(x,y)] = \frac{\alpha}{2} \exp[-\alpha|WZ(x,y) - SI(x,y)|], \qquad (9.1)$$

where $WZ(x,y)$ and $SI(x,y)$ are the pixel values at the location (x,y) in the WZ and SI frames, respectively, $p(\cdot)$ denotes the probability density function, and α is the Laplace distribution parameter defined as

$$\alpha = \sqrt{\frac{2}{\sigma^2}} \qquad (9.2)$$

Here, σ^2 is the variance of residuals between WZ and SI frames, where the granularity of residuals can be in the sequence, frame, block, and pixel levels. Moreover, the correlation parameter α can vary along both time and space, since the residual errors are usually large when there are high motions between frames or illumination changes within a frame.

Since the capability of correlation parameter estimation has strong impact on the WZ video coding efficiency, much research work has been done on improving correlation estimation in the literature. In the beginning, most of the LDPC and turbo coding-based WZ video-coding schemes assumed that the correlation noise statistics are stationary along both in time and space [67–69],

Distributed Source Coding: Theory and Practice, First Edition. Shuang Wang, Yong Fang, and Samuel Cheng.
© 2017 John Wiley & Sons Ltd. Published 2017 by John Wiley & Sons Ltd.
Companion Website: www.wiley.com/go/cheng/dsctheoryandpractice

where the correlation noise statistics could be obtained through training video sequences. However, the above assumption and estimation methods have many limitations, as the correlation statistics strongly depend on the video contents and may vary with time and space. To bridge this gap, nonstationary correlation models (e.g., on the pixel or block level) are studied in [58, 59, 70–72]. In general, the *online estimation* refers to the correlation estimation at the decoder side based on the decoded key frames. In contrast, the *offline estimation* implies the correlation estimation with the original WZ frame available. In practice, *offline estimation* is not feasible due to the requirement in DSC setups. Therefore, *offline estimation* is usually used as the benchmark reference in experiments. In the rest of this section, we will discuss the *offline correlation noise models* and *online correlation noise models* at the following granularity levels: sequence, frame, block, and pixel, based on [58].

9.1.2 Offline Correlation Estimation

Offline correlation estimation requires the original WZ frame and its corresponding SI frame to calculate the correlation parameter. Since the original WZ data is not available at the decoder side, the correlation value estimated through the offline correlation estimation scheme can only be treated as benchmark performance. Generally, the correlation is modeled by Laplace distribution, as shown in (9.1). In the pixel domain (PD) DVC, $WZ(x, y) - SI(x, y)$ corresponds to the residual of pixel values. Moreover, in the transform domain (TD) DVC, the residual corresponds to the differences of DCT coefficients. Moreover, since the SI quality varies with time and space, the choice of different granularities in correlation noise models has significant impact on decoding performance. The work in [58] studied a coarse-to-fine strategy to estimate the correlation parameter based on different granularities for both PD and TD DVC. The studies show that a fine-grained estimation level usually results in a better decoding performance in terms of rate-distortion performance. In the next subsection, we will review the details of offline correlation estimation methods at different granularities.

Pixel Domain Offline Correlation Estimation

To obtain the correlation parameter α for PD WZ video coding [58], four granularity levels are analyzed: sequence, frame, block, and pixel.

The general procedures for computing the correlation parameter α for PD WZ video coding can be summarized as follows [58]:

1) Residual frame generation:

$$R(x, y) = WZ(x, y) - SI(x, y), \tag{9.3}$$

where $R(x, y)$ denotes the residual frame and (x, y) represents the pixel position of a frame.

2) Variance computation: according to the definition, the variance of a random variable is $\sigma_Z^2 = E[Z^2] - (E[Z])^2$, where $E[\cdot]$ is the expectation operator, and the variance at different granularity levels can be obtained by averaging the variance on the corresponding levels.

a) Sequence level: since a whole video sequence is characterized by the same variance, variance will be averaged for the video sequence.

First, the variance for each residual frame is given by

$$\sigma_R^2 = E[R(x,y)^2] - (E[R(x,y)])^2$$

$$E[R(x,y)] = \frac{1}{H \times W} \sum_{x=1}^{H} \sum_{y=1}^{W} R(x,y)$$

$$E[R(x,y)^2] = \frac{1}{H \times W} \sum_{x=1}^{H} \sum_{y=1}^{W} [R(x,y)]^2 \qquad (9.4)$$

where H and W are the height and width of a frame.

Then the averaged frame variance at sequence level is:

$$\sigma_s^2 = \frac{\sum_{\text{F frames}} E_R[R^2]}{F} - \left(\frac{\sum_{\text{F frames}} E_R[R]}{F} \right)^2 \qquad (9.5)$$

where F is the total number of WZ frames in the coded video sequence.

b) Frame level: all the samples in one frame are characterized by the same variance. The average variance σ_R^2 of the residual frame is

$$\sigma_R^2 = E[R(x,y)^2] - (E[R(x,y)])^2 \qquad (9.6)$$

c) Block level: all the samples in each $m \times m$ block of a frame have the same variance.

First, the kth block in the residual frame can be written as

$$R_k(x,y) = WZ_k(x,y) - SI_k(x,y). \qquad (9.7)$$

Then the averaged variance of the kth block in the R frame can be calculated as

$$\sigma_{R_k}^2 = E_{R_k}[R_k(x,y)^2] - (E_{R_k}[R_k(x,y)])^2 \qquad (9.8)$$

where the expectations take over all pixels within the kth block.

d) Pixel level: the variance of each residual pixel is

$$\sigma_p^2 = (R(x,y))^2 = (WZ(x,y) - SI(x,y))^2. \qquad (9.9)$$

3) Correlation parameter α computation: according to the relationship between parameter α and variance in (9.2), the correlation parameter of different granularity levels can be written as

a) sequence level: $\alpha_s = \sqrt{\dfrac{2}{\sigma_s^2}}$

b) frame level: $\alpha_R = \sqrt{\frac{2}{\sigma_R^2}}$

c) block level: $\alpha_{R_k} = \begin{cases} \sqrt{2}, \sigma_{R_k}^2 \le 1 \\ \sqrt{\frac{2}{\sigma_{R_k}^2}}, \sigma_{R_k}^2 > 1. \end{cases}$

d) pixel level: $\alpha_p = \begin{cases} \sqrt{2}, |R(x,y)| \le 1 \\ \sqrt{\frac{2}{R(x,y)^2}}, |R(x,y)| > 1. \end{cases}$

Here, since the variances obtained from different granularity levels are based on the average of samples in different precisions, a variance value calculated with high precision (e.g., pixel, block levels) is more likely to be close to zero compared with low precision estimations (e.g., frame and sequence levels). In order to avoid numerical errors and maintain a reliable estimation, the maximum estimated correlation parameter is bounded by $\sqrt{2}$. Finally, the obtained correlation parameter of different granularities can be used to help the decoding of the WZ frame with different performance gains. In general, a higher estimation precision will result in a better decoding performance.

Transform Domain Offline Correlation Estimation
In contrast to PD offline correlation estimation, the estimation takes place on the DCT coefficients instead of pixel in TD DVC. For the offline correlation estimation in TD DVC [58], three granularity levels are studied: DCT band/sequence, DCT band/frame, and coefficient/frame.

The procedures for the computing correlation parameter α for TD WZ video coding are similar to the procedures for PD video coding, which can be summarized as follows [58]:

1) Residual frame generation: the residual frame $R(x, y)$ is generated as describe in (9.3).
2) Residual frame DCT transform: applying 4×4 block-based discrete cosine transform on the residual frame $R(x, y)$ will generate the DCT coefficients frame T, which consists of 16 DCT bands

$$T(u, v) = DCT[R(x, y)] \qquad (9.10)$$

where (u, v) denotes the DCT coefficient position of the T frame.
3) Variance computation: according to the definition of variance for a random variable as $\sigma_Z^2 = E[Z^2] - (E[Z])^2$, where $E[\cdot]$ is the expectation operator, the variance at different granularity levels can be obtained by averaging the variance on the corresponding levels:
   a) DCT band/sequence level: the same DCT band in a whole video sequence has the same variance, which can be obtained by averaging the corresponding DCT bands in the whole video sequence.

First, let T_b denote a set of coefficients at the bth DCT band of transformed frame T, where the length J of T_b is the ratio between the frame size and the number of total DCT bands, that is, $\frac{H \times W}{4 \times 4}$ in our setups. Then the variance of each DCT band within a given frame is given by

$$\sigma_b^2 = E_b[T_b^2] - (E_b[T_b])^2$$

$$E_b[T_b] = \frac{1}{J} \sum_{x=1}^{J} T_b(j)$$

$$E_b[T_b^2] = \frac{1}{J} \sum_{x=1}^{J} [T_b(j)]^2 \tag{9.11}$$

Then average variance at sequence level is:

$$\sigma_{b,s}^2 = \frac{\sum_{\text{F frames}} E_b[T_b^2]}{F} - \left(\frac{\sum_{\text{F frames}} E_b[T_b]}{F} \right)^2 \tag{9.12}$$

where F is the total number of WZ frames in the coded video sequence.

b) DCT band/frame level: all the samples in each DCT band of a transformed residual frame are characterized by the same variance. The average variance σ_b^2 of the DCT band b for a certain transformed residual frame is

$$\sigma_b^2 = E_b[T_b^2] - (E_b[T_b])^2 \tag{9.13}$$

c) Coefficient/frame level: the variance of each DCT coefficient has a different value, which can be written as

$$\sigma_c^2 = (T(u,c))^2. \tag{9.14}$$

4) Correlation parameter α computation: according to the relationship between parameter α and variance (9.2), the correlation parameter of different granularity levels can be written as

a) DCT band/sequence level: $\alpha_{b,s} = \sqrt{\dfrac{2}{\sigma_{b,s}^2}}$

b) DCT band/frame level: $\alpha_b = \sqrt{\dfrac{2}{\sigma_b^2}}$

c) Coefficient/frame level: $\alpha_c = \left\{ \begin{array}{l} \sqrt{2}, |T(u,v)| \leq 1 \\ \sqrt{\dfrac{2}{T(u,v)^2}}, |T(u,v)| > 1. \end{array} \right\}$

Here, the maximum the correlation parameter α is also bounded by $\sqrt{2}$ in case (c) for the same reasons discussed in the PD DVC.

9.1.3 Online Correlation Estimation

Due to the requirements of DSC setup, where each WZ frame needs to be encoded independently, the offline correlation parameter estimation is not feasible in practical DVC applications. As the original WZ frame is not available at the decoder, we should resort to other strategies to perform online correlation estimation at the decoder side. In [58], the online correlation estimation methods with three different granularities, that is, frame, block, pixel levels in PD and TD DVC, are studied. The strategies of online correlation parameter estimation at the decoder are similar to the offline case except that the residual is estimated in a different way as the original WZ frame is not explicitly available before decoding. The following sections review the *online correlation estimation* at different granularity levels in terms of PD and TD DVC.

Pixel Domain Online Correlation Estimation

The procedures for computing the correlation parameter α for PD DVC are summarized as follows [58]:

1) Residual frame generation: since the original WZ frame is not available at the decoder for the generation of the residual frame, the residual frame $R(x, y)$ will be approximated through the difference between motion compensated backward and forward frames X_B and X_F, which is given as follows:

$$R(x, y) = \frac{X_F(x + dx_f, y + dy_f) - X_B(x + dx_b, y + dy_b)}{2} \quad (9.15)$$

where $X_F(x + dx_f, y + dy_f)$ and $X_B(x + dx_b, y + dy_b)$ denote the forward and backward motion compensated frames, respectively, (x, y) represents the pixel position of a residual frame, and (dx_f, dy_f) and (dx_b, dy_b) represent the motion vectors for the X_F and X_B frames, respectively.

2) Variance computation: according to the variance definition of a random variable $\sigma_Z^2 = E[Z^2] - (E[Z])^2$, where $E[\cdot]$ is the expectation operator, the variance at different granularity levels can be obtained by averaging the variance on the corresponding levels:

a) Frame level: all the samples in each frame are characterized by the same variance. The estimated variance $\hat{\sigma}_R^2$ of the residual frame is

$$\hat{\sigma}_R^2 = E[R(x, y)^2] - (E[R(x, y)])^2 \quad (9.16)$$

b) Block level: all the samples in each $m \times m$ block of the R frame have the same variance.

First, the residual frame kth block can be obtained as

$$R_k(x, y) = \frac{X_{F_k}(x + dx_f, y + dy_f) - X_{B_k}(x + dx_b, y + dy_b)}{2} \quad (9.17)$$

Then the averaged variance of the R frame kth block is

$$\hat{\sigma}_{R_k}^2 = E_{R_k}[R_k(x, y)^2] - (E_{R_k}[R_k(x, y)])^2$$

$$E_{R_k}[R_k(x, y)] = \frac{1}{m \times m} \sum_{x=1}^{m} \sum_{y=1}^{m} R_k(x, y)$$

$$E_{R_k}[R_k(x,y)^2] = \frac{1}{m \times m} \sum_{x=1}^{m} \sum_{y=1}^{m} [R_k(x,y)]^2 \tag{9.18}$$

c) Pixel level: the variance of each residual pixel is

$$\hat{\sigma}_p^2 = (R(x,y))^2. \tag{9.19}$$

3) Correlation parameter α computation: according to the relationship between parameter α and variance (9.2), the correlation parameters of different granularity levels can be written as
   a) Frame level:

$$\hat{\alpha}_R = \sqrt{\frac{2}{\hat{\sigma}_R^2}} \tag{9.20}$$

b) Block level:

$$\hat{\alpha}_{R_k} = \begin{cases} \hat{\alpha}_R, & \hat{\sigma}_{R_k}^2 \leq \hat{\sigma}_R^2 \\ \sqrt{\dfrac{2}{\hat{\sigma}_{R_k}^2}}, & \sigma_{R_k}^2 > \hat{\sigma}_R^2, \end{cases} \tag{9.21}$$

In the above equation, for the case $\hat{\sigma}_{R_k}^2 \leq \hat{\sigma}_R^2$, we argue that the interpolated block has a higher quality than that averaged on the interpolated frame. However, to maintain the uncertainty between the WZ frame and the SI frame, the maximum value of $\hat{\alpha}_{R_k}$ is bounded by $\hat{\alpha}_R$, which can be obtained according to (9.16). When $\sigma_{R_k}^2 > \hat{\sigma}_R^2$, the interpolated block may have a low quality, which corresponds to a high residual error in the offline case. As a higher residual error implies a lower confidence about the similarity between the SI frame and the WZ frame, we will choose a smaller $\hat{\alpha}_{R_k}$ as $\sqrt{\dfrac{2}{\hat{\sigma}_{R_k}^2}}$.

c) Pixel level:

$$\hat{\alpha}_p = \begin{cases} \hat{\alpha}_R, & \sigma_{R_k}^2 \leq \hat{\sigma}_R^2 \\ \hat{\alpha}_{R_k}, & (\sigma_{R_k}^2 > \hat{\sigma}_R^2) \wedge (D_{R_k} \leq \hat{\sigma}_R^2) \\ \hat{\alpha}_{R_k}, & (\sigma_{R_k}^2 > \hat{\sigma}_R^2) \wedge (D_{R_k} > \hat{\sigma}_R^2) \wedge [R(x,y)]^2 \leq \hat{\sigma}_{R_k}^2 \\ \sqrt{\dfrac{2}{[R(x,y)]^2}}, & (\sigma_{R_k}^2 > \hat{\sigma}_R^2) \wedge (D_{R_k} > \hat{\sigma}_R^2) \wedge [R(x,y)]^2 > \hat{\sigma}_{R_k}^2 \end{cases} \tag{9.22}$$

where D_{R_k} is the distance between the average value of the R frame kth block and the average value of the R frame, which is written as $D_{R_k} = (E_{R_k}[R(x,y)] - E_R[R(x,y)])^2$. For case 1) $\sigma_{R_k}^2 \leq \hat{\sigma}_R^2$, since the interpolated block has a high quality, all the pixels of the kth block have the same confidence information as that on the frame level. For case 2) $D_{R_k} \leq \hat{\sigma}_R^2$ means that the kth block has similar proprieties as that in the frame level, but with low confidence, as $\sigma^2_{R_k} > \hat{\sigma}^2_R$. Similarly, for case 3), although both the kth block variance and block distance are higher than those in the frame level (i.e., $(\sigma^2_{R_k} > \hat{\sigma}^2_R) \wedge (D_{R_k} > \hat{\sigma}^2_R)$), the interpolated pixel itself has good quality as $[R(x, y)]^2 \le \hat{\sigma}^2_{R_k}$. Thus, such a pixel shares the correlation parameters as the kth block. Finally, for case 4), since the pixel is not well interpolated, it deserves a lower α value than that in the block level.

In summary, the above heuristic rules are based on the following guidelines:

1) If a finer grained estimation level (e.g., block or pixel levels) has equal or higher confidence of SI quality than that of a coarser estimation level (e.g., frame or block levels), which is also equivalent to $\sigma^2_{block} \le \sigma^2_{frame}$ or $\sigma^2_{pixel} \le \sigma^2_{block}$, the correlation parameter from the coarser level will be chosen as its correlation parameter. This scenario maintains a sufficient uncertainty between the WZ frame and the SI frame.

2) If a finer grained estimation level has a lower confidence of SI quality than that of a coarser estimation level, a smaller correlation parameter α will be calculated according to the formula (9.2).

Transform Domain Online Correlation Estimation

1) Residual frame generation: residual frame $R(x, y)$ is generated as described in (9.15).

2) Residual frame DCT transform: DCT transformed coefficients frame T is generated as in (9.10).

3) $|T|$ frame generation: take the absolute value of each element of T frame.

4) Variance computation: according to the variance definition of a random variable $\sigma^2_Z = E[Z^2] - (E[Z])^2$, where $E[\cdot]$ is the expectation operator, the variance at different granularity levels can be obtained by averaging the variance on the corresponding levels:

   a) $|T|$ frame DCT band/frame level: all the samples in the same DCT band of a transformed residual frame are characterized by the same variance. The average variance σ^2_b of the DCT band b for a certain transformed residual frame is

   $$\hat{\sigma}^2_b = E_b[|T|^2_b] - (E_b[|T|_b])^2 \qquad (9.23)$$

   b) Coefficient/frame level: the distance between a coefficient ($|T|_b(u, v)$) of the $|T|$ frame at band b, and the average value $\hat{\mu}_b$ of $|T|_b$ is computed as follows

   $$D_b(u, v) = |T|_b(u, v) - \hat{\mu}_b. \qquad (9.24)$$

5) Correlation parameter α computation: according to the relationship between parameter α and variance (9.2), the correlation parameter of different granularity levels can be written as

a) DCT band/frame level: $\hat{\alpha}_b = \sqrt{\frac{2}{\hat{\sigma}_b^2}}$

b) Coefficient/frame level:

$$\hat{\alpha}_c = \begin{cases} \hat{\alpha}_b, & [D_b(u,v)]^2 \leq \hat{\sigma}_b^2 \\ \sqrt{\frac{2}{[D_b(u,v)]^2}}, & [D_b(u,v)]^2 > \hat{\sigma}_b^2 \end{cases} \tag{9.25}$$

The heuristic rules introduced in the above equation follow the same guidelines as discussed in the PD DVC section. To avoid repetition, readers are referred to the previous section for details.

9.2 Recap of Belief Propagation and Particle Filter Algorithms

Since the decoding performance of DVC relies on the knowledge of correlation, the design of the correlation estimation scheme is essential in DVC application. The previous sections have reviewed the online estimation schemes proposed in [58], which are mainly based on the information extracted from adjacent motion-compensated frames as well as a few empirical rules. However, this failed to take into account the information from the received syndromes. Since the received syndromes contain important information on the original WZ frame, such additional information could potentially improve the estimation performance. In the next section we will introduce an on-the-fly (OTF) correlation estimation scheme, which can extract syndrome information for improving correlation estimation during the LDPC decoding. The OTF estimation scheme is based on the factor graph and Bayesian approximate inference (see Appendix C for more details). In this section, we will recap the belief propagation (BP) and particle filter algorithms before working through the OTF estimation scheme.

9.2.1 Belief Propagation Algorithm

The BP algorithm is an approximate technique for computing marginal probabilities by exchanging the message between neighboring nodes on a factor graph. Denote $m_{a \to i}(x_i)$ as the message sent from a factor node a to a variable node i, and $m_{i \to a}(x_i)$ as the message sent from a variable node i to a factor node a. Loosely speaking, $m_{a \to i}(x_i)$ and $m_{i \to a}(x_i)$ can be interpreted as the beliefs of node i taking the value x_i transmitting from node a to i and from node i to a, respectively. The message updating rules can be expressed as follows:

$$m_{i \to a}(x_i) \propto \prod_{c \in N(i) \backslash a} m_{c \to i}(x_i) \tag{9.26}$$

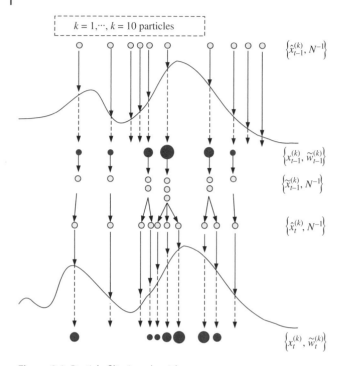

Figure 9.1 Particle filtering algorithm.

and

$$m_{a\to i}(x_i) \propto \sum_{\mathbf{x}_a \backslash x_i} \left(f_a(\mathbf{x}_a) \prod_{j\in N(a)\backslash i} m_{j\to a}(x_j) \right), \tag{9.27}$$

where $N(i)\backslash a$ denotes the set of all neighbors of node i excluding node a, f_a is the factor function for factor node a, and $\sum_{\mathbf{x}_a \backslash x_i}$ denotes a sum over all the variables in $\mathbf{x}_a$ that are arguments of f_a except x_i. Moreover, the BP algorithm approximates the belief of node i taking x_i as

$$b_i(x_i) \propto \prod_{a\in N(i)} m_{a\to i}(x_i). \tag{9.28}$$

9.2.2 Particle Filtering

Particle filters, also known as sequential Monte-Carlo methods, are sophisticated techniques for optimal numerical estimation when exact solutions cannot be analytically derived [73]. They are used to estimate posterior probability distributions of the unknown object states through a list of particles by estimating the state at one time from available measurements up to the next time [73].

The procedure for the particle filtering algorithm is shown as follows:

1) First, generate K number particles $x_{t-1}^{(k)}$ with weight $\frac{1}{N}$.

2) Then K new samples, $\tilde{x}_{t-1}^{(1)}, \cdots, \tilde{x}_{t-1}^{(K)}$, will be drawn with probabilities sampled from a weight distribution using systematic resampling [74]. As a result, some $x_{t-1}^{(k)}$ that have small probabilities will probably be discarded whereas those with high probability will be repeatedly drawn.

3) To maintain the diversity of the particles, the particle locations will be perturbed by a Metropolis–Wyner–Viz–Histing (MH) [75] based Gaussian random walk, which consists of two basic stages. First, let the proposed new K particles at each iteration be $\hat{x}_t^{(k)} = \tilde{x}_{t-1}^{(k)} + Z_r$, that is, the current value plus a Gaussian random variable $Z_r \sim N(0, \sigma_r^2)$. Second, decide whether the proposed values of new particles are rejected or retained by computing the acceptance probability $a\{\hat{x}_t^{(k)}, \tilde{x}_{t-1}^{(k)}\} = min\{1, \dfrac{p\left(\hat{x}_t^{(k)}\right)}{p\left(\tilde{x}_{t-1}^{(k)}\right)}\}$, where $\dfrac{p\left(\hat{x}_t^{(k)}\right)}{p\left(\tilde{x}_{t-1}^{(k)}\right)}$ is the ratio between the proposed particle value and the previous particle value. When the proposed value has a higher posterior probability than the current value $\tilde{x}_{t-1}^{(k)}$, it is always accepted, otherwise it is accepted with probability a.

4) Update weight by resetting to a uniform weight $\frac{1}{K}$ for each particle.

5) Iterate steps 2 to 4 unless the maximum number of iterations is reached or other exit condition is satisfied.

9.3 Correlation Estimation in DVC with Particle Filtering

In this section, we will introduce the OTF correlation estimation model, where the work flow is shown in Figure 9.2. Here, we first describe how a factor graph is designed for this problem, and then we explain the concept of adaptive graph-based decoding incorporating particle filtering for the OTF correlation estimation.

9.3.1 Factor Graph Construction

The construction of a factor graph capturing and connecting SW coding and correlation tracking will be described after identifying a set of variable and factor nodes. Variables nodes denote unknown variables such as coded bits and correlation variance, and factor nodes represent the connection among multiple variable nodes. We model the correlation between source and side information as Laplacian, and build a 3D graph to capture joint bit-plane

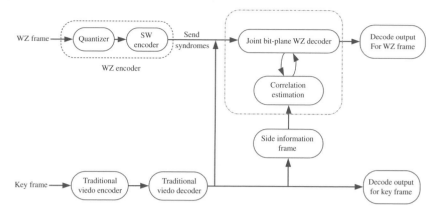

Figure 9.2 The workflow of a WZ decoder with OTF correlation estimation.

decoding, striving for efficient compression, accurate correlation modeling, and good performance.

For WZ coding, we carry out joint bit-plane coding by first quantizing an N-bit source x_i $(i = 1, \cdots, N)$ into $Q[x_i]$, using 2^q level Lloyd–Max quantization. We denote $x_i^1, x_i^2, \cdots, x_i^q$ as the binary format of the index $Q[x_i]$, and denote $\mathbf{B}_j = x_1^j, x_2^j, \cdots, x_N^j$ as the jth significant bit-plane. Each bit-plane is SW encoded independently by using q LDPC codes. Thus the jth SW encoder compresses $\mathbf{B}_j$ $(j = 1, \cdots, q)$ and computes the parity/syndrome bits $\mathbf{S}_j = s_1^j, s_2^j, \cdots, s_{M_j}^j$. This results in an $N{:}M_j$ SW compression ratio, where M_j denotes the number of output syndrome bits for the jth bit-plane. We thus define SW coding rate as $R = \frac{\sum_{j=1}^{q} M_j}{Nq}$, where $R = 1$ represents no compression.

The parity factor nodes $f_1^j, f_2^j, \cdots, f_{M_j}^j$ of the SW LDPC code take into account constraints imposed by the received parity bits. The variable nodes and factor nodes $f_a^j, a = 1, \cdots, q$ are represented as circle nodes and square nodes, respectively, in Region II of the factor graph illustrated in Figure 9.3, where $q = 3$. The q planes capturing SW parity factor nodes and source variable nodes represent the third dimension of the graph.

Let y_i $(i = 1, N)$ be the N-bit side information available at the decoder. To take into account the remaining correlation between quantized source $Q[x_i]$ and side information y_i, we additionally define correlation factor nodes g_i, $i = 1, 2, \cdots, N$ as in (9.29), where $P(\bullet)$ is the value of quantization partition at index "$\bullet$", that is, if a sampled source x_i satisfies $P(\bullet) \leq x_i < P(\bullet + 1)$, the quantization index $Q[x_i]$ of source x_i is equal to "$\bullet$". σ_z is the standard deviation of the correlation noise between x_i and y_i assuming an additive correlation model $x_i = y_i + z_i$, where z_i is Gaussian noise independent of y_i. Alternatively,

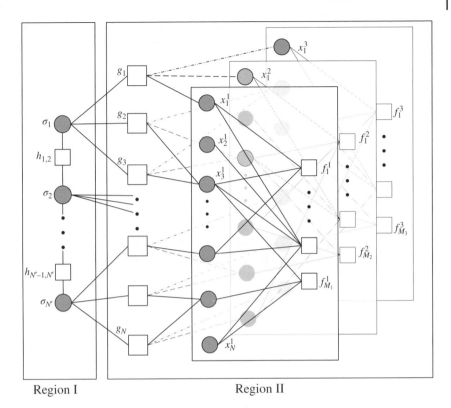

Region I   Region II

Figure 9.3 The WZ decoder with OTF correlation estimation.

the correlation factor node g_i can be expressed as (11.4) when we consider Laplacian noise.

$$f_i(Q[x_i], y_i, \sigma_z) = \int_{P(Q[x_i])}^{P(Q[x_i]+1)} \frac{1}{\sqrt{2\pi\sigma_z^2}} e^{-\left(\frac{x-y_i}{\sqrt{2}\sigma_z}\right)^2} dx$$

$$= \frac{1}{2}\mathrm{erfc}(-\frac{P(Q[x_i]+1)-y_i}{\sqrt{2}\sigma_z}) - \frac{1}{2}\mathrm{erfc}(-\frac{P(Q[x_i])-y_i}{\sqrt{2}\sigma_z}) \qquad (9.29)$$

$$g_i(Q[x_i], y_i, \alpha) = \int_{P(Q[x_i])}^{P(Q[x_i]+1)} \frac{\alpha}{2} e^{\alpha|x-y_i|} dx, \qquad (9.30)$$

where α is the scale parameter of the Laplacian distribution.

While in standard WZ decoding σ_z is assumed to be constant and known *apriori*, in practical applications such as DVC this is rarely the case. We assume that σ_z is unknown and varies slowly over time, typical for correlated frames in a video sequence. We now add extra correlation variable nodes $\sigma_1, \sigma_2, \cdots, \sigma_{N'}$ shown as circle nodes in Region I of the constructed factor graph in Figure 9.3.

Each factor node g_i in Region II is connected to an additional variable node corresponding to σ_l, $l = 1, 2, \cdots, N'$, where N' is the number of correlation variable nodes in Region I. The factor function $g_i(Q[x_i], y_i, \sigma_z)$ of g_i is the same as (9.29) or (11.4). The number of factor nodes g_i that each variable σ_l is connected to is defined as the connection ratio, which is three in the example shown in Figure 9.3. Since the assumption is that the correlation variance varies slowly temporally and spatially, it is expected that adjacent variable nodes σ_l will not differ much in value. This is represented in the graph by additional factor nodes $h_{l,l+1}$ that connect adjacent variable nodes σ_l and σ_{l+1}, as in (9.31), where λ is a hyper-prior, that is,

$$h_{l,l+1}(\sigma_l, \sigma_{l+1}) = \exp(-\frac{(\sigma_{l+1} - \sigma_l)^2}{\lambda}). \tag{9.31}$$

The final factor graph (Figure 9.3) comprises Region II, with a standard Tanner graph for bit-plane LDPC decoding, and Region I, with a bipartite graph capturing correlation variance σ_z^2, and factor nodes g_i, defined in (9.29) and (11.4) for Gaussian and Laplacian correlation, respectively, connecting the two regions.

9.3.2 Correlation Estimation in DVC with Particle Filtering

In this section, the message passing algorithm for efficient SW decoding and correlation estimation via the BP algorithm operating jointly with the particle filtering algorithm is described.

BP operates on the factor graph described in Section 9.2.1. Messages are passed iteratively between connected variable nodes and factor nodes in both regions of the graph until the algorithm converges or until a fixed number of iterations is reached. These messages (inferences or beliefs on source bits and correlation) will represent the influence that one variable has on another. We group these types of messages into the two connected regions identified previously, thus generalizing BP. Hence, the correlation estimation exploits variations in side information in each bit-plane, and hence pixel, and dynamically tracks spatial and temporal variations in correlation between source and side information.

Standard BP (the sum-product algorithm), generally used for SW decoding, can handle only discrete variables. The correlation variance, however, is not a discrete variable, since it varies continuously over time. We therefore resort to particle filtering [76], which is integrated within the standard BP algorithm in order to handle continuous variables. Particle filtering (PF) estimates the *aposteriori* probability distribution of the correlation variable node σ_l by sampling a list of random particles (tied to each correlation variable node) with associated weights.

Each variable node σ_l is modeled with N_p particles. The locations and corresponding weights of each particle σ_l^k, $k = 1, \cdots, N_p$, are adjusted with the updating of the BP algorithm. The belief $b(\sigma_l^k)$ of each particle is essentially the particle weight w_l^k, whose update is achieved by updating variable nodes using standard BP.

The first step of the particle-based BP algorithm is the initialization of each particle value $(\sigma_l^k)^2$ to $(\hat{\sigma}^2)$ and each particle weight w_l^k to $1/N_p$. $(\hat{\sigma}^2)$ is chosen with some prior knowledge of the source statistics. If the input codeword has the same parity as the received one in each bit-plane, the algorithm terminates, otherwise all variable nodes, factor nodes, and particles are updated iteratively.

Systematic resampling [74] is applied once all weights have been updated for all N_p particles in each variable node σ_l in Region I to discard particles with negligible weight and concentrate on particles with larger weights. However, after the resampling step, particles tend to congregate around values with large weight. To maintain the diversity of the particles, the new particle locations are perturbed by applying the random walk Metropolis–Hastings algorithm, which essentially adds Gaussian or Laplacian noise on the current value σ_l for each of the N particles. The weight of each particle is then reset to a uniform weight for each particle. A new codeword is generated at the end of each iteration until the BP algorithm finds a valid codeword or until it reaches a maximum number of iterations.

The message-passing schedule for the factor graph in Figure 9.3, incorporating PF, is summarized in Table 9.1.

Table 9.1 Message passing algorithm jointly updating inference on source and correlation variance.

1: Initialize the values of N_p variable particles in Region I, $\sigma_l = \hat{\sigma}$, for estimating the correlation noise and a uniform weight $1/N_p$.

2: Initialize messages sent from factor nodes $g_i(Q[x_i], y_i, l)$ connecting Regions I and II to variable nodes x_i in Region II as in (9.29), for each of the q bit-planes, where $\sigma_l = \hat{\sigma}$.

3: If the decoded estimate has the same syndrome as the received one or maximum number of iterations is reached, export the decoded codeword and finish. If not, go to Step 4.

4: Update variable nodes in Region II using standard BP (sum-product algorithm) for channel decoding.

5: Update particles in Region I by updating variable nodes using BP.

6: Compute the belief for each variable in Region II being $x_i \in \{0, 1\}$.

7: Compute the belief (= weight) of each particle for each variable node in Region I.

8: Systematic resampling of particles in Region I, followed by the Metropolis–Hastings algorithm and resetting the weight of particles to a uniform weight.

9: Update factor nodes firstly in Region II, Region I and finally those connecting the two regions.

10: Generate a new codeword based on the belief of variable nodes in Region II.

11: Go back to step 3.

9.3.3   Experimental Results

To verify the performance of OTF correlation tracking across WZ-encoded frames in a video sequence, we evaluated the above setup with many standard QCIF 15 Hz video sequences, "Soccer", "Coastguard", "Foreman", "Hall & Monitor", and "Salesman". These videos covered fast, median, and slow motion conditions. In the video sequences of our experiments, we considered two frames per group of picture (GOP) and a total of 149 frames for each video sequence. The odd frames are the key frames which are intraframe coded and reconstructed using H.264/AVC with the profile used in [77]. The even frames, between two intra-coded key frames, are WZ frames which are encoded using LDPCA codes and recovered through the joint bit-plane decoding with the correlation estimation described in previous section.

The test conditions for the WZ frame are described in the following. A 4×4 float DCT transform is performed for WZ frame first, and the uniform scalar quantizer with data range $[0, 2^{11})$ and a dead-zone quantizer with doubled zero interval and the dynamic data range $[-\text{MaxVal}_b, \text{MaxVal}_b)$ are applied for the DC and AC bands [77], respectively. At the decoder, the side-information frame is generated by the recovered frames from H.264/AVC decoding, where a search range with ±32 pixels is used for the forward motion estimation [77]. The following parameters are used in our simulation: number of particles

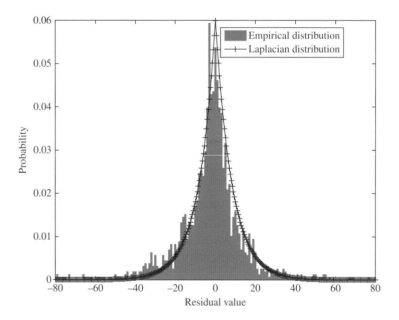

Figure 9.4   Residual histogram (DC band of Foreman at 15 Hz) in comparison with Laplacian distribution ($\alpha = 0.12$).

$N_p = 12$, connection ratio $C = 4$, and random walk step $\sigma_r = 0.005$. In addition, the PBP algorithm is used only after 50 BP iterations and then is performed every 20 BP iterations.

In Figure 9.4, we verified the Laplacian assumption of the correlation between the WZ frame and the side information frame. By setting $\alpha = 0.12$, the Laplace distribution provides an accurate approximation to the residual between the WZ frame and the side information frame.

Moreover, the Bjontegaard delta metric (BJM) [78] is used to illustrate the average difference between two rate-distortion curves in terms of PSNR or bitrate. The Bjontegaard delta measurements of the model and the DISCOVER model are given in Table 9.2, where the test is performed on all the sequences with different motion characteristics and difference quantization matrices. This metric shows that the model outperforms the previous method in terms of bitrate saving.

Results comparing the relative performance of DISCOVER (with the correlation estimator of [58]), joint bit-plane PBP (the transform-based codec with adaptive correlation estimation), joint bit-plane offline (joint bit-plane decoding using true correlation statistics estimated offline [58]) and joint bit-plane

Table 9.2 Average resulting Bjontegaard delta PSNR and corresponding bitrate for sequences Foreman, Soccer, Coastguard, Hall, and Salesman.

		Bjontegaard delta matix	
		Δ PSNR (dB)	Δ Rate (%)
Foreman	Q1	−0.1341	−6.0592
	Q3	−0.2327	−10.5428
	Q5	−0.2576	−14.1864
Soccer	Q1	−0.4253	−16.1156
	Q3	−0.0104	−14.6673
	Q5	−0.0315	−18.1785
Coastguard	Q1	−0.0384	−0.0271
	Q3	−0.7409	−11.7729
	Q5	−0.9804	−18.9212
Hall	Q1	−0.2010	−6.1717
	Q3	−0.2609	−7.5663
	Q5	−0.8541	−12.8799
Salesman	Q1	−0.0318	−5.1683
	Q3	−0.0971	−7.9557
	Q5	−0.3693	−10.5094

online (joint bit-plane decoding using correlation statistics estimated online [58]) codecs for the Soccer, Coastguard, Foreman, Hall & Monitor, and Salesman video sequences, respectively, are shown in Figures 9.5 to 9.9. Note that the offline estimation method of [58] models the correlation noise as Laplacian distributed variable whose true Laplacian parameter is calculated offline at the DCT-band/coefficient level for each frame using the residual between the WZ frame and the side information. This is impractical since in this case the encoder would need to perform side-information generation. On the other hand, the on-line estimation method of [58] models the correlation noise as Laplacian distributed whose Laplacian parameter is estimated using the difference between backward and forward motion compensated frames at the decoder.

As expected, the joint bit-plane decoder (offline) always achieves the best performance for all sequences (slow, median, and fast motion), since it knows the true correlation statistics between source and side information. Most importantly, the PBP-based codec consistently has the next best performance for all sequences, clearly outperforming the DISCOVER codec for all sequences since our PBP estimator iteratively refines the correlation statistics. We also note that the joint bit-plane setup also shows a better performance than that of the

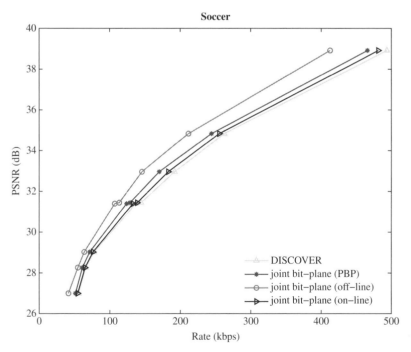

Figure 9.5 PSNR comparison of the PBP joint bit-plane DVC for the QCIF Soccer sequence, compressed at 15 fps.

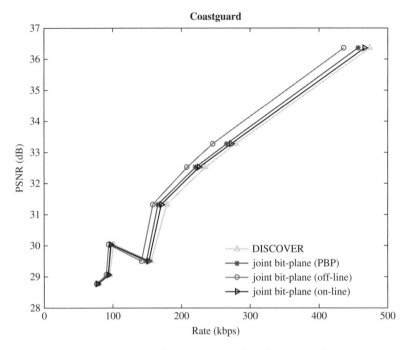

Figure 9.6 PSNR comparison of the PBP joint bit-plane DVC for the QCIF Coastguard sequence, compressed at 15 fps.

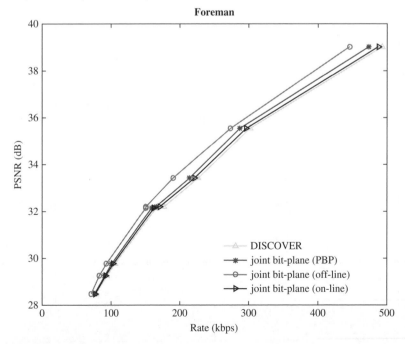

Figure 9.7 PSNR comparison of the PBP joint bit-plane DVC for the QCIF Foreman sequence, compressed at 15 fps.

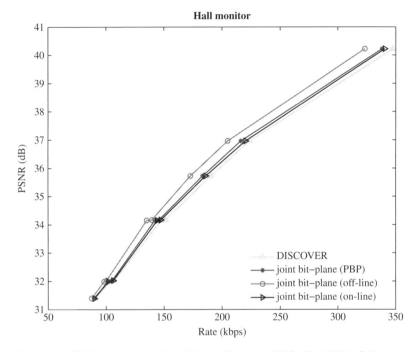

Figure 9.8 PSNR comparison of the PBP joint bit-plane DVC for the QCIF Hall Monitor sequence, compressed at 15 fps.

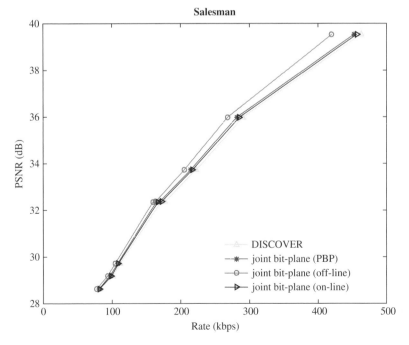

Figure 9.9 PSNR comparison of the PBP joint bit-plane DVC for the QCIF Salesman sequence, compressed at 15 fps.

Table 9.3 Execution time (full sequence in seconds) of joint bit-plane online, joint bit-plane PBP codec vs. DISCOVER codec.

Sequences	DISCOVER	Joint bit-plane online	Joint bit-plane PBP
Foreman Q1	347.9	359.9	1266.4
Q3	457.1	481.8	3159.1
Q5	508.5	631.7	6035.6
Soccer Q1	581.1	586.4	2527.3
Q3	615.0	623.4	4634.3
Q5	657.4	699.4	6847.5

DISCOVER codec since each code bit may obtain more information from its neighboring bit-planes than the traditional separate bit-plane decoding setup.[1] The estimation accuracy is studied in Figure 9.11. We can see that the PBP algorithm improves the online estimate [58], which also explains why the PBP algorithm outperforms DISCOVER codec. Further, the algorithm complexity in terms of execution time is compared in Table 9.3. The PBP algorithm needs longer execution time to achieve a better decoding performance in terms of bitrate saving. Please note that since the PBP algorithm is currently implemented by MATLAB incorporating JAVA, a shorter execution time is expected in the future by using more efficient programming language (e.g., C++).

Finally, a frame-by-frame PSNR variation for the Soccer sequence with quantization matrix Q8 is shown in Figure 9.10. We found that the PSNR variation across frames for soccer sequence is about 4.48 dB for the PBP codec and 2.81 dB for H.264 codec. Although the PSNR variation of the codec is slightly larger than that of H.264, the difference in average PSNR between H.264 and PBP codec is only 0.35 dB. Moreover, the result shows that the PSNR fluctuations of PBP and H.264 have similar trends and the maximum PSNR difference between them is about 1.33 dB.

9.3.4 Conclusion

This section discussed an OTF correlation estimation scheme for distributed video coding, where the correlation estimation step is embedded within the SW

---

1 Note that the LDPC code length in the joint bit-plane decoder is M_b times longer than that of the separate bit-plane decoder, where M_b is the quantization level of DCT band b. Since the decoding performance of LDPC code depends on the code length, the joint bit-plane decoder with longer code length is expected to outperform the separate bit-plane decoder in the simulation. Another reason for this behavior is that the factor g_i in the joint bit-plane decoder could obtain information from M_b variable nodes in Region III simultaneously, while such an advantage is not available in the separate bit-plane decoder.

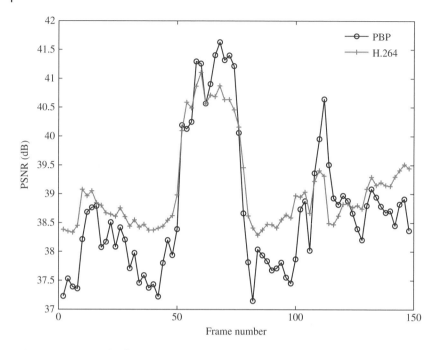

Figure 9.10 Frame-by-frame PSNR variance for Soccer sequence with quantization matrix Q8.

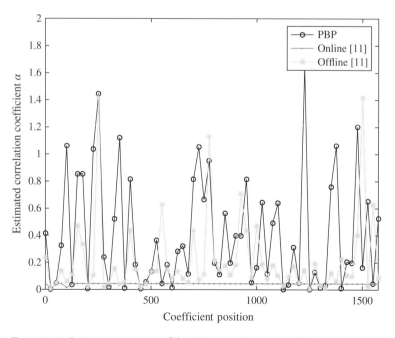

Figure 9.11 Estimation accuracy of the OTF estimation method for the AC band of Soccer sequence.

decoder itself to ensure dynamic tracking of the correlation estimation. This is achieved by augmenting the SW code factor graph with correlation parameter variable nodes together with additional factor nodes that connect the SW graph with the correlation variable nodes. Inference on the graph with multiple connected regions can then be achieved with standard belief propagation (sum product algorithm) together with particle filtering to handle the continuous variable. The results show that correlation statistics are accurately estimated through the OTF estimator, which also results in a better DVC decoding performance.

9.4 Low Complexity Correlation Estimation using Expectation Propagation

Although *on-the-fly correlation estimation* methods with particle filtering described in the previous section outperform the online estimation techniques, stochastic approximation methods usually introduce a large computational cost at the decoder. Since the deterministic approximation method (e.g., expectation propagation (EP)) provides low complexity and comparable estimation performance for modeling unimodal distribution, we will discuss the deterministic approximation method (e.g., EP) for the on-the-fly correlation estimation in DVC in this section.

9.4.1 System Architecture

To precisely catch the correlation between frames while recovering source frames, we introduce an adaptive DVC framework. In the Bayesian perspective, capturing correlation corresponds to estimating the posterior distribution of the correlation parameter. Since a factor graph, as a particular type of graphical model, enables efficient computation of marginal distributions through the message passing algorithm, the framework is carried out on the factor graph as shown in Figure 9.12. The key steps of the adaptive DVC framework can be outlined as follows: 1) factor graph construction: design a factor graph with appropriately defined factor functions to capture and connect SW coding and correlation tracking (see Section 9.4.2); 2) message-passing algorithm implementation: perform the message-passing algorithm on the constructed factor graph to calculate the posterior distribution of interested variables (see Section 9.4.3).

9.4.2 Factor Graph Construction

Compared with standard DVC, the factor graph (see Figure 9.12) of the adaptive DVC with correlation tracking consists of two regions, where Region I refers to the correlation parameter tracking and Region II corresponds to the traditional

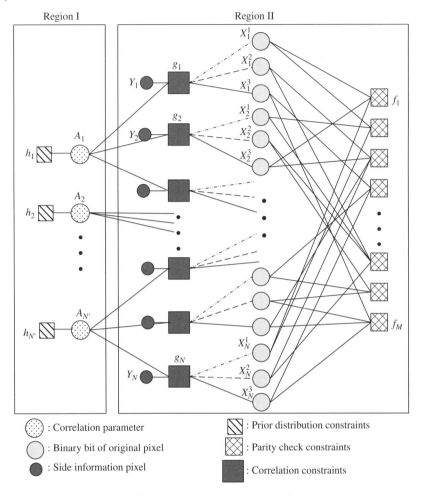

Region I   Region II

(dotted circle) : Correlation parameter   (hatched square ◺) : Prior distribution constraints

(light circle) : Binary bit of original pixel   (crossed square ⊠) : Parity check constraints

(dark circle) : Side information pixel   (filled square ■) : Correlation constraints

Figure 9.12 Factor graph of joint bit-plane SW decoding with correlation estimation.

WZ coding. In Figure 9.12, variable nodes (usually depicted by a circle) represent unknown variables that need to be estimated, such as coded bits and correlation parameter, and factor nodes (depicted by small squares) capture the relationship among the connected variable nodes.

Joint Bit-plane SW Coding (Region II)

WZ coding, also known as the lossy version of SW coding, is usually realized by quantization followed by SW coding of the quantized indices based on channel coding [79]. Here, for WZ coding, we carry out LDPC-based joint bit-plane SW coding after performing quantization, the factor graph of which is described as Region II in Figure 9.12.

Note that we suppose a N-length source sample x_i, $i = 1, \cdots, N$ is quantized into $Q[x_i]$ using 2^q levels of quantization, where $q = 3$ is taken as an example in Region II of Figure 9.12. We denote $x_i^1, x_i^2, \cdots, x_i^q$ as the binary format of the quantization index $Q[x_i]$, and denote $\mathbf{B} = x_1^1, x_1^2, \cdots, x_1^q, x_2^1, x_2^2, \cdots, x_N^q$ as the block which combines all the bit variables together. The block $\mathbf{B}$ is then encoded using LDPC-based SW codes and generates an M-length syndrome bits $\mathbf{S} = s_1, s_2, \cdots, s_M$, which results in a $qN{:}M$ SW compression ratio.

Similar to the standard LDPC decoding, the factor nodes $f_1, f_2, \cdots, f_M$ in Region II take into account the constraints imposed by the received syndrome bits. Thus, the factor function of factor node f_a, $a = 1, \cdots, M$ is defined as

$$f_a(\tilde{\mathbf{x}}_\mathbf{a}, s_a) = \begin{cases} 1, & \text{if } s_a \oplus \bigoplus \tilde{\mathbf{x}}_\mathbf{a} = 0, \\ 0, & \text{otherwise.} \end{cases} \tag{9.32}$$

where $\tilde{\mathbf{x}}_\mathbf{a}$ denotes the set of neighbors of factor node f_a, and $\bigoplus \tilde{\mathbf{x}}_\mathbf{a}$ denotes the binary sum of all elements of the set $\tilde{\mathbf{x}}_\mathbf{a}$.

Let an N-length sample y_i, $i = 1, \cdots, N$, the realizations of variable nodes Y_i, be the side information available at the decoder. The factor nodes g_i are introduced in the factor graph to capture the correlation constraints as shown in (9.1) between source x_i and side information y_i for SW decoding. Since source samples are first passed through a quantization process, the correlation constraints (i.e., the factor function of g_i) between the quantized indices $Q[x_i]$ and the side information y_i can be expressed as:

$$g_i(Q[x_i], y_i, \lambda) = \int_{P(Q[x_i])}^{P(Q[x_i]+1)} \frac{\lambda}{2} e^{\lambda |x - y_i|} dx, \tag{9.33}$$

where λ is the correlation parameter[2] of the Laplace distribution, $P(\bullet)$ denotes the lower boundary of quantization partition at index "$\bullet$", for example if a coefficient x_i satisfies $P(\bullet) \le x_i < P(\bullet + 1)$, the quantization index $Q[x_i]$ of coefficient x_i is equal to "$\bullet$". Actually, given a parameter λ, the factor node g_i plays the role of providing a predetermined likelihood $p(y_i | Q[x_i], \lambda)$ to variable node X_i^j, $j = 1, \cdots, q$ for LDPC-based SW decoding.

Correlation Parameter Tracking (Region I)

As described in the previous section, the correlation parameter, denoted by λ_l^t, can vary along both time and space, where the superscript t indicates the time dependence (i.e., frame) and the subscript l corresponds to the location of a single pixel or a group of pixels (i.e., block). In this section, for the simplicity of notation, we drop the superscript t and denote λ_l as the correlation of the

---

2 In this chapter, we use λ instead of α to denote the correlation parameter, where α is reserved for another purpose.

*l*th pixel block in a given frame, where each pixel block possesses C number of pixels. Then, the correlation parameter λ in (9.33) can be replaced by λ_l for a different pixel block.

Let us denote by N' the number of pixel blocks within a frame. Then, we introduce additional variable nodes A_l, $l = 1, 2, \cdots, N'$ to represent the correlation parameters λ_l in the factor graph (see, Region I of Figure 9.12). Since a block of C source samples (i.e., pixels) share the same correlation parameter, every C number of factor nodes g_i in Region II will be connected to the same variable node A_l, where we call C^3 the connection ratio. Moreover, to initialize a prior distribution for correlation parameter λ_l, additional factor nodes h_l, $l = 1, \cdots, N'$ are introduced, where Gamma distribution is assigned to each factor function $h_l(\lambda_l)$ for the mathematical convenience. Then, by implementing the message-passing rules introduced in the next subsection on the factor graph, each factor node g_i will periodically update the likelihood $p(y_i|Q[x_i], \lambda_l)$ for the corresponding bit variable nodes $X_i^1, X_i^2, \cdots, X_i^q$ when a new estimate of correlation parameter λ_l is available, instead of using a predetermined likelihood $p(y_i|Q[x_i], \lambda)$.

Consequently, by introducing correlation parameter estimation in Region I, the likelihood factor function in (9.33) will be updated as

$$g_i(Q[x_i], y_i, \lambda_l) = \int_{P(Q[x_i])}^{P(Q[x_i]+1)} \frac{\lambda_l}{2} e^{\lambda_l|x-y_i|} dx. \tag{9.34}$$

9.4.3 Message Passing on the Constructed Factor Graph

In Bayesian inference, the message-passing algorithm (e.g., BP) on a factor graph offers an very efficient way to calculate the marginal distributions (i.e., beliefs) of the unknown variables represented by their corresponding variable nodes. In the adaptive DVC factor graph (see Figure 9.12), we are interested in two unknown variables, which are represented by source variable nodes X_i^j in Region II and correlation parameter variable nodes A_l in Region I, respectively.

In Region II, without considering the connection to the Region I, the factor graph is identical to that of standard LDPC codes with discrete variables x_i^j. Hence, the posterior distribution of x_i^j can be calculated through the standard BP algorithm. However, in Region I the BP algorithm cannot be applied directly, as the correlation parameter λ_l represented by the variable node A_l is generally a nonGaussian continuous variable and the BP algorithm only handles discrete variables with small alphabet size or continuous variables with linear Gaussian distribution.

---

3 To estimate a stationary correlation parameter, we can set the connection ratio equal to the code length. Moreover, the connection ratio provides a trade-off between complexity and spatial variation.

To seek a workaround for this difficulty, let us start with the derivation of the posterior distribution of the correlation parameter λ_l. According to Bayes' rule and the message-passing rule, the posterior distribution of the correlation parameter λ_l can be expressed as:

$$p(\lambda_l | \mathbf{y}_l) = \frac{1}{Z_l} \prod_{i \in \mathcal{N}^{\setminus h_l}(A_l)} p(\lambda_l) p(y_i | \lambda_l)$$

$$= \frac{1}{Z_l} \prod_{i \in \mathcal{N}^{\setminus h_l}(A_l)} \int_{Q[x_i]} p(\lambda_l) p(Q[x_i]) p(y_i | Q[x_i]; \lambda_l)$$

$$= \frac{1}{Z_l} h(\lambda_l) \prod_{i \in \mathcal{N}^{\setminus h_l}(A_l)} \sum_{\mathbf{x}_i^q} g(y_i; Q[x_i], \lambda_l) \prod_{j \in 1,2,\cdots,q} m_{X_i^j \to g_i}(x_i^j)$$

$$= \frac{1}{Z_l} m_{h_l \to A_l}(\lambda_l) \prod_{i \in \mathcal{N}^{\setminus h_l}(A_l)} m_{g_i \to A_l}(\lambda_l), \qquad (9.35)$$

where Z_l is a normalization constant, $\sum_{\mathbf{x}_i^q}$ denotes a sum over all the bit variables in $\mathbf{x}_i^q$, the value of message $m_{X_i^j \to g_i}(x_i^j)$ is updated iteratively by variable node X_i^j in Region II according to the BP update rule, message $m_{h_l \to A_l}(\lambda_l) = h(\lambda_l)$ comes from prior factor node in Region I, and message $m_{g_i \to A_l}(\lambda_l) = \sum_{\mathbf{x}_i^q} g(y_i; Q[x_i], \lambda_l) \prod_{j \in 1,2,\cdots,q} m_{X_i^j \to g_i}(x_i^j)$ comes from likelihood factor node in Region II according to the BP update rule.

So far we have shown that the posterior distribution of correlation variable λ_l can be expressed as the product of all the incoming messages. In the rest of this subsection we investigate how to efficiently compute the posterior distribution using BP- and EP-based approximation algorithms.

Expectation Propagation

An approximate inference for solving the problem in (9.35) is to parametrize the variables through variational inference. Deterministic approximation schemes (e.g., EP [80]) provide some low-complexity alternatives based on the analytical approximations to the posterior distribution. For example, suppose that posterior distribution $p(\theta)$ of parameter θ is not feasible to be calculated directly. If the posterior can be factorized as $p(\theta) = \prod_k g_k(\theta)$, where each factor function $g_k(\theta)$ only depends a small subset of observations, EP solves this difficulty by replacing the true posterior distribution $p(\theta)$ with an approximate distribution $q(\theta) = \prod_k \tilde{g}_k(\theta)$ by sequentially computing each approximate term $\tilde{g}_k(\theta)$ for $g_k(\theta)$. The general workflow of the EP algorithm is listed in Table 9.4. In particular, for our problem in (9.35), EP is used to sequentially compute approximate messages $\tilde{m}_{h_l \to A_l}(\lambda_l)$ and $\tilde{m}_{g_i \to A_l}(\lambda_l)$ in place of true messages $m_{h_l \to A_l}(\lambda_l)$ and $m_{g_i \to A_l}(\lambda_l)$ in (9.35), then an approximate posterior on λ_l is obtained by combining these approximations. The details of correlation parameter estimation through EP for our problem will be discussed in the next section.

Table 9.4 Expectation propagation.

Initialize the term approximation $\tilde{g}_k(\theta)$ and $q(\theta) = \frac{1}{Z} \prod_{k=1}^{C} \tilde{g}_k(\theta)$, where $Z = \int_\theta \prod_{k=1}^{C} \tilde{g}_k(\theta)$

repeat

    for $k = 1, \cdots, C$ **do**

        Compute $q^{\backslash k}(\theta) \propto q(\theta)/\tilde{g}_k(\theta)$

        Minimize Kullback–Leibler (KL) divergence between $q(\theta)$ and $g_i(\theta)q^{\backslash k}(\theta)$ by performing moment matching

        Set approximate term $\tilde{g}_k(\theta) \propto q(\theta)/q^{\backslash k}(\theta)$

    end for

until parameters converged

9.4.4 Posterior Approximation of the Correlation Parameter using Expectation Propagation

In this section, we will derive the EP-based correlation estimator, which can provide a fast and accurate way to approximate the posterior distribution on the factor graph, as shown in Figure 9.12. The procedures of the EP algorithm have been detailed as follows:

1) Initialize the prior term

$$h_l(\lambda_l) = \text{Gamma}(\lambda_l, \alpha_l^0, \beta_l^0) = z_l^0 \lambda_l^{\alpha_l^0 - 1} \exp(-\beta_l^0 \lambda_l) \tag{9.36}$$

with $\alpha_l^0 = 2$, $\beta_l^0 = \frac{\alpha_l^0 - 1}{\lambda^0}$, $z_l^0 = \frac{\beta_l^{0\alpha_l^0}}{\Gamma(\alpha_l^0)}$, where λ^0 is the initial correlation parameter, and β_l^0 and α_l^0 are scale and shape parameters for Gamma distribution, respectively. The selection of the initial values for the above parameters guarantees the mode of prior distribution equal to the initial correlation λ^0.

2) Initialize the approximation term (uniform distribution)

$$\tilde{m}_{g_i \to A_l}(\lambda_l) = \text{Gamma}(\lambda_l, \alpha_{il}, \beta_{il}) = z_{il} \lambda_l^{\alpha_{il} - 1} \exp(-\beta_{il}\lambda_l) \tag{9.37}$$

with $\beta_{il} = 0$, $\alpha_{il} = 1$, $z_{il} = 1$.

3) Initialize α_l^{new} and β_l^{new} for approximate posterior $q(\lambda_l) = \text{Gamma}(\lambda_l, \alpha_l^{\text{new}}, \beta_l^{\text{new}})$, where $\alpha_l^{\text{new}} = \alpha_l^0 = 2$, and $\beta_l^{\text{new}} = \beta_l^0$.

4) For each variable node λ_l

    For each factor node g_i, where $g_i \in \mathcal{N}(\lambda_l)$

    a) Remove $\tilde{m}_{g_i \to A_l}(\lambda_l)$ from the posterior $q(\lambda_l)$, to get $q^{\backslash g_i}(\lambda_l) = \text{Gamma}(\alpha_l^{\text{tmp}}, \beta_l^{\text{tmp}})$

$$\alpha_l^{\text{tmp}} = \alpha_l^{\text{new}} - (\alpha_{il} - 1)$$
$$\beta_l^{\text{tmp}} = \beta_l^{\text{new}} - \beta_{il} \tag{9.38}$$

    b) Update $q^{\text{new}}(\lambda_l)$ by minimizing the Kullback–Leibler (KL) divergence $D(q^{\backslash g_i}(\lambda_l) m_{g_i \to A_l}(\lambda_l) \| q^{\text{new}}(\lambda_l))$ (i.e., performing moment matching (**Proj**)) (see Section 9.4.4 for details).

$$q^{\text{new}}(\lambda_l) = \frac{1}{Z_l}\mathbf{Proj}[q^{\backslash g_i}(\lambda_l)m_{g_i \to A_l}(\lambda_l)] \tag{9.39}$$

where $Z_l = \int_{\lambda_l} q^{\backslash g_i}(\lambda_l)m_{g_i \to A_l}(\lambda_l)$.

c) Set the approximated message

$$\alpha_{il} = \alpha_l^{\text{new}} - (\alpha_l^{\text{tmp}} - 1)$$

$$\beta_{il} = \beta_l^{\text{new}} - \beta_l^{\text{tmp}}$$

$$z_{il} = Z_l \frac{\beta_l^{\text{new}\alpha_l^{\text{new}}}}{\Gamma(\alpha_l^{\text{new}})} \left(\frac{\beta_l^{\text{tmp}\alpha_l^{\text{tmp}}}}{\Gamma(\alpha_l^{\text{tmp}})} \right)^{-1} \left(\frac{\beta_{il}^{\alpha_{il}}}{\Gamma(\alpha_{il})} \right)^{-1} \tag{9.40}$$

Moment Matching

Through moment matching, $q(\lambda_l)$ is obtained by matching the mean and variance of $q(\lambda_l)$ to those of $q^{\backslash g_i}(\lambda_l)m_{g_i \to A_l}(\lambda_l)$. Then, we get the updated α_l^{new} and β_l^{new}, the parameters of $q(\lambda_l)$, as follows,

$$\alpha_l^{\text{new}} = m_1 \beta_l^{\text{new}}$$

$$\beta_l^{\text{new}} = m_1/(m_2 - m_1^2), \tag{9.41}$$

where m_1 and m_2 are the first and second moments of the approximate distribution as shown below.

$$m_1 = \frac{1}{Z} \sum_{Q[x_i]} (\mathcal{F}_1(z_1) - \mathcal{F}_1(z_2)) \prod_{j=1}^{j=q} m_{X_i^j \to g_i}(x_i^j)$$

$$m_2 = \frac{1}{Z} \sum_{Q[x_i]} (\mathcal{F}_2(z_1) - \mathcal{F}_2(z_2)) \prod_{j=1}^{j=q} m_{X_i^j \to g_i}(x_i^j)$$

$$Z = \sum_{Q[x_i]} (\mathcal{F}_0(z_1) - \mathcal{F}_0(z_2)) \prod_{j=1}^{j=q} m_{X_i^j \to g_i}(x_i^j) \tag{9.42}$$

Here, Z is the normalization term and these unknown functions in (9.42) can be evaluated according to (9.43).

$$z_1 = P(Q[x_i] + 1)$$

$$z_2 = P(Q[x_i])$$

$$\mathcal{F}_0(z) = A(z) + B(z)$$

$$\mathcal{F}_1(z) = \frac{\alpha^{\text{tmp}}}{\beta^{\text{tmp}}} A(z) + \frac{\alpha^{\text{tmp}}}{\beta^{\text{tmp}} + |z - y_i|} B(z)$$

$$\mathcal{F}_2(z) = \frac{\alpha^{\text{tmp}}(\alpha^{\text{tmp}} + 1)}{(\beta^{\text{tmp}})^2} A(z) + \frac{\alpha^{\text{tmp}}(\alpha^{\text{tmp}} + 1)}{(\beta^{\text{tmp}} + |z - y_i|)^2} B(z)$$

$$A(z) = \frac{1}{2}(1 + \mathrm{sgn}(z - y_i))$$

$$B(z) = -\mathrm{sgn}(z - y_i)\frac{1}{2}\left(\frac{\beta^{\mathrm{tmp}}}{\beta^{\mathrm{tmp}} + |z - y_i|}\right)^{\alpha^{\mathrm{tmp}}} \tag{9.43}$$

9.4.5 Experimental Results

In this section, we employ a pixel-based DVC setup to demonstrate the benefit of the OTF correlation tracking. As in [11, 56, 58], the group of pictures (GOP) is equal to 2 in our study, where all even frames are treated as WZ frames and all odd frames are considered as key frames. The key frames are conventionally intra-coded, for example using the H.264 Advanced Video Coding (AVC) [81] intra-coding mode. WZ frames are first quantized pixel by pixel and then all bit-planes of the resulting quantization indices are combined together and compressed using an LDPCA code. At the decoder, side-information frame Y is generated using motion-compensated interpolation of the forward and backward key frames [11, 56]. Spatial smoothing [56], via vector-median filtering, is used to improve the result together with a half-pixel motion search. Each WZ frame is decoded by the EP-based OTF WZ decoder described in Section 9.4.1. Moreover, we incorporate LDPCA codes with a feedback channel for rate adaptive decoding.

To verify the effectiveness of correlation tracking across WZ-encoded frames in a video sequence, we tested the above setup with three standard QCIF (i.e., 176×144) video sequences, Carphone, Foreman, and Soccer, with different scene dynamics of low, medium, and high motions, respectively. All the results are based on the average of 50 WZ frames. The quantization parameters Q of H.264/AVC encoder for different video sequences with different WZ quantization bits q have been listed in Table 9.5. The selections of quantization parameters Q for different WZ quantization bits q make sure that both the decoded key and WZ frames have similar visual qualities in terms of PSNR. Moreover, we split each 176×144 WZ frame into 16 sub-frames with size 44×36 (i.e., $N = 1584$) for efficient coding purposes. Within each sub-frame, the block size for correlation estimation is equal to 4×6 (i.e., $C = 24$) for a total of $N' = 66$ blocks.

Results comparing the relative performance of pre-estimation in frame level DVC [58], pre-estimation in block level DVC [58], and the EP-based OTF DVC for the Carphone, Foreman, and Soccer video sequences, respectively, are shown in Figures 9.13, 9.14, and 9.15, where the implementation of standard DVC codec is based on the DISCOVER framework [58] with joint bit-plane setup.

The pre-estimation methods [58], either in frame or block levels, model the correlation as a Laplace distribution, whose correlation parameter is estimated using the difference between backward and forward motion compensated

Table 9.5 H.264/AVC quantization parameter Q for different video sequences.

Quantization bits	Carphone Q	Foreman Q	Soccer Q
2	46	46	44
3	36	36	34
4	28	28	26
5	22	21	19

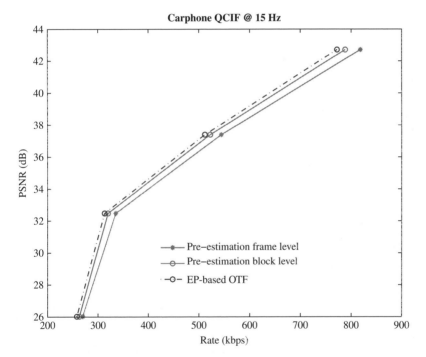

Figure 9.13 PSNR comparison of the EP-based OTF and pre-estimation DVC for the QCIF Carphone sequence, compressed at 15 fps.

frames/blocks at the decoder. The OTF estimator unifies the process of correlation estimation using EP and joint bit-plane decoding into a single joint process, where the updated decoding information can be used to improve the correlation estimation and vice versa. As expected, pre-estimation in the block level has better performance than that of the frame level in terms of bitrate saving, since block-level correlation offers a finer granularity than frame-level

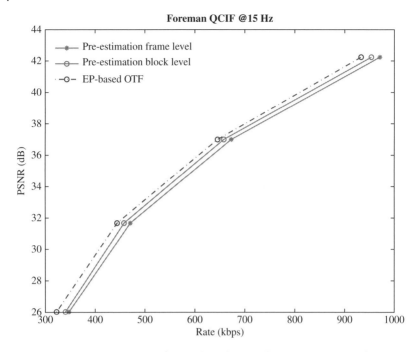

Figure 9.14 PSNR comparison of the EP-based OTF and pre-estimation DVC for the QCIF Foreman sequence, compressed at 15 fps.

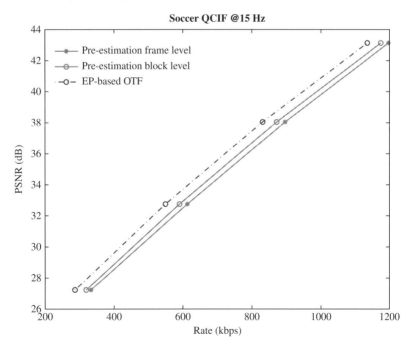

Figure 9.15 PSNR comparison of the EP-based OTF and pre-estimation DVC for the QCIF Soccer sequence, compressed at 15 fps.

correlation. More importantly, the EP-based OTF codec always achieves the best performance for all sequences (slow, medium, and fast motions), since the EP-based OTF estimator can iteratively refine the correlation statistics in each block.

In particular, for the Carphone sequence (slow motion), to obtain the same visual qualities (i.e., PNSRs), the EP-based OTF codec achieves about 10 kbps and 30 kbps saving compared to pre-estimation in the block and frame levels, respectively. For the Foreman sequence (medium motion), the average rate decrease of the EP-based OTF codec is 20 kbps for pre-estimation in the block level and 30 kbps for pre-estimation in the frame level. Moreover, for the Soccer sequence with fast motion, we again observe the superiority of the EP-based OTF codec over the pre-estimation codecs, where the EP-based OTF codec offers about 38.25 kbps and 56.5 kbps saving compared to pre-estimation in the block and frame levels, respectively. These results demonstrate that the EP-based OTF codec is more powerful for video sequences with fast motion.

A sub-frame-by-sub-frame (i.e., the first sub-frame of each WZ frame) rate variation for the soccer sequence with quantization bits equal to 3 is shown

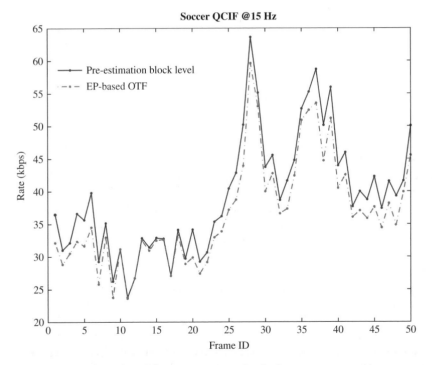

Figure 9.16 Subframe-by-subframe rate variance for the Soccer sequence with quantization bits equal to 3.

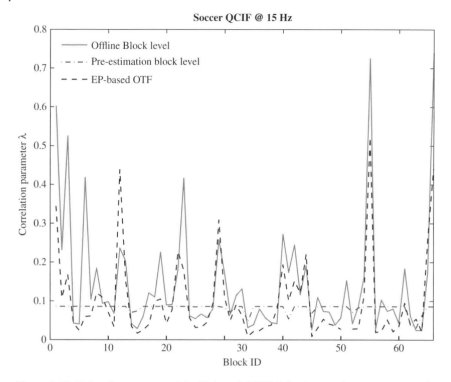

Figure 9.17 Estimation accuracy of the EP-based OTF DVC for the correlation parameter of the Soccer sequence.

in Figure 9.16. We found that the rate variation across frames is about 36.01 kbits for the EP-based OTF DVC codec and 40.01 kbits for the DVC codec with pre-estimation in the block level. Moreover, the result shows that the rate fluctuations of EP-based OTF and pre-estimation in block level DVC codecs have a similar trend and the EP-based OTF codec always has an equal or lower code rate than that of the pre-estimation in the block level DVC codec. The maximum difference in code rate between EP-based OTF and pre-estimation block level codecs is about −5.32 kbits. Similar results are obtained for other sub-frames in all three testing sequences.

The estimation accuracy of the correlation parameter is studied in Figure 9.17 for the soccer sequence. Here, we use the offline estimated correlation parameter as the benchmark, where the benchmark Laplacian parameter is calculated offline at the block level for each frame using the residual between the WZ frame and the side-information frame. We can see that the OTF correlation estimation scheme improves the estimates obtained through pre-estimation method [58], which also explains why the EP-based OTF DVC outperforms the pre-estimation-based DVC codec.

Finally, the EP-based OTF estimator offers a very low complexity overhead compared with the standard BP algorithm. The complexity of the estimator lies in the evaluation of equations (9.41), (9.42), and (9.43), as shown in Section 9.4.4. Roughly speaking, the EP-based OTF estimator introduces less than 10% computational overhead compared with the standard BP algorithm.

9.4.6 Conclusion

This section proposes an OTF correlation estimation scheme for distributed video coding using EP. Unlike previous work performing pre-estimation where estimation starts before decoding, the correlation estimation technique is embedded within the WZ decoder itself, thus ensuring dynamic estimation of correlation changes in the block level. This is achieved by augmenting the SW code factor graph to connect correlation parameter variable nodes together with additional factor nodes. Inference on the factor graph for continuous correlation parameter variables is achieved through EP-based deterministic approximation methods, which offer better trade-off between accuracy and complexity compared with other methods. The scheme boosts coding performance together with ease of integration with existing DVC codecs. We demonstrate the benefits of using the EP-based scheme via a pixel-based DVC setup. Simulation results show significant performance improvement due to correlation tracking for multiple video sequences with the Laplacian correlation model.

10

DSC for Solar Image Compression

Acquiring and processing astronomical images is becoming increasingly important for accurate space weather prediction and expanding our understanding about the Sun and the Universe. These images are often rich in content, large in size, and dynamic in range. Efficient, low-complexity compression solutions are essential to reduce onboard storage, processing, and communication resources. Distributed compression is a promising technique for onboard coding of solar images by exploiting correlation between successively acquired images. In this section, DSC and correlation estimation are applied to compress the stereo solar images captured by the twin satellites system of NASA's STEREO project, where an adaptive distributed compression solution using particle-based belief propagation (PBP) is used to track correlation, as well as perform disparity estimation, at the decoder side.

10.1 Background

Onboard data processing has always been a challenging and critical task in remote-sensing applications due to the severe computational and/or power limitations of onboard equipment. This is especially the case in deep-space applications, where mission spacecraft collect a vast amount of image data that is stored and/or communicated to the observation center. In such emerging applications, efficient low-complexity image compression is a must. While conventional solutions, such as JPEG, have been successfully used in many prior missions, demand for increasing image volume and resolution as well as increased space resolution and wide-swath imaging calls for greater coding efficiency at reduced encoding complexity.

NASA's STEREO (Solar TErrestrial RElations Observatory), launched in October 2006, has provided ground-breaking images of the Sun using two space-based observatories [82]. These images aim to reveal the processes in the solar surface (photosphere) through the transition region into the corona and provide the 3D structure of coronal mass ejections (CME). CMEs are violent

Distributed Source Coding: Theory and Practice, First Edition. Shuang Wang, Yong Fang, and Samuel Cheng.
© 2017 John Wiley & Sons Ltd. Published 2017 by John Wiley & Sons Ltd.
Companion Website: www.wiley.com/go/cheng/dsctheoryandpractice

eruptions of solar plasma into space. If the eruptions are directed towards the Earth they can have catastrophic effects on radio transmissions, satellites, power grids (which may result in large-scale and long-lasting power outages), and humans traveling in airplanes at high altitude.

The data streams that are transmitted 24 hours a day as weather beacon telemetry from each spacecraft have to be heavily compressed [82]. The reconstructed images are available online immediately after reception. Due to compression, many image artifacts have been spotted that have led to wrong conclusions (e.g., see [83]). Another scientific stream is recorded and transmitted daily using NASA Deep Space Network lightly compressed. These images become available 2–3 days after arrival in the Flexible Image Transport System (FITS) and/or JPEG format.

A variety of image compression tools are currently used in deep-space missions, ranging from Rice and lossy wavelet-based compression tools (used in PICARD mission by CNES2009), discrete cosine transform (DCT) + scalar quantization + Huffman coding (Clementine, NASA1994), ICER (a low-complexity wavelet-based progressive compression algorithm used in Mars missions, NASA 2003) to (12-bit) JPEG-baseline (Trace NASA1998, Solar-B JAXA2006) [84]. The compression algorithms have mainly been implemented in hardware (ASIC or FPGA implementation), but some of them run as software on DSP processors (e.g., ICER). The key characteristics of these algorithms are relatively low encoding power consumption, coding efficiency, and error resilience features. The latest Earth observation satellites usually employ JPEG2000 [85] or similar wavelet-based bit-plane coding methods implemented on FPGA, which might be too prohibitive for deep-space missions. Note that all current missions, including STEREO, use 2D mono-view image compression, trading-off computational cost and compression performance. Since STEREO images are essentially multi-view images, with high inter-view correlation, current compression tools do not provide an optimum approach. Thus, in this chapter we will discuss a distributed multi-view image compression (DMIC) scheme for such emerging remote-sensing setups.

When an encoder can access images from multiple views, a joint coding scheme [86] achieves higher compression performance than schemes with separate coding, since multi-view images are usually highly correlated. However, due to the limited computing and communication power of space imaging systems, it is not feasible to perform high-complexity, power-hungry onboard joint encoding of captured solar images. Although, intuitively, this restriction of separate encoding seems to compromise the compression performance of the system, DSC theory [55, 87] proves that distributed independent encoding can be designed as efficiently as joint encoding as long as joint decoding is allowed, making DSC an attractive low-complexity onboard source coding alternative.

The DMIC image codec is characterized by low-complexity image encoding and relatively more complex decoding meant to be performed on the ground. A joint bit-plane decoder is described that integrates particle filter (PF) with standard BP decoding to perform inference on a single joint 2D factor graph. We evaluated the lossy DMIC setup with grayscale stereo solar images obtained from NASA's STEREO mission [82] to demonstrate high compression efficiency with low encoding complexity and non-power-hungry onboard encoding, brought about by DSC. DSC has been used for onboard compression of multi-spectral and hyper-spectral images [88, 89], where DSC is used to exploit inter-band correlation efficiently. In [89], for example, a low-complexity solution that is robust to errors is proposed using scalar coset codes to encode the current band and the previous bands as decoder side information. The algorithms in [89] are implemented using FPGA, and simulations on AVIRIS images show promising results.

10.2 Related Work

Since the DMIC scheme intersects several research topics, we group prior work into three categories.

The first category relates to work in the area of low-complexity onboard or remote multi-view image coding. In [90], lossy compression of Earth orbital stereo imagery used for height detection with three or four views based on motion compensation and JPEG-2000 [85] and JPEG-LS is proposed. Note that motion compensation + JPEG-2000 might still be considered as power expensive for remote sensing, including deep-space missions. In [91], a modification of the mono-view ICER image coder, employed in the Mars Exploration mission, is proposed. The proposed coder optimizes a novel distortion metric that reflects better stereoscopic effects rather than conventional mean-square error (MSE) distortion. The results reported in [91] show improved stereo ranging quality despite the fact that correlation information between the left and right image pair was not exploited in any way or form (see also [92]).

The second relevant topic is correlation tracking in DSC applications. Most DSC designs, including DVC, (with few exceptions) usually simplify the problem by modeling correlation noise, that is, the difference between the source and side information, as Laplacian random variables and estimate the distribution parameters either based on training sequences or previously decoded data. This imposes a certain loss, especially for images or sequences that are very different or nonstationary. Nonstationarity of the scene has been dealt with mainly by estimating correlation noise (e.g., on the pixel or block level) from previously decoded data and different initial reliability is assigned to different pixels based on the amount of noise estimated both in pixel and transform domains [70, 59, 58, 71, 72]. In [93] the authors proposed an efficient way of estimating correlation between the source and side information for the pixel domain by tightly incorporating the process within the SW decoder via SW code factor graph augmentation to include correlation variable nodes with particles such that particle filtering is performed jointly with BP over the augmented factor graph during the SW decoding process. Note that the BP-based SW decoding and correlation statistics estimation are considered jointly. The proposed correlation estimation design was tested on a transform-domain DVC [77] with a channel, but with joint bit-plane coding. This work extends this result from mono-view to low-encoding complexity multi-view coding.

The third relevant topic is multi-view image coding using DSC principles. Despite the potential of DSC, attaining its ultimate performance relies on the assumption that both the correlation and the disparity among multi-view images are known a *priori* at the decoder. Direct measurement of the correlation and disparity at the encoder side is both expensive in terms of computation and impossible without communication among imaging sensors. Thus, estimating correlation and disparity at the decoder becomes the main challenge in DMIC. For disparity estimation in DMIC, the idea of motion compensation [69, 94] used in DVC offers a possible solution. However, these motion compensation methods usually require an excessive amount of computation. Thus, some low-complexity disparity learning schemes for DMIC have been proposed in the literature. In [95], Varodayan *et al.* developed an expectation maximization (EM) based algorithm at the decoder to learn block-based disparity for lossless compression [95] and then extended it to the lossy case [96]. In comparison with the system without disparity compensation, a better compression performance is observed when disparity compensation is employed at the decoder [95, 96].

Thus, knowing the correlation among multiview images is a key factor in determining the performance of a DMIC scheme. This correlation is generally nonstationary (spatially varying) and should be handled adaptively. For example, in [97] an edge-based correlation assignment method is proposed, where the correlation parameters of blocks with and without edges are assigned to different values. However, even the aforementioned work is based on the assumption that the correlation among images is known *a priori*. Similarly to disparity compensation, dynamic correlation estimation given at the decoder could also yield significant improvement in performance. However, most studies of correlation estimation in DSC focus on the correlation estimation of stationary binary sources [98, 99], which are not suitable for the nonbinary image sources in the DMIC case.

Several other approaches have been proposed [100–103], none of which uses correlation tracking. A review on multi-view video coding based on DSC principles can be found in [104, 105].

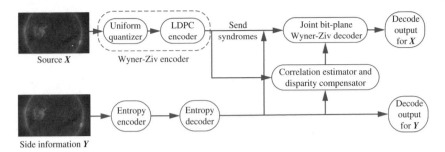

Figure 10.1 Lossy DMIC setup with disparity and correlation estimation.

10.3 Distributed Multi-view Image Coding

Let X and Y be a pair of correlated multi-view images with size M by N. Assuming that a horizontal disparity shift D exists between pixels of X and pixels of Y, the relationship between X and Y can be modeled as

$$X_{(x,y)} = Y_{(x-D_{(x,y)},y)} + Z_{(x,y)}, \tag{10.1}$$

where $x = 1, 2, \ldots, M$ and $y = 1, 2, \ldots, N$ denote the coordinates of pixels, and $Z_{(x,y)}$ satisfies a Laplace distribution $\mathcal{L}(Z_{(x,y)}|\sigma) = \frac{1}{2\sigma}\exp\left(-\frac{|Z_{(x,y)}|}{\sigma}\right)$.

Let us consider a lossy DMIC setup, as shown in Figure 10.1, where image Y, the side information, is perfectly known at the decoder through conventional image coding. At the encoder side, image X is first quantized into $Q[X_{(x,y)}]$ using 2^q level uniform nested scalar quantization (NSQ) [106] and then is encoded based on LDPC codes [16], where each bit-plane is independently encoded into syndrome bits of an LDPC code. Denote $X^1_{(x,y)}, X^2_{(x,y)}, \ldots, X^q_{(x,y)}$ as the binary format of the index $Q[X_{(x,y)}]$, and denote $\mathbf{B^j} = X^j_{(1,1)}, X^j_{(1,2)}, \ldots, X^j_{(M,N)}$ as the jth significant bit-plane, where the superscript $j = 1, \ldots, q$ is used to represent the jth quantized bit or the jth bit-plane in the rest of this chapter. At the LDPC decoder, the BP algorithm is employed to decode image X using the received syndrome bits, the given correlation, and the side information Y reordered by the given disparity information. Finally, when the BP algorithm converges, image X can be recovered based on the output belief for each pixel [16].

10.4 Adaptive Joint Bit-plane WZ Decoding of Multi-view Images with Disparity Estimation

10.4.1 Joint Bit-plane WZ Decoding

In popular layer WZ approaches such as [16], each bit-plane of the quantized source is recovered *sequentially* and this makes it difficult and inefficient for

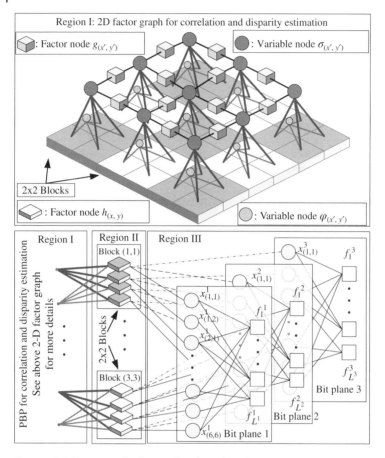

Figure 10.2 Factor graph of a joint bit-plane decoder with disparity and correlation estimation.

the decoder to perform the disparity and correlation estimation. In order to facilitate the disparity and the correlation estimation, let us introduce a joint bit-plane WZ decoding scheme, which can adaptively exploit the disparity and the correlation between a nonbinary source and side information during the decoding process. The main idea of the proposed joint bit-plane WZ decoding scheme is illustrated in Regions II and III of Figure 10.2, where all circle nodes denote variable nodes and all square nodes denote factor nodes. The encoder used in this chapter is the traditional syndrome-based approach using LDPC codes [16]. At the encoder side, a given bit-plane $\mathbf{B}^j$ is compressed into L^j number of syndrome bits, $\mathbf{S}^j = s_1^j, s_2^j \dots, s_{L^j}^j$, thus resulting in $(M \times N){:}L^j$ compression.

At the joint bit-plane decoder, the factor nodes $f_1^j, f_2^j \dots, f_{L^j}^j$ as shown in Region III of Figure 10.2, take into account the constraint imposed by the received syndrome bits. For a factor node $f_a^j, a = 1, \dots, L^j, j = 1, \dots, q$, the corresponding factor function is defined as

$$f_a^j(\tilde{\mathbf{X}}_a^j, s_a^j) = \begin{cases} 1, & \text{if } s_a^j \oplus \bigoplus \tilde{X}_a^j = 0, \\ 0, & \text{otherwise.} \end{cases} \qquad (10.2)$$

where $\tilde{\mathbf{X}}_a^j$ denotes the set of neighbors of the factor node f_a^j, $\oplus$ represents the bitwise addition, and $\bigoplus \tilde{X}_a^j$ denotes the bitwise sum of all elements of the set $\tilde{\mathbf{X}}_a^j$.

Then in Region II, the relationship among a candidate quantized source $Q[X_{(x,y)}]$, side information $Y_{(x,y)}$, and disparity compensation $D_{(x,y)}$ can be modeled by the factor function with

$$h_{(x,y)}(Q[X_{(x,y)}], Y_{(x,y)}, \sigma, D_{(x,y)}) =$$
$$\int_{P(Q[X_{(x,y)}])}^{P(Q[X_{(x,y)}]+1)} \frac{1}{2\sigma} \exp\left(\frac{|\mathcal{X} - Y_{(x-D_{(x,y)},y)}|}{\sigma}\right) d\mathcal{X}, \qquad (10.3)$$

where $P(Q)$ denotes the lower boundary of quantization partition at index "Q", for example if a pixel $X_{(x,y)}$ satisfies $P(Q) \leq X_{(x,y)} < P(Q+1)$, the quantization index $Q[X_{(x,y)}]$ of pixel $X_{(x,y)}$ is equal to Q. Then given the estimation of the correlation σ and the disparity $D_{(x,y)}$, standard BP can be used to perform joint bit-plane decoding based on the proposed factor graph (see Regions II and III in Figure 10.2) and the corresponding factor functions (10.2) and (10.3).

10.4.2 Joint Bit-plane WZ Decoding with Disparity Estimation

Let us assume that each block includes $n \times n$ pixels and shares the same disparity, which yields $\lceil \frac{M}{n} \rceil \times \lceil \frac{N}{n} \rceil$ number of blocks for an $M \times N$ image, where $\lceil \bullet \rceil$ represents the ceiling of $\bullet$ that rounds $\bullet$ toward positive infinity. Then the horizontal disparity field is a constant within a block and will be denoted $D_{(x',y')} \in \{-l, \dots, 0, \dots, l\}$, where $x' = 1, \dots, \lceil \frac{M}{n} \rceil$ and $y' = 1, \dots, \lceil \frac{N}{n} \rceil$ are the block indices. Thus, in the rest of this chapter we will use $D_{(x',y')}$ to represent the disparity $D_{(x,y)}$ of a pixel $X_{(x,y)}$ that lies inside the block (x', y'). For example, in the 2D factor graph of Figure 10.2, a 6×6 image is divided into 3×3 blocks with 2×2 pixels in each block.

In order to estimate the disparity between images, we introduce extra variable nodes $\phi_{(x',y')}$ in Region I (see the 2D factor graph in Figure 10.2). Each factor node $h_{(x,y)}$ in Region II is connected to an additional variable node $\phi_{(x',y')}$ in Region I. Here, we define the connection ratio as the number of factor nodes $h_{(x,y)}$ in Region II which each variable node $\phi_{(x',y')}$ is connected to, for example the connection ratio is equal to 4 in Figure 10.2. According to the BP update rules, the factor node update from Region II to the variable node

$\phi_{(x',y')}$ in Region I can be written as (10.4), where $\mathcal{N}(\phi_{(x',y')})/h_{(x,y)}$ denotes all the neighboring factor nodes of a variable node $\phi_{(x',y')}$ except $h_{(x,y)}$. Moreover, (10.4) can be interpreted as the E-step algorithm used in [95]. Similarly, the factor node update from Region II to Region III can be written as (10.5), where (10.5) can also be interpreted as the M-step algorithm used in [95].

$$m_{h_{(x,y)} \to \phi_{(x',y')}}(D_{(x',y')}) \propto$$

$$\sum_{Q[X_{(x,y)}] \in [0,2^q]} h_{(x,y)}(Q[X_{(x,y)}], Y_{(x,y)}, \sigma, D_{(x',y')}) \prod_{j=1}^{q} m_{X_{(x,y)}^j \to \phi_{(x',y')}}(X_{(x,y)}^j), \quad (10.4)$$

$$m_{h_{(x,y)} \to X_{(x,y)}^j}(X_{(x,y)}^j) \propto$$

$$\sum_{D_{(x',y')}} h_{(x,y)}(Q[X_{(x,y)}], Y_{(x,y)}, \sigma, D_{(x',y')}) m_{\phi_{(x',y')} \to h_{(x,y)}}(D_{(x',y')})$$

$$\prod_{j' \in [1,q]/j} m_{X_{(x,y)}^{j'} \to h_{(x,y)}}(X_{(x,y)}^{j'}), \quad (10.5)$$

10.4.3 Joint Bit-plane WZ Decoding with Correlation Estimation

To compress image X close to the WZ bound in the standard BP approach, the correlation parameter σ must be known a *priori*. However, in practice, the correlation between the colocated pixels of the pair of correlated images X and Y is unknown and, making the situation even more challenging, this correlation may vary over space. Thus, besides the proposed disparity estimation, we introduce an additional correlation estimation algorithm to perform online correlation tracking by extending the previous correlation estimation model [107] from 1D to 2D and from time varying to spatial varying. Moreover, the proposed framework is universal and can be applied to any parametric correlation model.

We assume that σ is unknown and varies block-by-block over space, with the same block-wise assumption also used in Section 10.4.2. To model this, we introduce another set of extra variable nodes $\sigma_{(x',y')}$ in Region I (see the 2D factor graph in Figure 10.2). Now, each factor node $h_{(x,y)}$ in Region II will be connected to an additional variable node $\sigma_{(x',y')}$ in Region I. Here the connection ratio used for correlation estimation is the same as that for disparity estimation in Section 10.4.2. Moreover, the correlation parameter σ used in the factor function $h_{x,y}(Q[X_{(x,y)}], Y_{(x,y)}, \sigma, D_{(x',y')})$ can be modified accordingly by replacing σ as $\sigma_{(x',y')}$, since we assume σ varies over space.

Further, the correlation changes among adjacent blocks may not be arbitrary [97]. The ability to capture correlation changes among adjacent blocks can significantly increase the stability of the correlation tracking of each block. To achieve this, we introduce additional factor nodes $g_{(x'',y'')}$ in Region I (see the 2D factor graph in Figure 10.2), where $x'' = 1, \ldots, \left\lceil \frac{M}{n} \right\rceil - 1$ and

$y'' = 1, \ldots, \left\lceil \frac{N}{n} \right\rceil - 1$ denote block indices just as x' and y'. The corresponding factor function can then be modeled as

$$g_{(x'',y'')}(\sigma_{(x'',y'')}, \sigma_{(x''+c,y''+d)}) = \exp\left(-\frac{(\sigma_{(x'',y'')} - \sigma_{(x''+c,y''+d)})^2}{\lambda}\right), \quad (10.6)$$

where the offset (c, d) is restricted to $\{(0, 1), (0, -1), (1, 0), (-1, 0)\}$ according to the defined 2D factor graph, and λ is a hyper-prior and can be chosen rather arbitrarily.

Since standard BP can only handle discrete variables with small alphabet sizes or continuous variables with a linear Gaussian model, it cannot be applied directly to estimating the continuous correlation parameters. However, by incorporating PF with BP, BP can be extended to handle continuous variables. Then the proposed factor graph model can be used to estimate the continuous correlation.

10.5 Results and Discussion

To verify the effect of correlation and disparity tracking for DMIC, let us evaluate the above setup with grayscale stereo solar images [82] captured by two satellites of the NASA's STEREO project, where the twin satellites are about 30 million miles apart, and the viewing angle is about $6 - 8$ degrees. For the purposes of illustrating accurate tracking of the correlation and the disparity, the simulations for the SW code use only a low-complexity regular LDPC code with variable node degree 5. More complex irregular codes would further improve the overall peak signal-to-noise ratio (PSNR) performance.

The following constant parameters are used in the simulations: image height $M = 128$ pixels, image width $N = 72$, maximum horizontal shift $l = 5$, block size $n = 4$, hyper-prior $\lambda = 10$, initial correlation for Laplace distribution $\sigma = 5$, and initial distribution of disparity

$$p(D_{(x',y')}) = \begin{cases} 0.75, & \text{if } D_{(x',y')} = 0; \\ 0.025, & \text{otherwise;} \end{cases}$$

where the selection of initial values follows [95]. Moreover, the two testing sets of solar images in the simulation results are labeled as solar image SET 1 and solar image SET 2, respectively.

First, let us verified the Laplacian assumption of the correlation between correlated images X and Y in Figure 10.3. By setting $\alpha = 0.23$, Laplace distribution provides an accurate approximate to the residual between images X and Y.

Then, let us examine the rate-distortion performance of the proposed adaptive DMIC scheme, where the PSNR of the reconstructed image is calculated as an indicator of the distortion. We consider the following five different setups.

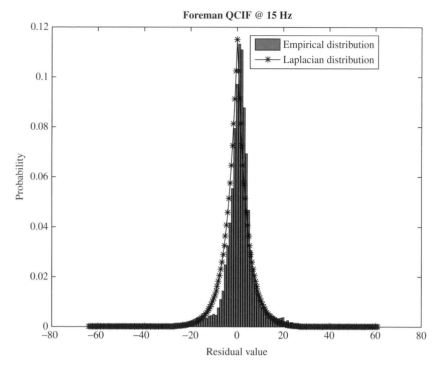

Figure 10.3 Empirical residual histogram for solar images in SET 1 in comparison of Laplace distribution ($\alpha = 0.23$).

a) Adaptive correlation DMIC with a known disparity, which is used as the benchmark performance.
b) Adaptive correlation and disparity DMIC, which is the proposed scheme in this chapter.
c) Adaptive disparity DMIC with a known fixed correlation, which corresponds to the setup used in [95, 96].
d) Non-adaptive joint bit-plane DMIC with known fixed correlation only, where the correlation and disparity estimators are not available at the decoder.
e) Non-adaptive separate bit-plane DMIC with known fixed correlation only, which corresponds to the setup used in [16].

In Figures 10.4 and 10.5 (corresponding to solar images in SET 1 and SET 2, respectively), as expected, the benchmark setup in case a) shows the best rate-distortion performance since the reference disparity is known before decoding. Comparing cases (b) and (c), we find a significant performance gain due to the improved knowledge of correlation statistics due to dynamic estimation. Moreover, all the adaptive DMIC schemes (cases (a), (b), and

Figure 10.4 Rate-distortion performance of the proposed adaptive DMIC scheme for solar images in SET 1.

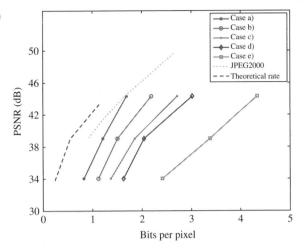

Figure 10.5 Rate-distortion performance of the proposed adaptive DMIC scheme for solar images in SET 2.

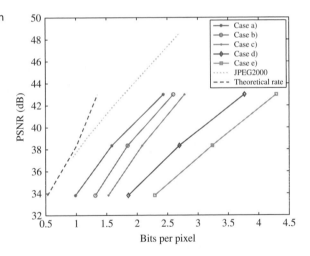

(c)) outperform the non-adaptive schemes (cases (d) and (e)). Besides, in the case without adaptive decoding, we find that the performance of joint bit-plane DMIC in case (d) is still better than separate bit-plane DMIC in case (e). One possible reason for this is that the joint bit-plane DMIC in case (d) can exploit the correlation between two nonbinary sources much better, since in case (e) each bit-plane is decoded separately.

The performance of JPEG2000 codec is also shown as references in Figures 10.4 and 10.5. One can see that JPEG2000 is still unreachable due to its high compression efficiency and uses arithmetic entropy coding at the cost of high encoding complexity. Note that in contrast to JPEG2000, the proposed

DMIC scheme does not employ any transform. In addition, we also plot the theoretical rate as in [16].

10.6 Summary

This chapter is motivated by the limited onboard processing and communications requirements of correlated images captured by different telescopes or satellites. Traditionally, these images are compressed independently using state-of-the-art, low-complexity compression algorithms such as JPEG without considering the spatial and temporal correlation among images captured by deep-space satellites. In order to exploit the correlation among the multiple views acquired from a solar event and enhance compression without jeopardizing the encoding onboard complexity and independent encoding process to minimize communication complexity, we discussed an adaptive DMIC algorithm, which can estimate the correlation and disparity between stereo images, and decode image sources simultaneously. To handle spatially varying correlation between stereo images, we extend the PBP algorithm [107] for correlation estimation to the 2D case. Moreover, the correlation and disparity estimation algorithms are all based on an augmented factor graph, which offers great flexibility for problem modeling in remote-sensing applications. Through the results, a significant decoding performance gain has been observed by using the proposed adaptive scheme, when comparing with the non-adaptive decoding scheme. While the proposed scheme performs worse than JPEG2000, the latter has significantly higher encoding complexity compared to the proposed method.

11

Secure Distributed Image Coding

11.1 Background

The introduction of the Virtual Lifetime Electronic Record (VLER) [108] by the Department of Veterans Affairs (VA) indicates a new era of electronic healthcare. VLER contains not only a veteran's administrative (i.e., personnel and benefits) information but also the complete medical record throughout his/her entire military career and beyond. The next phase of electronic healthcare intends to make patient information available securely, through the Nationwide Health Information Network (NHIN), to private and public healthcare providers for military officers [109]. This activity significantly improves the healthcare performance as well as increasing the flexibility and convenience for sharing patient diagnoses. Ultimately, the service is expected to be provided to all residents in the USA. President Obama proposed a massive effort to push the digital revolution in health records, and the FDA has approved an iPad/iPhone radiology app for mobile diagnoses, which means mobile applications may become a big potential market for cloud-based private data sharing and healthcare.

All these efforts have led to the explosion of medical data transmission over the network, which creates an urgent need to ensure both efficiency and confidentiality in the process. On one hand, since medical data contains sensitive information about personal privacy, privacy protection has become a crucial issue which requires that medical data always be safe. On the other hand, the transmission of medical data, especially medical imaging data, requires a significant amount of bandwidth. Efficient transmission has turned out to be another imperative concern, where promising compression algorithms with privacy-preserving capability are needed. However, privacy-preserving compression algorithms generally have to deal with the encrypted biomedical data directly, which creates big challenges for conventional compression

Distributed Source Coding: Theory and Practice, First Edition. Shuang Wang, Yong Fang, and Samuel Cheng.
© 2017 John Wiley & Sons Ltd. Published 2017 by John Wiley & Sons Ltd.
Companion Website: www.wiley.com/go/cheng/dsctheoryandpractice

techniques. This happens because the principle of compression is to remove redundancy through exploiting the structure of data, whereas encrypted data typically appear to be random, as they mask the structure of the original data (otherwise traces of the original data could be leaked out from any perceived structure of the encrypted data [110]). The "best" practice has been to reverse the order of these steps by compressing the data before encryption [111]. However, the conventional "best" practice obviously may not be suitable in medical-related applications with privacy concerns, since data hosts may not always be the data owners, and any manipulation (e.g., compression) of the unencrypted data initiated by the data host could lead to insecure exposure of private data.

To always keep data under encryption such that the privacy of the data and security of the system are well preserved, encryption followed by compression is essential but challenging. Although the encrypted data appears to be "completely" random, the fact that the encrypted data is not random conditioned on the cryptographic key offers some hope for tackling the impediment. More precisely, redundancy within the original data might be recovered by performing joint decompression and decryption at the decoder side given the cryptographic key, making direct compression on encrypted data feasible. Surprisingly, the aforementioned setup is nothing but a DSC problem, with side information only available at the decoder. More importantly, the theoretical result of DSC-based compression [55] guarantees that there is no performance loss even when performing compression without side information (e.g., the cryptographic key) at the encoder, which is identical to the compression of encrypted data. This essentially means that encrypted data can be compressed very efficiently, just like raw data, as long as the key is given to the decoder without sacrificing any security or privacy at the encoder side. Therefore, privacy-preserving compression opens up many possibilities, for example efficient medical data management, where the data hosts, who primarily focus on efficient data transmission/dissemination, and data owners, who are mainly concerned about data security and privacy, can be totally separated.

DSC offers a promising solution for this tricky problem (i.e., the compression of encrypted data), for which many secure compression algorithms have been investigated in [111–115]. For example, Johnson *et al.* [112] proved the feasibility of compressing encrypted data in theory. The encrypted data are as compressible as the original data when the cryptographic key is available at joint decompression and decryption [112]. Moreover, Schonberg *et al.* described a low-density parity-check (LDPC) codes-based secure compression scheme for binary images using a 1D Markov correlation structure [114] at the joint decoder. Since the 1D Markov correlation model has poor match performance for exploring 2D correlation within a binary image, a 2D Markov source model [113] is proposed to enhance the secure compression performance. However, all these efforts [113, 114] limit themselves to the scope of binary images. Recently, the work proposed by Schonberg *et al.* [111] described a way to losslessly compress encrypted gray-scale video sequences frame by frame. In this work, each pixel in a gray-scale frame is bitwisely decomposed into 8 binary bits with the outputs of 8 binary sub-frames for any given gray-scale frame. Moreover, by assuming that all above 8 binary sub-frames are equally significant in their gray-scale pixel representation, each binary sub-frame can be processed separately according to the aforementioned 2D Markov-based secure compression algorithm [113]. However, the correlation exploited within each independent binary sub-frame [113] cannot guarantee the ultimate compression efficiency, as the correlations among sub-frames are lost. Consequently, directly applying these models [114, 113] to practical gray-scale images/videos compression with privacy concerns involves many difficulties in coding efficiency.

In practical image/video coding, two of the most common techniques, discrete cosine transform (DCT) and localized predictor, are not feasible for dealing with encrypted images, since the exploitable structure or local information of original images are totally masked in the encrypted images. Fortunately, the temporal or spatial correlation among adjacent frames in video/image sequences offers a possible opportunity for frame-by-frame compression after encryption. In this chapter, primarily inspired by spatial correlation among adjacent CT image slices, we introduce a practical framework named SecUre Privacy-presERving Medical Image CompRessiOn (SUPERMICRO) to provide robust privacy-preserving compression on gray-scale medical images (i.e., CT image sequences).

11.2   System Architecture

The block diagram of the SUPERMICRO framework with compression of encrypted data is presented in Figure 11.1. Let $\tilde{X}$ and $\tilde{Y}$ denote the ith and $(i-1)$th original CT image slices, which are first encrypted into X and Y, respectively, through some standard encryption techniques (e.g., stream cipher). Further, the cryptographic keys used to encrypt these slices are sent to a joint decoder through secure channels. Then the encrypted slice X is compressed into S (i.e., so-called syndromes) using a standard DSC encoder without accessing the cryptographic key. At the decoder, the encrypted slice Y is treated as side information for helping the joint decompression and decryption given the received syndromes S and cryptographic keys of both X and Y, where $\hat{X}$ and $\hat{Y}$ denote the estimated slices by the decoder. The central process of the SUPERMICRO system consists of two key components as (1) compression of encrypted data (see Section 11.2.1) and (2) joint decompression and decryption (see Section 11.2.2). The following subsections present the details of each process.

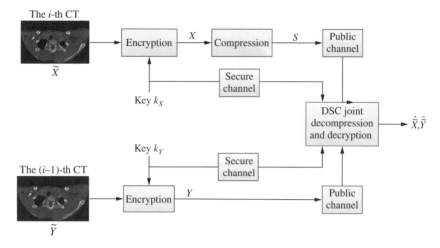

Figure 11.1 The workflow of the proposed framework.

11.2.1 Compression of Encrypted Data

In this module, each CT image slice is first encrypted with stream cipher and thus perfect secrecy is guaranteed [116]. This is illustrated in Figures 11.2(a) and 11.2(b). For example, the pixels of the original slice in Figure 11.2(a) have values in the range of 0 to 255. Therefore, each pixel can be represented by 8 bits such that each image consists of 8 bit-planes (from the most to the least significant), where each bit-plane includes all the bits with equal significance in the bitwise pixel representation. Then, eight randomly sampled keys with uniform distribution, where each of them corresponds to one bit-plane and has the same size as the original slice, will be generated. Encrypting the original slice in Figure 11.2(a) can be achieved by applying a bitwise exclusive-OR (XOR) between each bit-plane of the original slice and the generated key, where the encrypted slice is shown in Figure 11.2(b). Since the cryptographic key is i.i.d across all pixels and bit-planes, the correlation and redundancy among pixels after encryption will be totally destroyed and thus cannot be compressed in a traditional sense. Fortunately, given that the key is available at the decoder, we can compress the encrypted slice of CT image through the DSC principle.

The implementation of DSC is largely based on the idea of random code. Bits of different bit-planes and pixels are randomly intermixed and the resulting bitwise sum will be sent to the decoder as a "syndrome". The number of these 1-bit syndromes M will be smaller than the total number of encrypted bits and thus results in a compression, where each encrypted slice has N 8-bit pixels for total $8N$ encrypted bits and each pixel x_i is represented by $x_i^1, x_i^2, \cdots, x_i^8$ in its binary format. However, the decoding complexity in DSC depends on the code length, which means that direct compression and decompression on encrypted bits

Figure 11.2 Illustration of DSC-based compression of an encrypted CT image slice. The original slice (a) is encrypted into the slice shown in (b) using a stream cipher. The encrypted slice is then compressed using the DSC method into that shown in (c) with much smaller size.

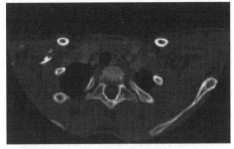

(a) Original slice

(b) Encrypted slice

(c) Compressed slice

with $8N$ bits may result in a significant computational burden for the decoder with limited computation power. One workaround to tackle this difficulty is to compress only the first q ($q \in [2, 8]$) most significant bit-planes instead of all 8 bits, which offers a great trade-off between compression performance and decompression complexity. The impact of different selections of q value on the compression performance will be studied in our experimental results.

Now, let us denote $\mathbf{B} = \{x_1^1, x_1^2, \cdots, x_1^q, \cdots, x_N^1, \cdots, x_N^q\}$ by the binary vector to be compressed. The remaining $(8 - q)N$ encrypted bits will be sent to the decoder directly without any compression. For LDPC codes based on SW coding (lossless DSC scheme [55]), the compressed syndromes are generated through $\mathbf{S} = \mathbf{H} \times \mathbf{B}^T$, where $\mathbf{H}$ is a parity check matrix of LDPC codes with size $M \times qN$ and $M < qN$. Thus, the LDPC-based SW codes result in a $R = ((8 - q)N + M) : 8N$ code rate.

11.2.2 Joint Decompression and Decryption Design

To make joint decompression and decryption for CT image sequences possible, a key factor is to be able to explore the spatial correlation between adjacent CT image slices at the decoder, which typically can be achieved by applying Bayesian inference on graphical models. The graphical model of the SUPERMICRO scheme for joint decompression and decryption of encrypted CT slices is shown in Figure 11.3, where variable nodes (usually depicted by circles) denote variables such as encrypted variables, encryption keys, syndromes (i.e., compressed bits) and unencrypted variables, and factor nodes (depicted by small squares) represent the relationship among the connected variable nodes.

The SUPERMICRO scheme is based on SW codes and only the first q significant bit-planes are compressed, as shown in Figure 11.3 for $q = 3$. On the right-hand side of Figure 11.3, x_i^l, the realization of variable nodes X_i^l, with $i = 1, \cdots, N, l = 1, \cdots, q$ represent the encrypted bits to be recovered, while x_i^l with $i = 1, \cdots, N, l = q + 1, \cdots, 8$ denote the received encrypted bits without any compression. In addition, s_j, the realization of variable nodes $S_j, j = 1, \cdots, M$, represent the received syndromes. The factor nodes $f_j, j = 1, \cdots, M$ connecting these variable nodes X_i^l and S_j take into account the parity check constraints. Since the received encrypted bits without any compression are directly available at the decoder side, we no longer need the factor nodes f_j to guarantee the parity check for the corresponding variable nodes. Actually, the right-hand side of Figure 11.3 without these uncompressed bits is identical to the factor graph of the standard LDPC codes. For the factor nodes $f_j, j = 1, \cdots, M$, the corresponding factor function can be expressed as

$$f_j(\mathbf{x}_{f_j}, s_j) = \begin{cases} 1, & \text{if } s_j \oplus \bigoplus \mathbf{x}_{f_j} = 0, \\ 0, & \text{otherwise.} \end{cases} \tag{11.1}$$

where $\mathbf{x}_{f_j}$ denotes the set of neighbors of factor node f_j, and $\bigoplus \mathbf{x}_{f_j}$ denotes the binary sum of all elements of the set $\mathbf{x}_{f_j}$.

Since pixels in the encrypted slices have approximately a uniform distribution, it is not feasible to directly obtain the correlation between original slices based on the encrypted versions, and the correlation information is essential for the inference algorithm to reconstruct the original slices. To find the correlation, one could resort to the cryptographic keys available at the decoder, where the factor nodes h_i^Y and h_{il}^X, $i = 1, \cdots, N$ is used to incorporate the constraints of cryptographic keys. The factor functions for factor nodes h_i^Y can be defined as

$$h_i^Y(y_i, \tilde{y}_i, \mathbf{k}_i^Y) = \begin{cases} 1, & \text{if } y_i^l \oplus \tilde{y}_i^l \oplus k_{il}^Y = 0, \\ & \text{for all } l = 1, \cdots, 8 \\ 0, & \text{otherwise} \end{cases} \tag{11.2}$$

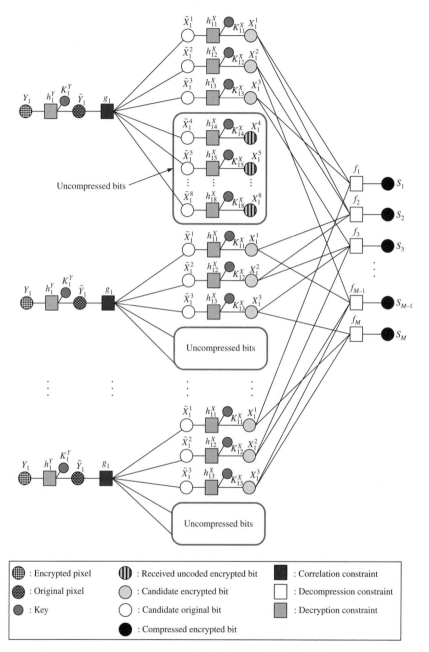

Figure 11.3 Factor graph for decompression of compressed encrypted data.

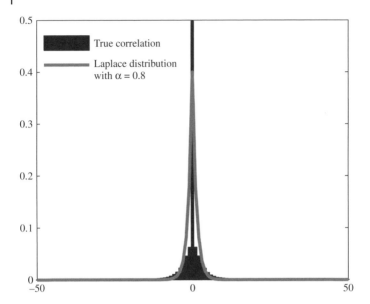

Figure 11.4 Residual histogram for slice sequences of a CT image, where the Laplace distribution with $\alpha = 0.8$ is shown as reference.

where y_i^l and $\tilde{y}_i^l$ represent the lth significant bit of pixel y_i and $\tilde{y}_i$ with $l = 1, \cdots, 8$, and key $\mathbf{k}_i^Y = \{k_{i1}^Y, \cdots, k_{i8}^Y\}$, respectively.

Similarly, the factor functions for factor nodes h_{il}^X can be expressed as

$$h_{il}^X(x_i^l, \tilde{x}_i^l, k_{il}^X) = 1 \oplus x_i^l \oplus \tilde{x}_i^l \oplus k_{il}^X \tag{11.3}$$

The definitions of the above factor functions guarantee that both side information pixels $\tilde{y}_i$ and any candidate source pixels $\tilde{x}_i$ will satisfy the encryption constraints. Then, we introduce additional factor nodes g_i to capture the correlation between $\tilde{y}_i$ and $\tilde{x}_i$. As identified by many previous studies [58, 117, 93], the statistical correlation between adjacent frames in a video sequence can be effectively modeled by a Laplace distribution. Here, for neighboring CT image slices, we observed that the inter-slice correlation also satisfies Laplace distribution, as depicted in Figure 11.4. Therefore, the factor function of factor node g_i is defined as

$$g_i(\tilde{x}_i, \tilde{y}_i, \alpha) = \int_{P(\tilde{x}_i)}^{P(\tilde{x}_i+1)} \frac{\alpha}{2} e^{-\alpha|x-\tilde{y}_i|} dx, \tag{11.4}$$

where α is the scale parameter of the Laplace distribution and $P(\tilde{x}_i)$ denotes the lower partition boundary at index $\tilde{x}_i$, for example whether a coefficient $\tilde{x}_i$ satisfies $P(\tilde{x}_i) \leq \tilde{x}_i < P(\tilde{x}_i + 1)$.

Based on the factor graph with factor function defined above, an estimate of the original slice $\hat{\hat{X}}$ can be decoded by performing the belief propagation (BP) [118] algorithm on the factor graph, which offers an efficient way to calculate the marginal distribution of unknown variables using Bayesian inference.

11.3   Practical Implementation Issues

In this section, we discuss some practical implementation issues of the SUPERMICRO system. The first issue is the side information availability at the decoder side. In this framework, the previous slice (i.e., the $(i-1)$th slice) is considered to be the side information slice of the ith source slice, as shown in Figure 11.1. Therefore, the most efficient way in terms of code rate saving is to send only the first encrypted slice uncompressed to the decoder as side information. Next, the second compressed slice can be jointly decoded with the help from the side information slice at the decoder and then it can serve as the side information for the third slice. We can continue this process for the rest of the compressed slices. The aforementioned setup yields high compression efficiency, as only one uncompressed slice (i.e., the key slice) is sent to the decoder. However, the cost for such a single key slice setup is an increased decoding latency, since to decode an interested slice one should go through all its previous compressed slices. One workaround for this drawback is to increase the number of key slices and evenly distribute them among all compressed slices with a slight loss of compression performance.

Then second practical implementation issue is an accurate estimate of the correlation parameter α for different source and side information slice pairs due to the dynamical changes of pixels within each slice. According to the assumption that the adjacent image slices are highly correlated, the correlation parameter α of the source and side information slice pair can be learnt from the residuals between the previous two decoded neighboring slices, as in [58]. It also implies that at least two uncompressed slices should be available at the decoder if we want to fully take advantage of correlation estimation. Hence, in this chapter we defined the group of slices (GOS) size as the number that represents how frequently two uncompressed slices will be distributed among the compressed slices. The impact of GOS size on compression performance is studied in the results section.

11.4   Experimental Results

In this section, we demonstrate the performance of the SUPERMICRO in privacy-preserving capability and compression efficiency through experimental results.

11.4.1 Experiment Setup

In our experiments, we used two different CT image sets with 100 slices in each set, giving a total of 200 slices. Figures 11.5(a),(b) and (c),(d) illustrate the first and last slices in CT image sets 1 and 2, respectively. As shown in Figures 11.5(a)–(d), great changes have taken place from the first to the last slice in each CT image set, therefore the image sets offer diversely experimental conditions for verifying the robustness of the SUPERMICRO framework. The size of each slice is 484 × 396 for all CT image sets. For the sake of coding efficiency, each slice is partitioned into sub-slices of size 44 × 36, which results in a total of 11 × 11 = 121 sub-slices for each slice, as depicted in Figure 11.6(a). Further, the slice partitioning technique can be used to easily fit a CT image slice with any size in practice. Then, in each sub-slice the number of pixels is equal to $N = 44 \times 36 = 1584$. Hence, the length of possible encode bits qN varies from 3168 to 12672 for $q \in [2, 8]$ in our experiments, where LDPCA codes with corresponding lengths were used for rate adaptive decoding. In the rest of this section, the GOS size is equal to 100 for all experiments, except where explicitly indicated otherwise.

CT image set 1   CT image set 2

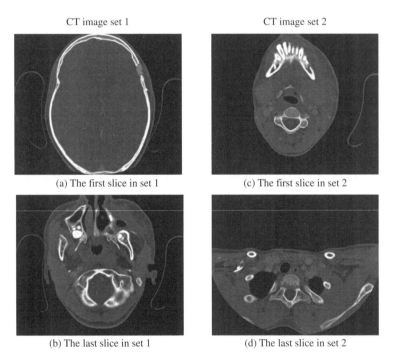

(a) The first slice in set 1   (c) The first slice in set 2

(b) The last slice in set 1   (d) The last slice in set 2

Figure 11.5 The first and the last slices in CT image set 1 ((a) and (b), respectively) and set 2 ((c) and (d), respectively).

11.4.2   Security and Privacy Protection

The secrecy of a privacy-preserving compression scheme is essential, where the strong security [111] is achieved by using the stream cipher technique in our experiment. In this framework, each bit-plane of the original CT image slice is encrypted separately by using its own encryption key with the same size. For example, Figure 11.6(b) shows the encryption keys of all 8 bit-planes in decimal representation. By applying each encryption key on the corresponding bit-plane, the encrypted slice can be found in Figure 11.6(c). Here, the grids in Figures 11.6(a), (b), and (c) partition the whole slice into sub-slices. In Figure 11.6(c) we can see that the correlation of pixels among the original slice has been totally destroyed. Therefore, without knowing the encryption key, the stream cipher-based encryption technique offers strong security and privacy protection. Besides security, compression performance is another important criterion for the privacy-preserving compression scheme, which will be studied in the next subsection.

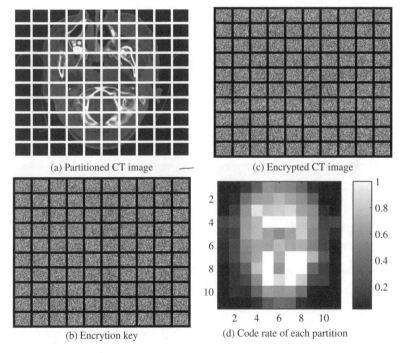

(a) Partitioned CT image   —   (c) Encrypted CT image

(b) Encrytion key   (d) Code rate of each partition

Figure 11.6 Examples of (a) partitioned CT image slice, (b) encryption key, (c) encrypted CT image, and (d) code rates of each partition using the SUPERMICRO framework, where the grids in (a), (b), and (c) partition the whole slices into sub-slices and the average code rate in (d) is $R = 0.38$ for the given slice.

11.4.3 Compression Performance

First, we investigate how compression efficiency varies partition by partition. Figure 11.6(d) depicts the code rate of each partition after applying the SUPERMICRO framework, where a lower code rate refers to a higher compression efficiency and the average code rate of the given slice is $R = 0.38$. In Figure 11.6(d) we can see that a partition (i.e., a sub-slice) with homogeneous background (e.g., the boundary regions with dark background) is usually associated with the lowest code rate, while a partition with irregularly illuminating changes (e.g., the central regions) normally results in a higher code rate. This is because the achievable code rate of a partition largely depends on how well the source and the side information partitions are correlated, where a source partition with a homogeneous background usually has a higher correlation with its side-information partition.

As illustrated in Section 11.2.1, the decoding complexity is directly proportional to the number of encoded bits. Second, we are interested in the impact of the number of encoded bits on the compression performance in terms of code rate, which offers a trade-off between decoding complexity and compression performance. Figure 11.7 shows the average code rates with error bars for the two CT image sets based on all encoded slices, where the code rates for both CT image sets decrease as the number of encoded bits increases. Moreover, the lowest achievable code rates of CT image sets 1 and 2 are 0.41 and 0.32, respectively, when the number of encoded bits is $q = 8$.

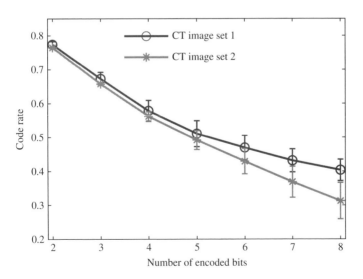

Figure 11.7 Code rate versus the different number of encoded bits for both CT image set 1 (i.e., black dash-circle line) and set 2 (i.e., gray dash-dot line) using the SUPERMICRO system.

Image set 1

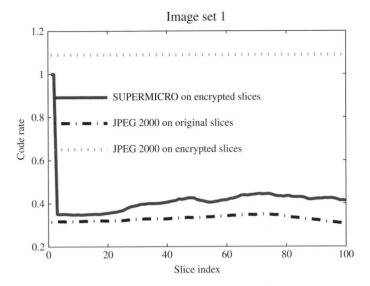

Figure 11.8 Code rate versus different CT image slices for image set 1, which compared three different setups: the SUPERMICRO on encrypted slices, JPEG 2000 lossless compression on both original slices, and encrypted slices.

Next, we study the varieties of each individual slice's code rate with three different setups: the SUPERMICRO on encrypted slices, JPEG 2000 lossless compression on both original slices, and encrypted slices in Figures 11.8 and 11.9, where the first two slices are sent to the decoder uncompressed in our SUPERMICRO framework. Here, the state-of-the-art JPEG 2000 lossless compression on original slices serves as the baseline performance to show how well a slice could be compressed without any encryption. Ideally, the closer the performance of the SUPERMICRO on encrypted slices can approach the baseline performance, the higher the compression performance of this framework gets. Moreover, the setup of JPEG 2000 lossless compression on encrypted slices depicts an intuitive sense on how difficult the encrypted slice could be compressed through traditional compression techniques. In both Figures 11.8 and 11.9 we can see that the performance of the SUPERMICRO framework on encrypted slices is very close to the baseline performance in terms of code rate, where the minimum differences are 0.029 and 0.074 for CT image sets 1 and 2, respectively. In addition, the maximum differences are bounded by 0.11 and 0.095 for sets 1 and 2, respectively. Interestingly, the average code rates of JPEG 2000 lossless compression on encrypted slices for both set 1 and set 2 are 1.0889 and 1.0888, respectively, which means that JPEG 2000 lossless "compression" on encrypted slices will result in data size increase instead of data size reduction after compression. Such observations verify our previous statement that direct compression on encrypted data is usually not feasible. It is even worse than not

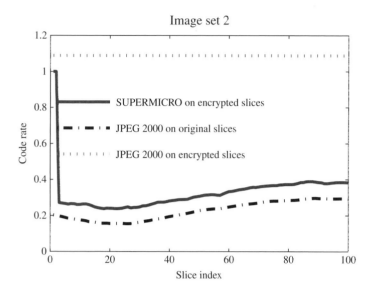

Figure 11.9 Code rate versus different CT image slices for image set 2, which compared three different setups: the SUPERMICRO on encrypted slices, JPEG 2000 lossless compression on both original slices, and encrypted slices.

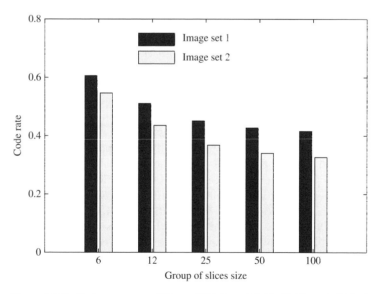

Figure 11.10 Code rate versus different GOS sizes of 6, 12, 25, 50, and 100 for image sets 1 and 2.

compressing the image, as the traditional compression technique (e.g., JPEG 2000) needs to introduce additional overheads for compression.

Finally, we investigate how the average code rate changes with different GOS sizes. In Section 11.3 we showed that a smaller GOS size could bring a shorter decoding latency, as the SUPERMICRO framework requires that the decoding must start from the immediate neighboring slice of the key slice. In other words a smaller GOS size offers more key slices, which are evenly distributed among the encoded slices, therefore an interested encoded slice can be decoded through discovery of the shortest pathway from its nearest neighboring key slice. However, the cost of using a smaller GOS size is that the overall code rate is also increased, as key slices cannot be compressed. As expected, Figure 11.10 shows that the overall code rates of both image sets does indeed increase as the GOS size decreases.

11.5   Discussion

In this chapter we have discussed the compression of an encrypted medical image sequences framework based on the DSC principle, called Secure Privacy-presERving Medical Image CompRessiOn (SUPERMICRO), which makes the compression of the encrypted data possible without compromising security and compression efficiency. SUPERMICRO also guarantees the data transmission in a privacy-preserving manner. Compared to the previous work performing decompression on each bit-plane (i.e., sub-frame) separately with a 2D Markov-based symmetric correlation model, the proposed SUPERMICRO framework employs a joint bit-plane decoder based on a factor graph which can handle the inter-pixel correlation between adjacent CT image slices and uses a more realistic Laplacian correlation model. Moreover, the SUPERMI-CRO framework incorporates LDPCA codes for rate adaptive decoding. The experimental results based on two CT image sets with 200 slices show that the proposed system provides high-level security and privacy protection, as well as significant compression performance, even when compared with that of the state-of-the-art JPEG 2000 lossless compression on unencrypted slices.

12

Secure Biometric Authentication Using DSC

12.1 Background

Biometrics have been widely used as the technology for identification and authentication in systems that provide secure private access to specific users. Compared to conventional access keys (e.g., badges, access cards, passwords, etc.), biometrics (e.g., fingerprints, retina/iris scan, face, etc.) offer an effective and convenient way to authenticate individuals given their universality and distinctiveness. Biometrics are not likely to be lost or left at home. There are moves toward electronic medical information. Combining biometric identifiers with electronic health records (EHR) offers a promising solution for accurate and secure data access and linkage among different EHR databases [119, 120]. Moreover, the federal Drug Enforcement Administration (DEA) is pressuring states to implement biometric authentication for e-prescribing of controlled substances [121]. As biometric systems become increasingly ubiquitous, the security and privacy concerns (closely related with biometric traits) get more attention from the public. These are critical issues to be tackled for public trust and practical deployment of biometric systems.

The most common concerns of biometrics-related security and privacy risks are the following. First, biometric systems may reveal more than just the information needed for identification, and incidentally disclose information about a personal health or medical history. For example, retina/iris scans can leak medical condition of a user's eye (e.g., glaucoma [122]) and ethnic group. There is strong evidence that unusual fingerprint patterns have high probabilities of being associated with genetic diseases, such as Down or Turner syndrome [123, 124]. Therefore, given biometric identifiers, people can infer significant health information from the owner. Second, unlike conventional password or access keys, biometric identifiers are usually irrevocable when they are disclosed [125], since biometric traits are permanent (i.e., nonrenewable resources) for individuals. These shortcomings of biometric systems have motivated numerous

Distributed Source Coding: Theory and Practice, First Edition. Shuang Wang, Yong Fang, and Samuel Cheng.
© 2017 John Wiley & Sons Ltd. Published 2017 by John Wiley & Sons Ltd.
Companion Website: www.wiley.com/go/cheng/dsctheoryandpractice

studies of secure biometric strategies [126–131] to alleviate the potential security and privacy risks of biometric traits.

A typical biometric recognition system comprises enrollment and authentication components. In enrollment, the raw biometrics of an individual are captured and the features depicting the characteristics (or traits) of the biometrics are extracted and stored in a database. In authentication, similar procedures are applied to extract biometric features, which are compared with those stored in a database. However, such a system may create security holes and serious privacy risks. As reported by Ratha *et al.* [132–134], there are several vulnerable points within a fingerprint-based biometric system (see Figure 12.1). If an attacker gains access to a biometric system through one of these eight possible attack points the entire system might be compromised. The situation may become even worse as stolen biometrics could lead to significant security and privacy problems, and they cannot be revoked or easily replaced. Therefore, a number of efforts [126–131] have been made to build robust secure biometric systems in the past decade.

Regarding attack points 1, 2, and 3, as shown in the dashed block of Figure 12.1, a malicious user can gain fingerprint information through observing the fingerprint residue on a contact-based sensor (attack point 1), eavesdropping on the communication channel at attack point 2, or hacking into the feature extraction module at attack point 3. The security can be enforced by moving fingerprint capture and feature extraction modules into integrated secure hardware [126] (e.g., a smartphone) that can only be accessed by genuine users. Note that in this chapter we restrict our discussion to biometric systems with true/false output. Attack point 8 is out of the scope of this chapter since it is a problem only associated with a biometric system with matching score outputs, and the workarounds of such a problem have been studied in [135]. Attack points 4 and 7 correspond to the secure communication of features among the biometric database, the authentication server, and users, which could be enforced by transmitting encrypted features instead of raw features [136]. It turns out that privacy-preserving authentication (attack point

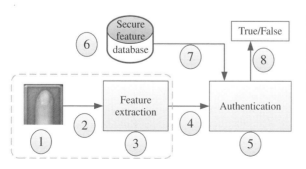

Figure 12.1 Example of several possible attack points of a fingerprint-based biometric system, as identified by Ratha *et al.* [132–134].

5) and maintaining a secure feature database (attack point 6) are critical tasks in building a secure biometric system.

An efficient way to mitigate the privacy and security problems in attack points 5 and 6 is to encrypt the biometric features before storing them in the database. However, encrypting these features alone cannot guarantee either the privacy of the biometric owner or the security of the system [129] since these features have to be decrypted when the comparison step (at attack point 5) is performed. An ideal way to enhance data privacy and security is to process the encrypted data directly without going through the decryption step. While processing encrypted data can enhance the security and privacy of the biometric, it also poses significant challenges. Due to the nonstatic characteristic of biometric identifiers (e.g., the elastic deformation of fingerprints), variability between the features captured during enrollment and authentication is inevitable in practice. Thus, making a direct comparison of encrypted features in a biometric system seems a difficult task. In the past decade there were only a few efforts (e.g., feature transformation [137] and biometric crypto-systems [129, 130]) to tackle this difficulty. In the next section we provide a detailed review of previous work and point out the shortcomings of existing solutions.

12.2   Related Work

Existing techniques for secure biometric systems [119, 138, 137, 126, 139] can be categorized into two classes: feature transformation approaches and biometric crypto-systems. Feature transformation techniques [137] focus on transforming the extracted biometric features through a non-invertible transformation function, which is revocable in case of a system compromise. However, the security of such techniques is usually not provable, which plagues their implementation in practice. The key idea of a biometric crypto-system is to bind a secret key with the original biometrics to generate a low privacy risk alternative (i.e., secure template) of the original biometrics, where the secure template can tolerate the changes (e.g., elastic deformation) in biometrics introduced by different capturing environments. For example, the biometric crypto-systems in [138, 126, 139] used error-correcting codes (ECC). The changes in biometrics are captured by the error-correcting capability of the code and the security of such a system is provable based on information theoretic security. However, earlier studies also showed that traditional ECC-based biometric crypto-systems experience a poor matching performance [138, 126, 139].

DSC [140, 55] is a communication technology that studies separate compression of correlated sources in a distributive manner. It has emerged as a promising technology in solving the difficulty of privacy-preserving biometric authentication. More specifically, asymmetric SW coding [55, 141], a special case of DSC that compresses only one source with another source as side information, provides a model to allow secure biometric storage and can accommodate a slight variability in biometrics [129, 130].

In a nutshell, the captured features during enrollment are compressed into some SW coded bits (also called **syndromes**) and at the same time the features are processed by a hash function. Both compressed bits and hashed values are stored in the database. During authentication, the system will try to decompress the stored SW coded bits using the features captured during authentication as side information. If the subject is an intruder, the captured features at enrollment cannot be decoded successfully. Thus, they will not have the same hashed values as the stored ones. Otherwise, the hashed values of the decoded features will match the stored ones and one can be sure that the subject is an authorized user. Note that in the former case, the compressed features are never decoded to the original features without a genuine fingerprint as side information and therefore a malicious user with illicit biometrics will be unable to steal the protected features during authentication even if he is able to hack into the server. This asymmetric SW-based secure biometric system successfully solved the security hole at attack point 6 and partially tackled the privacy risk at attack point 5, shown in Figure 12.1. A number of asymmetric SW-based biometric systems have been proposed to improve matching performance and individual privacy security [129–131, 119].

However, these asymmetric SW-based secure biometric systems are not yet applicable to practical scenarios in which fingerprint capture devices, and database and authentication servers are deployed at different locations. The feature transmission procedures among devices result in additional vulnerability attack points 4 and 7 (see Figure 12.1). The shortcomings of the asymmetric SW-based biometric system are the following:

- A malicious user can easily gain a certain amount of information with the intercepted side information feature through attack point 4, as the traditional asymmetric SW-based schemes miss protection on side information.
- An even worse situation arises when the encrypted feature is no longer secure at attack points 6 and 7 given that the side information feature is available at the malicious user side. Traditional DSC-based authentication techniques have no controls on who can carry out the authentication step. In other words, a malicious user can build a replica of the authentication server to recover the original features with encrypted features and side information features included.
- To verify any input, the authentication server must first decode an estimated feature based on the input (i.e., the side information). Then, a genuine user would be confirmed if and only if the hashed value of the estimated feature is exactly the same as that of the original feature. Such a procedure significantly increases the privacy risk at attack point 5, as the estimated feature of

a genuine user is identical to the original feature and could be eavesdropped by a malicious user.

In this work we introduce a novel and practical framework, the PRivacy-Oriented Medical Information Secure biomEtric system (PROMISE), which aims to resolve the aforementioned security holes within the traditional asymmetric SW-based biometric system. The main advantages of the PROMISE framework are the following:

- Compared to the previous work on asymmetric SW code models, the model based on non-asymmetric SW code[1] technology [142, 29] is more general, namely, it can provide protection for biometric template storage and ensure high security for the transmission and verification of biometric features.
- A traditional DSC-based authentication server lacks a secure mechanism, so a malicious user can bypass the authentication step by using a replica. To tackle this difficulty, we propose a secure distributed source coding (SeDSC) scheme that incorporates a pre-encryption layer (e.g., stream cipher) before implementing traditional DSC. Therefore, without knowing a specific decoding key, the authentication step is not reproducible. Although the SeDSC scheme is applicable to both asymmetric and non-asymmetric SW setups, SeDSC always refers to the non-asymmetric SW setup in the rest of this chapter except where specifically indicated as an asymmetric SW setup.
- During authentication, SeDSC only recovers the pre-encrypted features instead of the raw features with the specific decoding key. Hence, the comparisons among hashed values can be directly carried out on the encrypted features rather than the raw features. We show that decoding a key-based SeDSC can achieve the same coding efficiency as the traditional DSC without sacrificing individual security and privacy.

The rest of this chapter is organized as follows. In Section 12.3 we introduce the PROMISE framework. In Sections 12.4 and 12.5 we present the SeDSC scheme in detail, focusing on encoder/encrypter and decoder/decrypter,[2] respectively. The experimental results and concluding remarks can be found in Sections 12.6 and 12.7.

12.3 System Architecture

The basic workflow of the framework is shown in Figure 12.2. The central process of the PROMISE framework consists of four steps: (1) feature extraction

---

1 In contrast to asymmetric SW codes, where only the enrollment feature will be encoded, non-asymmetric codes imply that both the enrollment and verification features can be encoded through syndrome coding with different code rates.
2 The terms encoder and encryptor are equivalent in this chapter, as are the terms decoder and decrypter.

(see Section 12.3.1): the original minutiae points are transformed into a binary feature vector through a thresholding mechanism [131]; (2) feature pre-encryption (see Section 12.3.2): binary feature vectors extracted from both enrollment and verification sides are pre-encrypted with a stream cipher [143] and the hashed value of each pre-encrypted feature is generated, stored in the hashed feature database or sent to the authentication server (see Figure 12.2); (3) the SeDSC encrypter/decrypter (see Section 12.3.3): the pre-encrypted feature vectors from both the enrollment and verification steps are passed through the SeDSC encrypter, where SeDSC encrypted features will be treated as secure biometric features, stored in a database or transmitted to the authentication server (given the cipher keys and secure biometric features from enrollment and verification sides as inputs, the SeDSC decrypter decodes the corresponding estimates of the original secure biometric features); (4) privacy-preserving authentication (see Section 12.3.4): given the estimates of the original pre-encrypted features from the SeDSC decrypter, the authentication server can generate corresponding hashed values, to be compared with the original hashed values for authentication purposes. The following subsections present the details of each step in the PROMISE framework. For the sake of clarity, the symbols used in this chapter are listed in Table 12.1.

12.3.1 Feature Extraction

The extracted biometric features used for authentication should facilitate the matching procedure while protecting individuals' privacy. Minutiae points, which denote the startings, endings or bifurcations of the fingerprint ridge map with spatial locations (a, b) and orientation θ in three dimensions, are the most common extractions for capturing fingerprint characteristics and comparing for authentication purposes in a traditional biometric system. Minutiae points-based features cannot easily fit into the ECC/DSC-based biometric crypto-systems as these features cannot satisfy the i.i.d. and ergodic

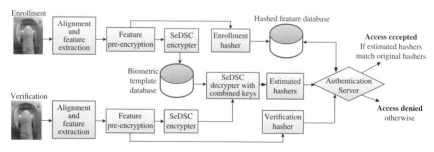

Figure 12.2 The workflow of the PROMISE framework. Four subsequent steps in PROMISE are: (1) alignment and feature extraction, (2) feature pre-encryption, (3) SeDSC encrypter/decrypter, and (4) privacy-preserving authentication.

Table 12.1 List of symbols.

H, H_1 and H_2	Parity check matrices
P, P_1 and P_2	Partitioned matrices of H
I_M	Identity matrix of size $M \times M$
$0_{N_1 \times N_2}$	All zeros matrix of size $N_1 \times N_2$
$\mathbf{x}$ and $\mathbf{y}$	Binary feature vectors in enrollment and verification
$\mathbf{u}$ and $\mathbf{v}$	Encrypted feature vectors of $\mathbf{x}$ and $\mathbf{y}$, respectively
$\mathbf{u}^1$ and $\mathbf{u}^2$	Superscripts of a vector, indicating its partitions
$\mathbf{t}_1, \mathbf{t}_2$ and $\mathbf{t}'$	Reconstructed code-bit vectors at the decrypter
$\hat{\mathbf{x}}, \hat{\mathbf{y}}$ and $\hat{\mathbf{t}}'$	Decrypted feature vectors of $\mathbf{x}$, $\mathbf{y}$ and $\mathbf{t}'$, respectively
$\mathbf{h}_x$ and $\mathbf{h}_y$	Hashed values of $\mathbf{x}$ and $\mathbf{y}$, respectively
R_x and R_y	Code rates for encrypting $\mathbf{x}$ and $\mathbf{y}$, respectively
N	Length of binary feature vectors
N_1 and N_2	Lengths of uncoded bits for $\mathbf{x}$ and $\mathbf{y}$, respectively
$\mathbf{k}_x$ and $\mathbf{k}_y$	Stream cipher keys for $\mathbf{x}$ and $\mathbf{y}$, respectively
$\mathbf{k}_c$	The combined key of $\mathbf{k}_x$ and $\mathbf{k}_y$

statistics. There are some studies [129–131] that tackle this difficulty of binary feature extraction for ECC/DSC-based biometric crypto-systems. These binary features are extracted using the methods in [129–131] and offer lower spatial resolutions compared to minutiae-based features, therefore providing better privacy protection. In this chapter we adopt the feature extraction method proposed by Nagar *et al.* [131] as it shows excellent performance in terms of distinctiveness compared with other existing methods. For completeness, we will briefly summarize Nagar's feature extraction method in this subsection.

1) **Feature aggregation:** For minutiae-based features with spatial locations (a, b, θ), n cuboids, represented by the boundaries [a_i^{min}, a_i^{max}, b_i^{min}, b_i^{max}, θ_i^{min}, θ_i^{max}] are first randomly selected, where [$0 < a_i^{min}, a_i^{max} < h$; $0 < b_i^{min}, b_i^{max} < w$; $0 < \theta_i^{min}, \theta_i^{max} < 360$] and $i = 1, \cdots, n$. Here, h and w are the height and width of the input fingerprint image, respectively. Then, for the ith cuboid C_i of the jth fingerprint, the average minutiae ($\bar{a}_{ij}, \bar{b}_{ij}, \bar{\theta}_{ij}$) and the corresponding standard deviation ($\sigma_{a_{ij}}, \sigma_{b_{ij}}, \sigma_{\theta_{ij}}$) are calculated over all the minutiae points of the jth fingerprint within the cuboid C_i [131]. Therefore, each fingerprint can be represented by $6n$ aggregated minutiae features.

2) **Binary feature extraction:** Suppose we have m training fingerprints. For each cuboid C_i, the median of these m fingerprints is used as the threshold to binarize the aggregated minutiae features. For example, the binarized feature $\bar{a}_{ij}^{bin} = 1$, if $\bar{a}_{ij} > \bar{a}_i^{threshold}$, otherwise, $\bar{a}_{ij}^{bin} = 0$, where the threshold

$\overline{a}_i^{\text{threshold}}$ is calculated by the median of $\overline{a}_{i1}, \cdots, \overline{a}_{im}$. Hence, each fingerprint can be represented by $6n$ binary features.

3) **Feature selection:** To achieve the best matching performance, the following feature selection steps are involved: (1) eliminate redundant cuboids/features, where highly overlapped cuboids or highly correlated features will be removed [130] and (2) select highly discriminative features, where the discriminability is quantified by the intra-user similarity and inter-user unlikeness, see [131] for more details. After this step, each fingerprint can be represented by the N most discriminative features selected from its $6n$ features.

12.3.2 Feature Pre-encryption

Compared to minutiae-based features, the binary features extracted in Section 12.3.1 have reduced privacy risk through aggregation. However, to minimize the privacy risk existing in the feature storage, transmission and comparison/authentication, the extracted binary features should be protected by encryption.

In the system the binary features are protected through a dual encryption protocol, that is, stream cipher-based pre-encryption followed by SeDSC-based post-encryption. The advantages of the dual encryption protocol will be discussed in the next section. The implementation of stream cipher-based pre-encryption is straightforward, where a ciphertext is obtained by merging a pseudorandom keystream with the plaintext bits (i.e., binary feature) through a bitwise exclusive-or (XOR) operation. The pseudorandom keystream is generated from a short key through a keystream generator.

12.3.3 SeDSC Encrypter/decrypter

The key benefit of applying a DSC-based approach as encryption technology for biometric features is that the output of a DSC encoder appears to be completely random and thus it cannot be used to deduce the original input. However, a simple scheme offered by the asymmetric SW-based DSC [129, 130] would fail in many ways. For example, the unprotected side information feature may leak a certain amount of a user's private information. Besides that, given the encrypted biometric template and side information, it is easy to build a replica of the authentication server to recover the original biometric features. To authorize a genuine user, the original biometric features must be recovered and hashed for verification purposes.

To ensure secure biometric features so that no raw features will be stored, transmitted or observed at the decoder side, we propose a novel SeDSC scheme which will take the pre-encrypted features as the inputs of non-asymmetric SW codes. Here, non-asymmetric SW codes, as a special case of DSC, offer the possibility of encoding/encrypting both enrollment and verification features, which satisfies our requirement for protecting both enrollment and verification features. In this chapter the non-asymmetric SW codes are developed using a code-partitioning approach, which will be introduced in Section 12.4.

As the SeDSC scheme takes the pre-encrypted features as the inputs of non-asymmetric SW encoder, the encoder outputs (i.e., syndromes) are highly secure. In other words, both the SeDSC encrypter and decrypter operate directly on the encrypted features instead of the raw features. The SeDSC decrypter requires the combined cipher key of both enrollment and verification features to explore the correlation between raw features. Hence, a malicious user could not easily replicate the decrypter procedure without knowing the specific combined cipher key.[3] Given the combined cipher key, there is no way to recover the original biometric features. Therefore, SeDSC guarantees the privacy and security of the authentication server. The design of the SeDSC decrypter is based on the factor graph model, which is described in Section 12.4.

12.3.4 Privacy-preserving Authentication

The last step of DSC-based authentication is the comparison of the hashed values of the input features of the DSC encoder and those of the output features of the DSC decoder. Note that DSC theory guarantees that correct identification (i.e., the estimates of the DSC decoder are identical to the original inputs of the DSC encoder) will happen if and only if the features from enrollment and verification are highly correlated (i.e., enrollment and verification features are from the same user). Hence, the hashed values, as security keys, offer a way of efficiently verifying the identity between the estimates and the original features. SeDSC advanced the security of this step by generating hashed values of pre-encrypted features instead of the raw features, as it can work on the encrypted features directly.

12.4   SeDSC Encrypter Design

Let $\tilde{\mathbf{x}}$ and $\tilde{\mathbf{y}}$ denote a binary feature pair extracted from the enrollment and verification steps, respectively, as described in Section 12.3.1. Then both $\tilde{\mathbf{x}}$ and $\tilde{\mathbf{y}}$ will be pre-encrypted into $\mathbf{x} = \tilde{\mathbf{x}} + \mathbf{k}_x$ and $\mathbf{y} = \tilde{\mathbf{y}} + \mathbf{k}_y$ through stream cipher keys $\mathbf{k}_x$ and $\mathbf{k}_y$, respectively (refer to Section 12.3.2). The SeDSC encrypter based on the non-asymmetric SW setup uses pre-encrypted features as input, where pre-encrypted features from both enrollment and verification will be further encoded into syndromes through $\mathbf{S}_x = H_1\mathbf{x}$ and $\mathbf{S}_y = H_2\mathbf{y}$, respectively. There are several efforts to implement non-asymmetric SW codes, including

---

3 The security of cipher keys corresponds to secure key management approaches [143], which are out of the scope of this chapter.

time-sharing, source splitting [24], and code partitioning [144, 27]. The code-partitioning approach implemented using irregular repeated accumulate (IRA) codes [145] has been reported by Stanković *et al.* to have the best performance for non-asymmetric SW coding. The following subsections describe the code-partitioning approach in the context of the SeDSC scheme.

12.4.1 Non-asymmetric SW Codes with Code Partitioning

The idea of the code-partitioning approach is to convert a non-asymmetric SW coding problem into a channel coding problem, where a main code defined by the parity check matrix H is partitioned into two channel subcodes defined by the parity check matrices H_1 and H_2. Before walking through the implementation details, let us prove the following Lemma of code partitioning.

Lemma 12.1 (Code partitioning). The codes defined by H_1 and H_2 partition the code defined by H.

Proof. The parity check matrix of a systematic linear block code has the form $H = [P|I_M] = [P_1 P_2|I_M]$, where I_M is an identity matrix of size $M \times M$, and the matrix P is split horizontally into two matrices P_1 and P_2 with widths N_1 and N_2, respectively. Therefore, H is of size $M \times N$, where $N = N_1 + N_2 + M$.

Now we define the parity check matrices of the first and second encoders,

$$H_1 = \begin{bmatrix} I_{N_1} & 0_{N_1 \times N_2} & 0_{N_1 \times M} \\ 0_{M \times N_1} & P_2 & I_M \end{bmatrix}, H_2 = \begin{bmatrix} 0_{N_2 \times N_1} & I_{N_2} & 0_{N_2 \times M} \\ P_1 & 0_{M \times N_2} & I_M \end{bmatrix}, \quad (12.1)$$

where $0_{k \times l}$ is all zero matrix of size $k \times l$. Then, given H_1 and H_2, input features with N binary bits of the two encoders can be encoded into $N_1 + M$ and $N_2 + M$ syndromes, respectively.

Let

$$G_1 = [0_{N_2 \times N_1}|I_{N_2}|P_2^T], G_2 = [I_{N_1}|0_{N_1 \times N_2}|P_1^T] \quad (12.2)$$

be the possible generator matrices of H_1 and H_2, such that $H_1 G_1^T = 0_{(N_1+M) \times N_2}$ and $H_2 G_2^T = 0_{(N_2+M) \times N_1}$[4]. Then, it is straightforward to show that

$$H \begin{bmatrix} G_2 \\ G_1 \end{bmatrix}^T = 0_{M \times (N_1+N_2)}, \quad (12.3)$$

and so $G \triangleq \begin{bmatrix} G_2 \\ G_1 \end{bmatrix}$ is a generator matrix of H. Moreover, since the row spaces of G_1 and G_2 are linearly independent, the codes defined by H_1 and H_2 partition the code defined by H. □

---

4 In this chapter all additions involved in the matrix or vector operations are bitwise XOR.

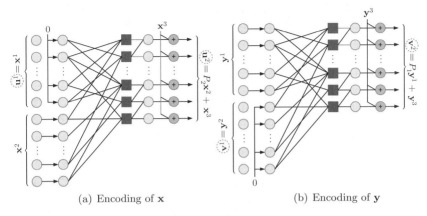

(a) Encoding of **x**   (b) Encoding of **y**

Figure 12.3 Non-asymmetric SW for binary features **x** and **y** using an IRA code, where the encrypted (compressed) bits **u** and **v** are circled.

Given the parity check matrices H_1 and H_2 as defined in (12.1) at the enrollment and verification encoders, the pre-encrypted feature vectors **x** and **y** with N bits each extracted from enrollment and verification will be encrypted into syndromes (i.e., post-encrypted features) $\mathbf{u} = H_1\mathbf{x}$ and $\mathbf{v} = H_2\mathbf{y}$, respectively.

12.4.2 Implementation of SeDSC Encrypter using IRA Codes

In practice, a straightforward implementation of the aforementioned code-partitioning scheme using channel codes (e.g., LDPC codes) may lead to poor performance. This is due to the fact that the identity part in the systematic parity check matrix H corresponds to the variable nodes of degree one. Such nodes are difficult to estimate correctly using the BP algorithm due to the lack of diversified information. Systematic IRA codes turn out to be better alternatives as their performances are almost the same as that of LDPC codes in a nonsystematic setup.

Figure 12.3 illustrates the encoding of pre-encrypted features **x** and **y** using IRA codes with parity check matrix $H = [P|I_M]$ and $P = [T_M H'_{M \times (N_1 + N_2)}]$, where H' is a nonsystematic parity check matrix, and T is a matrix with all 0s in the upper triangle region and all 1s in the remaining region, for example $T_3 = \begin{pmatrix} 1 & 0 & 0 \\ 1 & 1 & 0 \\ 1 & 1 & 1 \end{pmatrix}$. For ease of explanation, let us split **x** into $\mathbf{x} = \begin{bmatrix} \mathbf{x}^1 \\ \mathbf{x}^2 \\ \mathbf{x}^3 \end{bmatrix}$, where the lengths of each partition are N_1, N_2 and M, respectively, and split **u** into $\mathbf{u} = [\begin{smallmatrix} \mathbf{u}^1 \\ \mathbf{u}^2 \end{smallmatrix}]$ with partition lengths N_1 and M, respectively. Therefore, we have $\mathbf{u}^1 = \mathbf{x}^1$ and $\mathbf{u}^2 = P_2\mathbf{x}^2 + \mathbf{x}^3$. Similarly, **y** is split into $\mathbf{y}^1$, $\mathbf{y}^2$ and $\mathbf{y}^3$, and **v** is split into $\mathbf{v}^1$ and $\mathbf{v}^2$, which gives us $\mathbf{v}^1 = \mathbf{y}^2$ and $\mathbf{v}^2 = P_1\mathbf{y}^1 + \mathbf{y}^3$.

12.5 SeDSC Decrypter Design

At the decoder, the received bits of $\mathbf{u}$ and $\mathbf{v}$ will be rearranged and padded with zeros into

$$
\mathbf{t}_1 = \begin{bmatrix} \mathbf{u}^1 \\ \mathbf{0}_{N_2 \times 1} \\ \mathbf{u}^2 \end{bmatrix}, \quad \mathbf{t}_2 = \begin{bmatrix} \mathbf{0}_{N_1 \times 1} \\ \mathbf{v}^1 \\ \mathbf{v}^2 \end{bmatrix} \tag{12.4}
$$

The following lemma shows that a non-asymmetric SW coding problem is equivalent to a channel coding problem.

Lemma 12.2 *(Virtual noise channel).* The code $\mathbf{t}' = \mathbf{t}_1 + \mathbf{t}_2 + \mathbf{x} + \mathbf{y} = \begin{bmatrix} \mathbf{y}^1 \\ \mathbf{x}^2 \\ P_2\mathbf{x}^2 + P_1\mathbf{y}^1 \end{bmatrix}$ is a valid codeword defined by the parity check matrix H, such that $\mathbf{t} = \mathbf{t}_1 + \mathbf{t}_2$ and $\mathbf{z} = \mathbf{x} + \mathbf{y}$ can be considered as the noise channel output and the channel noise, respectively.

Proof. Note that the encoding steps ensure $H[\mathbf{t}_1 + \mathbf{x}] = H \begin{bmatrix} \mathbf{0}_{N_1 \times 1} \\ \mathbf{x}^2 \\ P_2\mathbf{x}^2 \end{bmatrix} = \mathbf{0}_{M\times 1}$ and

$H[\mathbf{t}_2 + \mathbf{y}] = H[\mathbf{0}_{N_2 \times 1}\ \begin{smallmatrix}\mathbf{y}^1\\ \\P_1\mathbf{y}^1\end{smallmatrix}] = \mathbf{0}_{M\times 1}$. Thus, $H([\mathbf{t}_1 + \mathbf{x}] + [\mathbf{t}_2 + \mathbf{y}]) = \mathbf{0}_{M\times 1}$ and $\mathbf{t}' = \mathbf{t}_1 +$

$\mathbf{t}_2 + \mathbf{x} + \mathbf{y} = \begin{bmatrix} \mathbf{y}^1 \\ \mathbf{x}^2 \\ P_2\mathbf{x}^2 + P_1\mathbf{y}^1 \end{bmatrix}$ is a codeword defined by parity check matrix H. Let us denote by $\mathbf{t} = \mathbf{t}' + \mathbf{z}$ the noise channel output with channel noise $\mathbf{z} = \mathbf{x} + \mathbf{y}$, then it is easy to show that $\mathbf{t} = \mathbf{t}_1 + \mathbf{t}_2$.

According to Lemma 12.2, given noise channel output $\mathbf{t}$ at the decoder, the original codeword $\mathbf{t}'$ could be recovered by taking $\mathbf{t}$ as a corrupted codeword passing through a virtual channel with noise $\mathbf{z}$. For ease of explanation, let us split the noise channel output $\mathbf{t}$ and virtual noise $\mathbf{z}$ into $\mathbf{t} = \begin{bmatrix} \mathbf{t}^1 \\ \mathbf{t}^2 \\ \mathbf{t}^3 \end{bmatrix}$ and $\mathbf{z} = \begin{bmatrix} \mathbf{z}^1 \\ \mathbf{z}^2 \\ \mathbf{z}^3 \end{bmatrix}$ with partition lengths N_1, N_2 and M, respectively. Then, $\mathbf{t}^1 = \mathbf{u}^1 = \mathbf{x}^1 = (\mathbf{x}^1 + \mathbf{y}^1) + \mathbf{y}^1 = \mathbf{z}^1 + \mathbf{y}^1$ can be considered as the noise version of $\mathbf{y}^1$. Similarly, $\mathbf{t}^2 = \mathbf{z}^2 + \mathbf{x}^2$ and $\mathbf{t}^3 = \mathbf{z}^3 + P_2\mathbf{x}^2 + P_1\mathbf{y}^1$ can be considered as the noise versions of $\mathbf{x}^2$ and $P_2\mathbf{x}^2 + P_1\mathbf{y}^1$, respectively.

Supposing that $\hat{\mathbf{t}}'$ is the estimate of $\mathbf{t}'$ given $\mathbf{t}$ at the decoder, we can directly read out $\hat{\mathbf{y}}^1$ and $\hat{\mathbf{x}}^2$ as the estimates of $\mathbf{y}^1$ and $\mathbf{x}^2$ from the first $N_1 + N_2$ bits of $\hat{\mathbf{t}}'$. Then, the estimates $\hat{\mathbf{x}}^3$ and $\hat{\mathbf{y}}^3$ can be obtained according to the relationship

$\mathbf{u}_2 = P_2\mathbf{x}^2 + \mathbf{x}^3$ and $\mathbf{v}_2 = P_1\mathbf{y}^1 + \mathbf{y}^3$ by knowing the estimates $\hat{\mathbf{x}}^2$ and $\hat{\mathbf{y}}^1$. Finally, $\mathbf{x}^1$ and $\mathbf{y}^2$ can be read out directly from the first $N_1 + N_2$ bits of the received noise channel output $\mathbf{t}$. By combining all above information, the estimates $\hat{\mathbf{x}}$ and $\hat{\mathbf{y}}$ can be expressed as $\hat{\mathbf{x}} = \begin{bmatrix} \mathbf{x}^1 \\ \hat{\mathbf{x}}^2 \\ \hat{\mathbf{x}}^3 \end{bmatrix}$ and $\hat{\mathbf{y}} = \begin{bmatrix} \hat{\mathbf{y}}^1 \\ \mathbf{y}^2 \\ \hat{\mathbf{y}}^3 \end{bmatrix}$, respectively.

Given the estimates $\hat{\mathbf{x}}$ and $\hat{\mathbf{y}}$, the authentication server will accept a genuine user if and only if

$$\begin{cases} \mathbf{u} == H_1\hat{\mathbf{x}} \quad \& \quad \mathbf{v} == H_1\hat{\mathbf{y}} \\ f_{hash}(\hat{\mathbf{x}}) == h_{\mathbf{x}} \quad \& \quad f_{hash}(\hat{\mathbf{y}}) == h_{\mathbf{y}} \end{cases} \tag{12.5}$$

where the double equals sign "==" denotes a logical operator to test equality of values, $h_{\mathbf{x}}$ and $h_{\mathbf{y}}$ are the hashed values of original pre-encrypted features $\mathbf{x}$ and $\mathbf{y}$ during enrollment and verification, respectively, and $f_{hash}(\cdot)$ is the hash function to generate $h_{\mathbf{x}}$ and $h_{\mathbf{y}}$. It is worth mentioning that all aforementioned procedures are carried out directly on the pre-encrypted features instead of the raw feature (i.e., one of the contributions of the framework), where the security and privacy of an individual during the authentication are protected by the stream cipher-based pre-encryption. In the rest of this section, we will explain how SeDSC can directly handle the pre-encrypted features without exposing any information on the original features.

Let us denote by $\mathcal{X}$ and $\mathcal{Y}$ two i.i.d. random binary feature vectors with uniform distribution,[5] then the SW [55] theorem shows that the theoretical bounds for the lossless coding are as follows,

$$R_{\mathcal{X}} = \frac{N_1 + M}{N} \geq \mathcal{H}(\mathcal{X}|\mathcal{Y}) = 1 - I(\mathcal{X}; \mathcal{Y})$$

$$R_{\mathcal{Y}} = \frac{N_2 + M}{N} \geq \mathcal{H}(\mathcal{Y}|\mathcal{X}) = 1 - I(\mathcal{X}; \mathcal{Y})$$

$$R_{\mathcal{X}} + R_{\mathcal{Y}} \geq \mathcal{H}(\mathcal{X}, \mathcal{Y}) = 2 - I(\mathcal{X}; \mathcal{Y}), \tag{12.6}$$

where $\mathcal{H}(\mathcal{X}|\mathcal{Y})$, $\mathcal{H}(\mathcal{X}, \mathcal{Y})$, and $I(\mathcal{X}; \mathcal{Y})$ are the conditional entropy, the joint entropy, and the mutual information, respectively. For example, Figure 12.4 shows the box plots of the mutual information of 100 sampled sequence pairs ($\tilde{\mathbf{x}}$ and $\tilde{\mathbf{y}}$) with length $N = 1000$ before and after performing pre-encryption, where the correlation between $\tilde{\mathbf{x}}$ and $\tilde{\mathbf{y}}$ is modeled by a binary symmetric channel (BSC) model with crossover probability $p = 0.05$, which is the probability that a bit flips to its opposite (i.e., 1 to 0 or 1 to 0). Thus, the mutual information can be evaluated as $I(\tilde{\mathbf{x}}; \tilde{\mathbf{y}}) = 1 - \mathcal{H}(p_{\tilde{\mathbf{x}}, \tilde{\mathbf{y}}})$, where $p_{\tilde{\mathbf{x}}, \tilde{\mathbf{y}}} = E[\tilde{\mathbf{x}} + \tilde{\mathbf{y}}]$ and $E[\cdot]$ is the expectation operator.

In Figure 12.4 we can see that $\tilde{\mathbf{x}}$ and $\tilde{\mathbf{y}}$ are highly correlated (i.e., a large $I(\tilde{\mathbf{x}}; \tilde{\mathbf{y}})$). Therefore, the code rate $R_{\tilde{\mathbf{x}}}$ and $R_{\tilde{\mathbf{y}}}$ can be rather small according to

---

5 The statistical analysis of feature vector distribution can be found in [130].

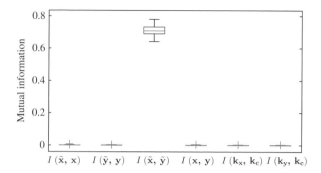

Figure 12.4 The mutual information of original feature $\tilde{x}$ and pre-encrypted feature x (i.e., $I(\tilde{x}; x)$), original feature $\tilde{y}$ and pre-encrypted feature y (i.e., $I(\tilde{y}; y)$), original feature pair (i.e., $I(\tilde{x}; \tilde{y})$), pre-encrypted feature pair (i.e., $I(x; y)$), cipher key of feature x and combined cipher key (i.e., $I(k_x; k_c)$), and cipher key of feature y and combined cipher key (i.e., $I(k_y; k_c)$) in a box plot based on 100 samples.

(12.6), where a smaller code rate refers to a higher information theoretical security [129–131]. However, after pre-encryption, the correlation between the pre-encrypted feature pair x and y has been totally obliterated (i.e., $I(x; y) \approx 0$). Moreover, the correlations between the pre-encrypted features and the original features also vanish after pre-encryption (i.e., $I(\tilde{x}; x) \approx 0$ and $I(\tilde{y}; y) \approx 0$). Although the above-mentioned decorrelation results in a strong protection of original features offered by the pre-encryption procedure, this is bad for the traditional DSC-based coding due to $R_x \approx 1$ and $R_y \approx 1$ (i.e., no encryption/compression). This disobeys our original aim of dual encryption, as the traditional DSC fails to handle decorrelated features. More importantly, pre-encryption alone cannot guarantee perfect protection, that is, performing authentication on encrypted features directly is not sufficient. The following lemma shows that the SeDSC decoder can efficiently tackle this difficulty without sacrificing any system security or privacy.

Lemma 12.3 *(SeDSC decoder).* Given combined cipher key $k_c = k_x + k_y$, the SeDSC decoder can achieve the same decoding performance as that of a traditional DSC decoder without sacrificing any security or privacy.

Proof. As shown in Lemma 12.2, the correlation between the virtual noise channel output t and input t' is modeled by the BSC model with noise $z = x + y$. The factor function that captures the correlation in traditional DSC can be expressed as

$$f(t_i, \hat{t}'_i) = p_z^{t_i \oplus \hat{t}'_i}(1 - p_z)^{1 \oplus t_i \oplus \hat{t}'_i} \tag{12.7}$$

where $\oplus$ denotes binary summation, t_i and $\hat{t}'_i$ are the ith bit of the channel output t and the estimate $\hat{t}'$ through the BP algorithm, respectively, with

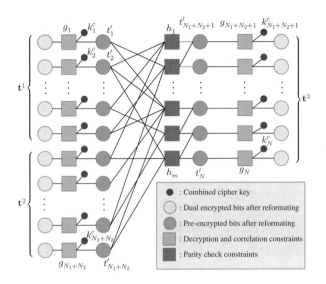

Figure 12.5 Factor graph for non-asymmetric SW decoding.

$i = 1, \cdots, N$, and $p_z = E[z] = E[x + y]$ is the crossover probability in the BSC model. Since x and y are independent after pre-encryption (i.e., $I(x; y) \approx 0$ as shown in Figure 12.4), it is easy to show that $p_z \approx 0.5$. Hence, in this case, the BP algorithm will fail to recover the input t', as the factor function $f(t_i, \hat{t}'_i)$ cannot provide any helpful information.

To alleviate this problem, in the SeDSC decoder we introduce the combined key $k_c = k_x + k_y$ to our factor graph design, as shown in Figure 12.5. Then the corresponding factor function is given by

$$g_i(t_i, \hat{t}'_i, k_i^c) = p_{\tilde{z}}^{k_i^c \oplus t_i \oplus \hat{t}'_i}(1 - p_{\tilde{z}})^{1 \oplus k_i^c \oplus t_i \oplus \hat{t}'_i} \tag{12.8}$$

where k_i^c is the ith bit of combined key k_c, and $p_{\tilde{z}}$ is the combined-key-informed correlation. Given the combined key k_c at the SeDSC decoder, it is easy to show that $\tilde{z} = t + t' + k_c = \tilde{x} + \tilde{y}$ and $p_{\tilde{z}} = E[\tilde{x} + \tilde{y}]$, which is the same as the correlation between $\tilde{x}$ and $\tilde{y}$ before pre-encryption. Therefore, given the combined key k_c, the SeDSC with ciphertext as inputs can achieve the same decoding performances as that of the traditional DSC decoder with plaintext as inputs.

For security and privacy protection, to recover the original features $\tilde{x}$ and $\tilde{y}$ with the estimate $\hat{t}'$, the cipher keys k_x and k_y are required [143]. However, for a malicious user, it is impossible to learn the required cipher keys from the combined key k_c due to "pair-wise" independence among k_c, k_x and k_y. Note that $I(k_c; k_x) \overset{(a)}{=} H(k_c) - H(k_c|k_x) \overset{(b)}{=} 1 - H(k_c|k_x) \overset{(c)}{=} 1 - H(k_y|k_x) \overset{(d)}{=} 1 -$

$H(\mathbf{k}_y) \overset{(e)}{=} 1 - 1 = 0$, where (a) is from the definition of mutual information, (b) is due to the fact that $\mathbf{k}_c$ is uniformly distributed, (c) is because $\mathbf{k}_c = \mathbf{k}_x + \mathbf{k}_y$ and $\mathbf{k}_x$ has no effect on the value of the conditional entropy when $\mathbf{k}_x$ is given, (d) is due to the fact that $\mathbf{k}_x$ and $\mathbf{k}_y$ are independent, and (e) is because k_y is uniformly distributed. Similarly, we have $I(\mathbf{k}_c; \mathbf{k}_y) = 0$.

So far, we have restricted our discussion of SeDSC to the non-asymmetric setup. It is worth mentioning that the same trick (i.e., combined-key-informed decoder) is also applicable to the asymmetric SW setup. However, the asymmetric SW-based SeDSC lacks dual encryption protection on the side information feature (i.e., verification feature). The comparison of authentication accuracy between the asymmetric SW and the non-asymmetric SW-based SeDSC setups will be shown in the next section.

12.6 Experiments

12.6.1 Dataset and Experimental Setup

In our experiments, all fingerprints were taken from the fingerprint verification competition (FVC) 2002 database-2a [146], which contains 100 different fingers with eight fingerprint samples captured from each finger. Similar to [129, 130], all these fingerprints are pre-aligned in our experiments. As suggested in [131], the first and second fingerprint samples of each finger in the database are used for enrollment and verification, respectively. For each finger, the remaining six fingerprint samples are treated as the genuine queries in the training set, and the second fingerprint samples of another 20 randomly selected different users are treated as the intruder queries in the training set. Based on the training data, the N most discriminable binary feature vectors representing each fingerprint can be extracted using the procedures mentioned in Section 12.3.1.

In our experiments, we first selected 750 3D cuboids after reducing the redundancy through removal of overlapped cuboids, where each cuboid contains six minutia features including three average minutia and three deviation minutia. Then, the $750 \times 6 = 4500$ minutiae-based features for each fingerprint were binarized into corresponding binary features through the median value threshold obtained from the training dataset. Finally, we selected a different number of the most discriminable bits, such as $N = 100, 300, 500, 700, 900,$ and 1100 from the total 4500 binary features to verify the performance of the PROMISE framework. In the testing phase, the database contains one genuine impression and 99 fake impressions for each input query (i.e., verification fingerprint sample). There are 100 total users, with 100 genuine matches and 9900 (i.e., 99×100) false matches in the test data.

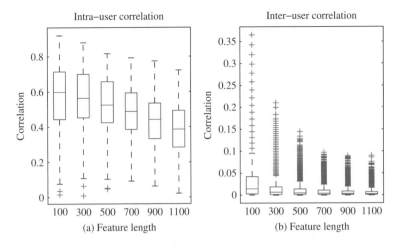

Figure 12.6 Correlations in terms of crossover probability between enrollment (**x**) and verification (**y**) features versus different feature length: (a) intra-user correlation and (b) inter-user correlation.

12.6.2 Feature Length Selection

Figure 12.6(a) and (b) show the box plots of intra-user and inter-user correlations in terms of mutual information $I(\mathbf{x}, \mathbf{y})$ between enrollment (**x**) and verification (**y**) features with varying feature length. In Figure 12.6(a) we can see that the intra-user correlation is much higher than the inter-user correlation, as shown in Figure 12.6(b). This result demonstrates that the extracted binary features are highly discriminable. Moreover, Figure 12.6(a) and (b) also shows that as the feature length increases, both the average intra-user correlations and the variance of the inter-user correlations decrease. In other words, a shorter feature length results in a higher intra-user likeness but in a lower inter-user distinctiveness and vice versa. This means that there is a clear trade-off between intra-user and inter-user correlation/discriminability in feature length selection. Since a lower correlation refers to a larger crossover probability, Figure 12.6(a) also indicates that a larger initial crossover probability should be used for a larger feature length in the SeDSC decoder to achieve a better decoding performance.

12.6.3 Authentication Accuracy

Authentication Performances on Small Feature Length (i.e., $N = 100$)
Figure 12.7 illustrates a portion of the receiver operating characteristic (ROC) curves of the non-asymmetric and asymmetric SeDSC setups with the top 100 most discriminable features, where the crossover probability is equal to 0.1 for the BP-based SeDSC decoder with a maximum of 50 BP iterations. In our experiment, the genuine accept rate (GAR) and the false accept rate (FAR) are

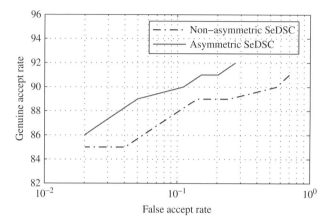

Figure 12.7 ROC curves of non-asymmetric SeDSC and asymmetric SeDSC setups with the top 100 most discriminable features.

used to quantify the authentication accuracy, where different GAR and FAR were obtained by varying code rate. Ideally, a robust authentication system should provide a high GAR, while maintaining a very low FAR. In Figure 12.7, we can see that both the non-asymmetric and the asymmetric SeDSC setups achieve similar authentication accuracy with GARs of 85% and 86% at the same FAR of 0.02%, respectively. Moreover, with a feature length of 100, we can see that the non-asymmetric SW-based SeDSC results in a slightly higher FAR than that of the asymmetric SW-based SeDSC given the same GAR, and the reasons for this are two-fold. On the one hand, a feature length of 100 usually shows a much lower inter-user distinctiveness (see Figure 12.6(a)). On the other hand, the IRA codes (originally designed for channel coding) offer a very strong error correction capability, which is better for channel coding, but undesirable in authentication with short feature lengths and low inter-user distinctiveness. However, as it will be shown in our next study of large feature length (i.e., Figure 12.8), the performance of non-asymmetric SW-based SeDSC becomes better than that of asymmetric SW-based SeDSC as long as a reasonably large feature length is selected.

Performances on Large Feature Lengths (i.e., $N \geq 300$)
Figure 12.8 depicts the impact of feature length and total code rate R on the GAR and FAR for both non-asymmetric and asymmetric SW-based SeDSCs, where the feature length varies from 300 to 1100, and the individual code rates are $R_x = R_y = \frac{R}{2}$ in the non-asymmetric case. For the BP-based SeDSC decoder, the crossover probability and the maximum BP iterations are set to 0.25 and 100, respectively. Here, the aforementioned parameters in the current studies are larger than those in the previous study of small feature length, and

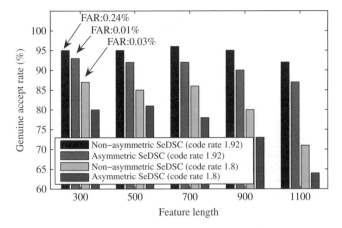

Figure 12.8 Authentication performances (i.e., GAR) of both non-asymmetric and asymmetric SW-based SeDSCs with different feature length and code rate, where GAR points with nonzero FARs have been specifically indicated.

the purpose is to seek a better decoding performance, as discussed in Section 12.6.2.

In Figure 12.8, the corresponding FAR of each GAR point is zero except where specifically indicated. For example, when the feature length is equal to 300, although non-asymmetric SeDSC setups show better GAR performances than those of asymmetric SeDSC setups, they also show worse FAR performances (i.e., 0.24% and 0.01% at the code rates 1.92 and 1.8, respectively) than those of the asymmetric SeDSC setups (i.e., 0.03% and 0% at the same given code rates). However, when the feature length is equal to or larger than 500, the FAR comes to zero for all setups in our experiments, which coincides with our previous discussions in Section 12.6.2 and observations in Figure 12.6(b).

Figure 12.8 also shows that a smaller code rate will result in a lower GAR for both the non-asymmetric and the asymmetric setups, where a smaller code rate corresponds to a higher information theoretic security [129]. Finally, given the same code rate, the GARs of the non-asymmetric setup always outperform these of the asymmetric setup, where the non-asymmetric SeDSC achieves its best GAR of 96% at a feature length of 700 and a code rate of 1.92.

12.6.4 Privacy and Security

We provide comprehensive protection to cover attack points 1–7 in Figure 12.1. Thanks to DSC-based techniques, PROMISE only operates on encrypted binary features to allow a high level of privacy and security protection during enrollment. Regarding authentication, as indicated by Figure 12.9, the diagonal dominant pre-encrypted correlation matrix becomes flat after applying PROMISE, which makes our system highly confidential.

(a) Correlation before DSC encryption

Figure 12.9 Intra-user and inter-user correlation before non-asymmetric SW encryption (a) and after non-asymmetric SW encryption (b).

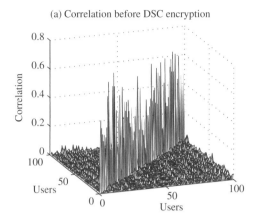

(b) Correlation after DSC encryption

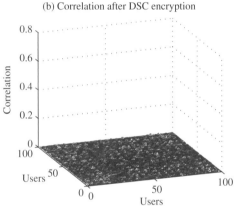

The syndrome encryption (see Figure 12.3) through non-asymmetric SW encoding offers high secrecy, as corresponding syndromes from each pair are statistically independent of one another and can only be reverted when both encrypted features are from the same user based on the DSC principle. Figures 12.9(a) and (b) show the intra-user and inter-user correlation (i.e., the mutual information between feature vectors) before and after applying the non-asymmetric SW-based encryption scheme, respectively. We can see that the original feature vectors have high intra-user correlation, as shown in the diagonal direction of Figure 12.9(a), but the encrypted feature vectors, as shown in Figure 12.9(b), have been totally decorrelated, as all correlations approach zero. From an information-theoretic perspective, the complexity of recovering encrypted syndrome features without any side information is proportional to $2^{(n-m)}$, where n and m are the length of binary features and the corresponding encrypted features, respectively.

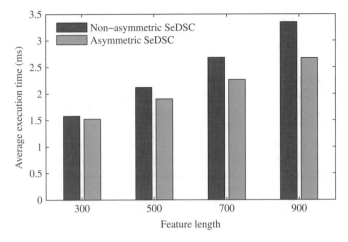

Figure 12.10 Comparison of complexities between non-asymmetric and asymmetric SW-based SeDSCs in terms of average execution time per match in milliseconds (ms).

12.6.5 Complexity Analysis

Figure 12.10 illustrates the complexities of both non-asymmetric and asymmetric SW-based SeDSCs in terms of average execution time per match in millisecond (ms), where all results are obtained by averaging $10,000$ independent matches. The algorithms are implemented in MATLAB/JAVA and tested on a dual-processors (Intel 3.6 GHz) workstation. In Figure 12.10, as expected, the average execution time of both algorithms linearly increases as the feature length increases. Moreover, the average execution time of non-asymmetric SW-based SeDSC is slightly higher than that of asymmetric SW-based SeDSC, as the former requires additional operations for feature vector reconstruction during the authentication phase (see Section 12.5). However, we can see that both non-asymmetric and asymmetric SW-based SeDSCs offer ultra-fast authentication throughputs (i.e., a few milliseconds per match).

12.7 Discussion

We presented a PRivacy-Oriented Medical Information Secure biomEtric system (PROMISE) based on state-of-the-art DSC principles. The work focuses on extending the asymmetric SW codes based the DSC model to a novel stream cipher-based non-asymmetric SW model (i.e., SeDSC) that ensures high security for biometric feature storage, transmission and authentication. By using a carefully designed factor graph model, all operations in the SeDSC framework are carried out directly on the encrypted features rather than the raw biometric features, which significantly improves the system security. Experimental results show that the system provides much stronger privacy protection than existing systems as well as a high authentication accuracy (i.e., 96%). Compared to the previous asymmetric SW-based models [129–131], we conclude that the system is more applicable in practice, where different components (e.g., feature capture, storage and authentication) are developed in a distributed manner.

Similar to [129, 130], the PROMISE framework requires the pre-alignments of all the fingerprints, which limit its flexibility in practice. The secure fingerprint alignment method proposed in [131] can be used to release such a requirement of pre-alignment. In this chapter we only considered three minutiae-point-based features (i.e., the spatial locations (a, b) and orientation θ) in the feature extraction step. Previous work [131] indicates that other features (e.g., ridge orientation and ridge wavelength features) can be used to improve the authentication accuracy.

A

Basic Information Theory

A.1 Information Measures

A.1.1 Entropy

The notion of entropy in information theory is borrowed from the same terminology used in statistical physics. In the latter field, entropy is a quantity that measures the amount of randomness of a system. On the other hand, entropy in information theory quantifies the amount of information contained in a statistically random event. When we equate the two, we have the conclusion that the more randomness an event has, the more information it contains. It may seem unnatural at first sight. However, if we interpret the entropy of a discrete random variable as the amount of information revealed when the outcome of the variable is uncovered, then it makes perfect sense! The more random the event is, the more information we obtain after we know the outcome. On the other hand, if the event is not random at all (completely deterministic), it means that we already know the outcome and so there is no more information revealed when someone tells us what we already know!

A few words on notation should be added here. We will write the probability of a drawn sequence x^n, $Pr(X^n = x^n)$, as $p(x^n)$ or $p(x_1, x_2, \cdots, x_n)$. Mathematically, the entropy of a random variable is defined as follows.

Definition A.1 *(Entropy).* Let X be a discrete random variable that takes values from the alphabet $\mathcal{X}$, and has probability mass function[1] $p(x) = p_X(x) = Pr(X = x), x \in \mathcal{X}$. Then, the entropy of a discrete random variable X, denoted by $H(X)$, is defined as

$$H(X) = - \sum_{x \in \mathcal{X}} p(x) \log p(x). \tag{A.1}$$

---

1 For the rest of the book, we will misuse the terminology of "distribution" slightly and call $p(x)$ the distribution of X.

Distributed Source Coding: Theory and Practice, First Edition. Shuang Wang, Yong Fang, and Samuel Cheng.
© 2017 John Wiley & Sons Ltd. Published 2017 by John Wiley & Sons Ltd.
Companion Website: www.wiley.com/go/cheng/dsctheoryandpractice

Even though $H(X)$ is written like a function of X, it really depends (and only depends) on the distribution of X. Therefore, it may be more proper to interpret $H(X)$ as $H(p_X)$ instead. Moreover, when X is binary, $p_X(0) = p$ and $p_X(1) = 1 - p$. With a slight misuse of notation, we also often write $H(X) = H(p) \triangleq -p\log_2(p) - (1-p)\log_2(1-p)$.

If the logarithm is of base 2 as above, then the entropy is expressed in bits; if it is of base e, then the entropy is expressed in nats. In this book we will always use base-2 logarithms, and hence express entropies in bits.

Example A.1 *(Discrete binary source).* A discrete binary source taking values 0 or 1 has entropy 1 when $Pr(X = 0) = Pr(X = 1) = 0.5$ and entropy 0 when $Pr(X = 0) = 1$ or $Pr(X = 1) = 1$. Other values of $Pr(X = 0)$ are shown in Figure A.1.

From this example we can see that the more we are uncertain about the source, the higher the entropy. When we are absolutely certain of the value of the random source, that is, $Pr(X = x) = 1$ for some $x \in \mathcal{X}$, $H(X)$ will be 0. In this case, the source is no longer random.

Since $H(X)$ describes the information content of X, we expect that $H(X) \geq 0$. Moreover, $H(X) \leq \log_2|\mathcal{X}|$. The latter inequality is also self-evident since the entropy is equal to the amount of information that a variable contains. For example, if we are given 5 bits, these 5 bits should be able to represent all variables with alphabet sizes equal to $2^5 = 32$ regardless of their distributions (by mapping each 5-bit pattern as one symbol in the alphabet).

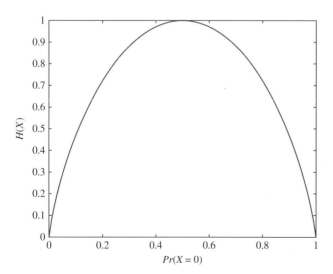

Figure A.1 The entropy of a binary source.

Therefore, the information content of these variables should be less than $5 = \log_2(32) = \log_2|\mathcal{X}|$. Thus, we have the following lemma.

Lemma A.1 *(Range of entropy).* For any random variable X,

$$0 \leq H(X) \leq \log_2|\mathcal{X}|. \tag{A.2}$$

For the first inequality, we will defer the proof to later in this chapter (see Lemma A.8). For the second inequality, we will use a simple fact as follows.

Fact A.1. *For any real a,*

$$\ln(a) \leq a - 1. \tag{A.3}$$

Moreover, the equality only holds when a=1.

Partial proof of Lemma A.1. We will use Fact A.1 to prove the second inequality first.

$$
\begin{aligned}
H(X) - \log_2|\mathcal{X}| &\overset{(a)}{=} \sum_{x \in \mathcal{X}} p(x) \log_2 \frac{1}{p(x)} - \sum_{x \in \mathcal{X}} p(x) \log_2|\mathcal{X}| \\
&= \sum_{x \in \mathcal{X}} \frac{p(x)}{\ln 2} \ln \frac{1/|\mathcal{X}|}{p(x)} \\
&\overset{(b)}{\leq} \sum_{x \in \mathcal{X}} \frac{p(x)}{\ln 2} \left(\frac{1/|\mathcal{X}|}{p(x)} - 1 \right) \\
&= \frac{1}{\ln 2} \sum_{x \in \mathcal{X}} \left(\frac{1}{|\mathcal{X}|} - p(x) \right) \\
&= \frac{1}{\ln 2} (1 - 1) = 0,
\end{aligned}
$$

where (a) is due to $\sum_{x \in \mathcal{X}} p(x) = 1$ and (b) is from Fact A.1. $\qquad\square$

Now, consider two discrete random variables, X and Y. The joint entropy of X and Y specifies the total amount of information revealed when both outcomes of X and Y are uncovered. Similar to the definition of entropy, we have the following definition.

Definition A.2 *(Joint entropy).* The joint entropy $H(X, Y)$ of a pair of discrete random variables X and Y, with joint probability distribution $p(x, y)$, is defined as

$$H(X, Y) = - \sum_{(x,y) \in \mathcal{X} \times \mathcal{Y}} p(x, y) \log_2 p(x, y). \tag{A.4}$$

Similarly, for n random variables $X_1, X_2, \cdots, X_n$ with joint probability distribution $p(x_1, x_2, \cdots, x_n)$, the joint entropy is given by

$$H(X_1, X_2, \cdots, X_n) = - \sum_{(x_1, \cdots, x_n) \in \mathcal{X}_1 \times \cdots \times \mathcal{X}_n} p(x_1, \cdots, x_n) \log_2 p(x_1, \cdots, x_n). \quad (A.5)$$

Lemma A.1 can be naturally extended to the following lemma.

Lemma A.2 *(Range of joint entropy).* For random variables $X_1, X_2, \cdots, X_n$,

$$0 \le H(X_1, X_2, \cdots, X_n) \le \log_2 |\mathcal{X}_1 \times \mathcal{X}_2 \times \cdots \times \mathcal{X}_n|$$
$$= \log_2 |\mathcal{X}_1| + \log_2 |\mathcal{X}_2| + \cdots + \log_2 |\mathcal{X}_n|.$$

Intuitively, $H(X, Y)$ is uncertainty in (X, Y), or the information we get once we reveal X and Y. Therefore, if we know Y beforehand, the remaining uncertainty in X is $H(X, Y) - H(Y)$, which intuitively should be smaller than or equal to $H(X)$ – the information that we would get from X if we did not know Y before. This remaining uncertainty in X, if we know Y, is called the conditional entropy of X given Y and is formally defined next.

Definition A.3 *(Conditional entropy).* The conditional entropy of X given Y, denoted by $H(X|Y)$, is defined as

$$H(X|Y) = - \sum_{(x,y) \in (\mathcal{X}, \mathcal{Y})} p(x, y) \log_2 p(x|y). \quad (A.6)$$

From the above definitions we get mathematically what we concluded earlier intuitively: $H(X|Y) = H(X, Y) - H(Y)$. Moreover, from our previous discussion (which we will show more formally in Corollary A.1), $H(X) \ge H(X|Y)$. Therefore, we have $H(X, Y) = H(X|Y) + H(Y) \le H(X) + H(Y)$.

Note that $H(X, X) = H(X) + H(X|X)$. Also, from Definition A.2, it follows $H(X, X) = H(X)$. Thus, $H(X|X) = 0$. This is expected indeed, since if we know X, there is zero uncertainty in X remaining.

It can also be verified easily that $H(g(X)|X) = 0$ if $g(\cdot)$ is a deterministic function. This is reasonable as there should be no more information remaining from $g(X)$ if X is known.

We can extend $H(X, Y) = H(X) + H(Y|X)$ to more variables. This is usually known as the chain rule for entropy. It is obtained immediately from the entropy definitions, hence the proof is omitted.

Lemma A.3 *(Chain rule for entropy).*

$$H(X_1, X_2, \cdots, X_n) = H(X_1) + H(X_2|X_1) + H(X_3|X_1, X_2) + \cdots$$
$$+ H(X_n|X_1, X_2, \cdots, X_{n-1}). \quad (A.7)$$

A.1.2 Relative Entropy

Despite its name, relative entropy or Kullback–Leiber (KL) divergence considers some very different concepts from entropy. In a nutshell, relative entropy measures the difference between two probability distributions. Consider two distributions $p(x)$ and $q(x)$. Their relative entropy denoted by $D(p(x) \| q(x))$ is defined as

$$D(p(x) \| q(x)) \triangleq \sum_{x \in \mathcal{X}} p(x) \log_2 \frac{p(x)}{q(x)}. \tag{A.8}$$

If the two distributions are identical, their relative entropy will be zero since $\log_2 \frac{p(x)}{q(x)} = 1$ for all x. Actually, we can say more, as follows.

Lemma A.4   (Relative entropy).   The relative entropy has to be positive and will only be zero when the two input distributions are identical.

Proof.

$$D(p(x) \| q(x)) = \sum_{x \in \mathcal{X}} p(x) \log_2 \frac{p(x)}{q(x)}$$

$$= -\sum_{x \in \mathcal{X}} \frac{p(x)}{\ln 2} \ln \frac{q(x)}{p(x)}$$

$$\overset{(a)}{\geq} -\frac{1}{\ln 2} \sum_{x \in \mathcal{X}} p(x) \left(\frac{q(x)}{p(x)} - 1 \right)$$

$$= \frac{1}{\ln 2} \sum_{x \in \mathcal{X}} (p(x) - q(x)) = 0,$$

where the inequality in (a) is from Fact A.1 and equality only holds when $\frac{q(x)}{p(x)} = 1$ for all x with nonzero $p(x)$. Since both $p(x)$ and $q(x)$ are probability distributions that will be summed to 1, these two conditions force $p(x)$ and $q(x)$ to be practically the same for all x.   □

The above lemma gives relative entropy its "distance" property. However, unlike a real distance measure, relative entropy is not symmetric in the sense that $D(p(x) \| q(x)) \neq D(q(x) \| p(x))$. It turns out that many information measures can be expressed by relative entropies. Moreover, as shown later in the chapter, it has many uses in other disciplines.

Now, let us use Lemma A.4 to show a useful inequality widely used in information theory.

Lemma A.5   (Log-sum   inequality).   For any $a_1, \cdots, a_n \geq 0$ and $b_1, \cdots, b_n \geq 0$, we have

$$\sum_i a_i \log_2 \frac{a_i}{b_i} \geq \sum_i a_i \log_2 \frac{\sum_i a_i}{\sum_i b_i}. \tag{A.9}$$

Proof. We can define two distributions $p(x)$ and $q(x)$ with $p(x_i) = \frac{a_i}{\sum_i a_i}$ and $q(x_i) = \frac{b_i}{\sum_i b_i}$. Since $p(x)$ and $q(x)$ are both non-negative and sum to 1, they are indeed valid probability mass functions. Now consider the non-negative property of the relative entropy. We have

$$D(p(x) \| q(x)) = \sum_i p(x_i) \log_2 \frac{p(x_i)}{q(x_i)}$$

$$= \sum_i \frac{a_i}{\sum_i a_i} \left(\log_2 \frac{a_i}{b_i} - \log_2 \frac{\sum_i a_i}{\sum_i b_i} \right) \geq 0.$$

Rearranging the terms we obtain the lemma. □

A.1.3 Mutual Information

As $H(X)$ is equivalent to the information revealed by X and $H(X|Y)$ the remaining information of X knowing Y, we expect that $H(X) - H(X|Y)$ is the information of X shared by Y, or the amount of information that Y reveals about X. This quantity is called the mutual information of X and Y, and is formally defined next.

Definition A.4 *(Mutual information).* The mutual information of X and Y, denoted by $I(X; Y)$, is given by

$$I(X; Y) = H(X) - H(X|Y)$$

$$= - \sum_{(x,y) \in \mathcal{X} \times \mathcal{Y}} p(x, y) \log_2 \frac{p(x)}{p(x|y)}$$

$$= \sum_{(x,y) \in \mathcal{X} \times \mathcal{Y}} p(x, y) \log_2 \frac{p(x, y)}{p(x)p(y)}.$$

$$= H(Y) - H(Y|X). \tag{A.10}$$

From (A.10) we see that the definition of mutual information is symmetric, thus $I(X; Y) = I(Y; X)$. In other words, the information of X shared by Y is equal to that of Y shared by X, or information that Y tells us about X is equal to information that X tells us about Y. We can also see from (A.10) $I(X; Y) = D(p(x, y) \| p(x)p(y))$, that is, the mutual information between two random variables is equivalent to the relative entropy between the joint distribution of the two variables and the product of their marginal distributions. As relative entropy has to be non-negative, we can immediately conclude that mutual information has to be non-negative also.

Note that $I(X; X) = H(X) - H(X|X) = H(X)$. This actually should be within our expectation since the information that X reveals about itself (X) should be nothing but its entropy.

Extending the definition of mutual information, we can define the "conditioned" form of mutual information as follows.

Definition A.5 *(Conditional mutual information).* The conditional mutual information of X and Y given Z, denoted by $I(X; Y|Z)$, is defined as

$$I(X; Y|Z) = H(X|Z) - H(X|Y, Z)$$

$$= \sum_{(x,y,z) \in \mathcal{X} \times \mathcal{Y} \times \mathcal{Z}} p(x, y, z) \log_2 \frac{p(x, y|z)}{p(x|z)p(y|z)}$$

$$= H(Y|Z) - H(Y|X, Z).$$

We can think about this quantity as information that Y tells us about X, if we know already Z.

Similar to the chain rule for entropy, we can easily verify the chain rule for mutual information as below. The proof can be readily obtained from the definition and will be left as an exercise for the reader.

Lemma A.6 *(Chain rule for mutual information).*

$$I(X_1, X_2, \cdots, X_n; Y) = I(X_1; Y) + I(X_2; Y|X_1) + \cdots$$
$$+ I(X_n; Y|X_1, X_2, \cdots, X_{n-1}). \tag{A.11}$$

A.1.4 Entropy Rate

In our previous discussion we used entropy to quantify the amount of information contained in a random variable or, equivalently, the level of uncertainty in a random variable. For a discrete stochastic process, which is a sequence of random variables, we use *entropy rate* instead.

Definition A.6 *(Entropy rate).* For a discrete stochastic process, $X_1, X_2, \cdots, X_n, \cdots$, the entropy rate is defined as

$$H(\mathcal{X}) \triangleq \lim_{n \to \infty} \frac{1}{n} H(X_1, X_2, \cdots, X_n) \tag{A.12}$$

when the limit exists.

For an i.i.d. source $H(X_1, X_2, \cdots, X_n) = nH(X)$, as will be shown in (A.15), hence we have $H(\mathcal{X}) = H(X)$. Essentially in this case, $H(X)$ is sufficient to describe the information content of the process.

Alternatively, we can quantify the random process using conditional entropy

$$H'(\mathcal{X}) \triangleq \lim_{n \to \infty} H(X_n|X_1, \cdots, X_{n-1}). \tag{A.13}$$

For a stationary stochastic process, where the joint distribution of any subset of the sequence of random variables is invariant with respect to index shifts, $H'(\mathcal{X})$ exists since $H(X_n|X_1,\cdots,X_{n-1})$ is a monotonic decreasing sequence (from Corollary A.1) bounded below by 0. Moreover, we have the following theorem.

Theorem A.1 *(Equivalent notions of entropy rate for a stationary process).*

$$H'(\mathcal{X}) = H(\mathcal{X}).\tag{A.14}$$

We will need the following lemma to prove Theorem A.1. The lemma is rather intuitive and we will skip a proof here. The interested reader may refer to [12] for a proof.

Lemma A.7 *(Cesáro mean).* Consider a sequence $a_1, a_2, \cdots, a_n, \cdots$ and let $b_n = \frac{1}{n} \sum_{i=1}^{n} a_i$ be another sequence constructed by sequence a_n. If the limit of the sequence a_n exists and equals a, then the limit of the sequence b_n exists and equals a also.

Proof of Theorem A.1. By the chain rule,

$$\frac{1}{n}H(X_1,\cdots,X_n) = \frac{1}{n} \sum_{i=1}^{n} H(X_i|X_1,\cdots,X_{i-1}).$$

Note that $H(X_i|X_1,\cdots,X_{i-1})$ converges to $H'(\mathcal{X})$ and hence by Lemma A.7 $\frac{1}{n}H(X_1,\cdots,X_n)$ also converges to $H'(\mathcal{X})$. On the other hand, by Definition A.6, $\frac{1}{n}H(X_1,\cdots,X_n)$ converges to $H(\mathcal{X})$. Hence $H(\mathcal{X}) = H'(\mathcal{X})$. ☐

A.2 Independence and Mutual Information

From basic probability theory we know that two variables X and Y are independent if their joint probability distribution $p(x, y)$ is equal to the product of their marginal distributions $p(x)p(y)$. Intuitively, we expect that the two random variables are independent if they share no mutual information, in other words $I(X; Y) = 0$. The following lemma shows that this is indeed the case.

Lemma A.8 For two random variables X and Y, mutual information $I(X; Y) \geq 0$, where equality holds if and only if X and Y are independent.

Proof. From the definitions of mutual information and relative entropy we can write $I(X; Y)$ as $D(p(x, y) \| p(x)p(y))$. As mentioned previously, the mutual information has to be non-negative since relative entropies cannot be negative. Moreover, a relative entropy is zero only if the two input distributions are identical, that is, $p(x, y) = p(x)p(y)$, which is the precise definition of X and Y being independent.

Note that $I(X; Y) = 0$ implies $H(X) = H(X|Y)$, that is, if X and Y are independent, knowing Y will not vary the amount of information that can be revealed by X. This indeed captures the idea of the independence of X and Y. In particular, consider the entropy of a sequence coming out of an i.i.d. source X. By the chain rule, we have

$$H(X_1, X_2, \cdots, X_n) = H(X_1) + H(X_2|X_1) + \cdots + H(X_n|X_1, \cdots, X_{n-1})$$

$$\overset{(a)}{=} H(X_1) + H(X_2) + \cdots + H(X_n)$$

$$\overset{(b)}{=} nH(X) \tag{A.15}$$

where (a) comes from the fact that the source samples at different time instances are independent and (b) is due to the fact that the source is identically distributed. Therefore, as expected, the total information of a sequence of n samples is equal to n times of the information of individual sample!

Moreover, since $I(X; Y) = H(X) - H(X|Y)$, we have the important corollary that we have already mentioned when we define conditional entropy.

Corollary A.1 *(Conditioning reduces entropy).* For any two random variables X and Y,

$$H(X) \geq H(X|Y). \tag{A.16}$$

Moreover, since $I(X; X) = H(X) - H(X|X) = H(X)$, we have $H(X) = I(X; X) \geq 0$. This completes the proof of Lemma A.1 that we have promised.

Now consider an additional random variable Z. From basic probability theory we know that two variables X and Y are conditionally independent given Z if and only if $p(x, y|z) = p(x|z)p(y|z)$. By similar reckoning we have the following lemma.

Lemma A.9 *(Conditionally independent variables).* Two random variables X and Y are conditionally independent given another random variable Z if $I(X; Y|Z) = 0$.

The proof is very similar to the proof of Lemma A.8 and hence will be left as an exercise for the reader. Moreover, we have the following lemma and corollary similar to Lemma A.8 and Corollary A.1.

Lemma A.10   *(Non-negativity of conditional mutual information).*   For any random variables X, Y, and Z,

$$I(X; Y|Z) \geq 0. \tag{A.17}$$

Corollary A.2   *(Conditioning reduces entropy).*   For any random variables X, Y, and Z,

$$H(X|Y) \geq H(X|Y, Z). \tag{A.18}$$

An alternative interpretation of conditional independence is by a Markov chain. When X and Y are conditionally independent given Z, we say X, Z, and Y form a Markov chain. It is usually denoted by $X \to Z \to Y$.

Definition A.7   *(Markov chain).*   Random variables X, Z, and Y form a Markov chain $X \to Z \to Y$ if $p(x|z) = p(x|z, y)$. In other words, given Z, the value of Y will not have an effect on the probability of X.

Note that if $X \to Z \to Y$, we have $p(x, y|z) = p(y|z)p(x|y, z) = p(y|z)p(x|z)$, where the last equality is from the definition of $X \to Z \to Y$ and $p(x, y|z) = p(x|z)p(y|z)$ means that X and Y are conditionally independent given Z. Therefore, the definition above is indeed consistent with the definition of conditional independence.

Since the concept of conditional independence is symmetric, $X \to Z \to Y$ is equivalent to $Y \to Z \to X$. To emphasize the symmetry, in this book we will use notation $X \leftrightarrow Y \leftrightarrow Z$ to denote a Markov chain between random variables X, Y, and Z.

If $I(X; Y|Z) = 0$, we have $H(X|Z) = H(X|Y, Z)$. In other words, given Z, the amount of information revealed by X does not depend on whether we also know Y. This indeed captures the notion of the Markov property and thus conditional independence.

Finally, it is worth mentioning that the definition of a Markov chain can be extended to more variables.

Definition A.8   *(General Markov chain).*   Random variables $X_1, X_2, \cdots, X_n$ form a Markov chain denoted by $X_1 \leftrightarrow X_2 \leftrightarrow \cdots \leftrightarrow X_n$ if $p(x_l|x_{l-1}) = p(x_l|x^{l-1})$, for $l = 1, 2, \cdots, n$, where x^{l-1} is the shorthand notation for $x_1, x_2, \cdots, x_{l-1}$.

An important inequality using the Markov property is the data-processing inequality, which is stated as follows.

Lemma A.11   *(Data-processing inequality).*   If random variables X, Y, and Z satisfy $X \leftrightarrow Y \leftrightarrow Z$, then

$$I(X; Y) \geq I(X; Z).$$

For an important special case, we can imagine Y being a processed output of X, where Z in turn is a processed output of Y. The data-processing inequality basically states that the mutual information decreases as the number of processes increases. The proof is basically one line:

$$I(X;Y) = I(X;Y,Z) - I(X;Z|Y) \stackrel{(a)}{=} I(X;Y,Z)$$

$$= I(X;Z) + I(X;Y|Z) \geq I(X;Z),$$

where (a) is due to $I(X;Z|Y) = 0$ as $X \leftrightarrow Y \leftrightarrow Z$.

A.3   Venn Diagram Interpretation

Loosely speaking, if we denote S_X as the "set" of information out of X, we expect the amount of information out of X (or the entropy of X) to be equal to the cardinality of the set, that is, $H(X) = |S_X|$. Moreover, one will expect the set of total information out of X and Y to be equal to the union of the sets of information out of X and out of Y, that is, $S_{X,Y} = S_X \cup S_Y = S_X + (S_Y \backslash S_X)$. Here, we use $+$ instead of $\cup$ for the set union operation in order to emphasize that the operands S_X and $S_Y \backslash S_X$ are disjoint. Therefore, we have $|S_{X,Y}| = |S_X| + |S_Y \backslash S_X|$. If we compare this with $H(X,Y) = H(X) + H(Y|X)$, we conclude that $H(Y|X) = |S_Y \backslash S_X|$. Moreover, from $S_{X,Y} = (S_X \backslash S_Y) + (S_X \cap S_Y) + (S_Y \backslash S_X)$ and $H(X,Y) = H(X|Y) + I(X;Y) + H(Y|X)$, we can conclude that $I(X;Y) = |S_X \cap S_Y|$. Therefore, one may interpret the relationship between entropies and mutual information for two random variables with a Venn diagram as depicted in Figure A.2. This idea can be extended to more variables. For three variables X, Y, and Z,

$$H(X|Y,Z) = |S_X \backslash S_{Y,Z}| = |S_X \backslash (S_Y \cup S_Z)|. \tag{A.19}$$

On the other hand,

$$H(X|Y) = |S_X \backslash S_Y|$$

$$= |S_X \cap (U \backslash S_Y)|$$

$$= |S_X \cap ((U \backslash (S_Y \cup S_Z)) + (S_Z \backslash Y))|$$

$$= |S_X \cap (U \backslash (S_Y \cup S_Z))| + |S_X \cap S_Z \backslash S_Y|$$

$$= |S_X \backslash (S_Y \cup S_Z)| + |S_X \cap S_Z \backslash S_Y|,$$

where U denotes the universal set, essentially representing the "set" containing all available information. Since $I(X;Z|Y) = H(X|Y) - H(X|Y,Z)$, we can conclude that $I(X;Z|Y) = |S_X \cap S_Z \backslash S_Y|$. To summarize, we have

$$H(X) = |S_X|,$$

$$H(X|Y) = |S_X \backslash S_Y|,$$

$$I(X;Y) = |S_X \cap S_Y|,$$

$$I(X;Y|Z) = |S_X \cap S_Y \backslash S_Z|.$$

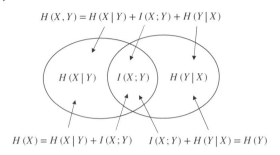

$$H(X,Y) = H(X \mid Y) + I(X;Y) + H(Y \mid X)$$

$$H(X) = H(X \mid Y) + I(X;Y) \qquad I(X;Y) + H(Y \mid X) = H(Y)$$

Figure A.2 Relationship between entropies and mutual information.

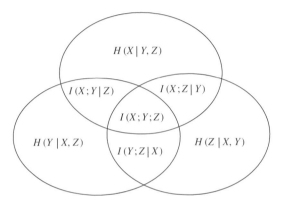

Figure A.3 Venn diagram with relationships of information measures among three variables.

Note that the above four equations include all possibilities since X, Y, and Z above can be interpreted as multiple variables themselves. For example, $I(X, V; Y \mid W, Z) = |S_{X,V} \cap S_Y \backslash S_{W,Z}| = |(S_X \cup S_V) \cap S_Y \backslash (S_W \cup S_Z)|$.

The Venn diagram interpretation is intuitively appealing but needs to be used with caution. For example, the relationship of information measures for three variables X, Y, and Z is shown in Figure A.3. The overlapping region of the three sets is not defined in information-theoretical sense. We will just denote it with $I(X;Y;Z)$ here. From the diagram it is tempting to conclude that $I(X;Y) = I(X;Y \mid Z) + I(X;Y;Z) \geq I(X;Y \mid Z)$, but this is not true in general. The problem is that unlike all other information measures for discrete variables described in this chapter, $I(X;Y;Z)$ need not be positive.

A.4 Convexity and Jensen's Inequality

We call a function $f(\mathbf{x})$ convex if, for any $\lambda \in [0, 1]$, we have $f(\lambda \mathbf{x}_1 + (1 - \lambda)\mathbf{x}_2) \leq \lambda f(\mathbf{x}_1) + (1 - \lambda)f(\mathbf{x}_2)$. Geometrically, a typical convex function is like a smooth bowl and a global minimum exists when $f(\mathbf{x})$ is bounded below. We call a function $g(\mathbf{x})$ concave if $-g(\mathbf{x})$ is convex.

For a convex function $f(\mathbf{x})$, the Jensen's inequality states that for any real vector random variable $\mathbf{X}$,

$$E[f(\mathbf{X})] \geq f(E[\mathbf{X}]). \tag{A.20}$$

The inequality is rather intuitive. Let us consider the case when X only takes two outcomes $\mathbf{x}_1$ and $\mathbf{x}_2$ with probability p and $1 - p$. Then

$$E[f(\mathbf{X})] = pf(\mathbf{x}_1) + (1 - p)f(\mathbf{x}_2) \overset{(a)}{\geq} f(p\mathbf{x}_1 + (1 - p)\mathbf{x}_2) = f(E[\mathbf{X}]),$$

where (a) is from the convexity of $f(\mathbf{x})$. The argument can easily be extended to X with more than two outcomes [12].

Many of the information quantities described here are either convex or concave. Let us list several examples here.

Lemma A.12 (Convexity of relative entropy). Consider four distributions $p_1(\cdot)$, $p_2(\cdot)$, $q_1(\cdot)$, and $q_2(\cdot)$ on a set of outcomes $\mathcal{X}$:

$$\lambda_1 D(p_1 \| q_1) + \lambda_2 D(p_2 \| q_2) \geq D(\lambda_1 p_1 + \lambda_2 p_2 \| \lambda_1 q_1 + \lambda_2 q_2),$$

where $\lambda_1, \lambda_2 \geq 0$ and $\lambda_1 + \lambda_2 = 1$.

Proof.

$$\lambda_1 D(p_1 \| q_1) + \lambda_2 D(p_2 \| q_2)$$

$$= \lambda_1 \sum_{x \in \mathcal{X}} p_1(x) \log \frac{p_1(x)}{q_1(x)} + \lambda_2 \sum_{x \in \mathcal{X}} p_2(x) \log \frac{p_2(x)}{q_2(x)}$$

$$= \sum_{x \in \mathcal{X}} \lambda_1 p_1(x) \log \frac{\lambda_1 p_1(x)}{\lambda_1 q_1(x)} + \lambda_2 p_2(x) \log \frac{\lambda_2 p_2(x)}{\lambda_2 q_2(x)}$$

$$\overset{(a)}{\geq} \sum_{x \in \mathcal{X}} (\lambda_1 p_1(x) + \lambda_2 p_2(x)) \log \frac{\lambda_1 p_1(x) + \lambda_2 p_2(x)}{\lambda_1 q_1(x) + \lambda_2 q_2(x)}$$

$$= D(\lambda_1 p_1 + \lambda_2 p_2 \| \lambda_1 q_1 + \lambda_2 q_2),$$

where (a) is from the log-sum inequality (Lemma A.5). □

Lemma A.13 (Concavity of entropy). The entropy function $H(X)$ is a concave function of $p(x)$.

Proof.

$$H(X) = \sum_x p(x) \log \frac{1}{p(x)} = \sum_x p(x) \log \frac{1/|\mathcal{X}|}{p(x)} + \sum_x p(x) \log |\mathcal{X}|$$

$$= -D(p(x) \| 1/|\mathcal{X}|) + \log |\mathcal{X}|.$$

Since $D(p \| q)$ is a convex function over (p, q), $-D(p \| q)$ is a concave function over (p, q). Specifically, $-D(p \| 1/|\mathcal{X}|)$ is a concave function over $p(x)$ with $1/|\mathcal{X}|$ being a constant. Hence $H(p)$ is a concave function of $p(x)$. □

Lemma A.14 *(Concavity of mutual information with respect to marginal distribution).* For any random variables X and Y, $I(X;Y)$ is a concave function of $p(x)$ for a fixed $p(y|x)$

Proof.

$$I(X;Y) = \sum_{x,y} p(x,y) \log \frac{p(x,y)}{p(x)p(y)} = \sum_{x,y} p(x)p(y|x) \log \frac{p(x)p(y|x)}{p(x)p(y)}$$

$$= \sum_{x,y} p(x)p(y|x) \log p(y|x) + H\left(\sum_{x} p(x)p(y|x)\right)$$

Note that the first term is a linear function of $p(x)$ whereas from Lemma A.13 $H(p(y))$ is a concave function of $p(y)$, which in turn is a concave function of $p(x)$ as $p(y)$ is a linear function of $p(x)$. □

Lemma A.15 *(Convexity of mutual information with respect to conditional distribution).* For any random variables X and Y, $I(X;Y)$ is a convex function of $p(y|x)$ for a fixed $p(x)$.

Proof. Let us write

$$I(X;Y) = D(p(x,y) \| p(x)p(y))$$

$$= D\left(p(x)p(y|x) \| p(x) \sum_{x} p(x)p(y|x)\right) \triangleq f(p(y|x)).$$

It is easy to see that $I(X;Y)$ is convex with respect to $p(y|x)$ since

$$\lambda f(p_1(y|x)) + (1-\lambda)f(p_2(y|x))$$

$$= \lambda D\left(p(x)p_1(y|x) \| p(x) \sum_{x} p(x)p_1(y|x)\right)$$

$$+ (1-\lambda)D\left(p(x)p_2(y|x) \| p(x) \sum_{x} p(x)p_2(y|x)\right)$$

$$\overset{(a)}{\geq} D\left(\lambda p(x)p_1(y|x) + (1-\lambda)p(x)p_2(y|x) \| \lambda p(x) \sum_{x} p(x)p_1(y|x)\right.$$

$$\left. + (1-\lambda)p(x)\sum_{x} p(x)p_2(y|x)\right)$$

$$= D\left(p(x)[\lambda p_1(y|x)+(1-\lambda)p_2(y|x)] \| p(x)\sum_{x} p(x)[\lambda p_1(y|x)+(1-\lambda)p_2(y|x)]\right)$$

$$= f(\lambda p_1(y|x) + (1-\lambda)p_2(y|x)),$$

where (a) is due to Lemma A.12. □

A.5 Differential Entropy

Definition A.1 is targeted to a discrete source. The entropy measures the average amount of information of the random variable. If the variable is continuous, apparently the entropy described by Definition A.1 will be unbounded since we need an infinite number of bits to represent a continuous random variable in general. In such a case we may require an alternative information measure, as defined below.

Definition A.9 (*Differential entropy*). The differential entropy of a continuous random variable X with probability density function $p(x)$ is given by

$$h(X) \triangleq - \int p(x) \log p(x) dx,$$

where the integration is taken over the support set of X.

Consider a scalar continuous random variable X and let X_q be the quantized X with step size q. We can relate the entropy of X_q to the differential entropy of X as follows.

$$h(X) = - \int p_X(x) \log p_X(x) dx \approx - \sum p_X(x_q) \log p_X(x_q) q$$

$$\overset{(a)}{\approx} - \sum \frac{p_{X_q}(x_q)}{q} \log \frac{p_{X_q}(x_q)}{q} q$$

$$= - \sum p_{X_q}(x_q) \log p_{X_q}(x_q) + \log q = H(X_q) + \log q,$$

where (a) is due to $p_{X_q}(x_q) \approx p_X(x_q) q$ for small q.

We can define the conditional differential entropy in a similar manner to differential entropy, namely,

$$h(X|Y) \triangleq - \int p(x, y) \log p(x|y) dx dy.$$

Consequently, we can define the mutual information between two continuous variables X and Y as

$$I(X; Y) = h(X) - h(X|Y) = h(Y) - h(Y|X).$$

An interesting property of mutual information is that the quantity generally preserves after quantization. Let X_q and Y_q be the corresponding quantized X and Y. For a fine quantization with a small step size,

$$I(X_q; Y_q) \approx H(X_q) - H(X_q|Y_q) \approx H(X_q) - H(X_q|Y)$$

$$\approx h(X) - \log q - (h(X|Y) - \log q) = I(X; Y).$$

A.5.1 Gaussian Random Variables

For scalar Gaussian variable X with mean μ and variance σ_X^2, the differential entropy defined in Definition A.9 can be computed by

$$h(X) = \frac{1}{2} \log(2\pi e \sigma_X^2). \tag{A.21}$$

Note that the quantity does not depend on μ as one may expect. The derivation is straightforward and we will leave it as an exercise for the reader [12]. As for a vector variable $\mathbf{X}$ with covariance matrix $\Sigma_{\mathbf{X}}$, the differential entropy is given by

$$h(\mathbf{X}) = \frac{1}{2} \log |det(2\pi e \Sigma_{\mathbf{X}})|. \tag{A.22}$$

Just as in the scalar case, the quantity does not depend on the mean of the variable.

A.5.2 Entropy Power Inequality

From A.21 we know that the differential entropy of a scalar Gaussian random variable X is given by $h(X) = \frac{1}{2} \log(2\pi e \sigma_X^2)$. This gives us the variance of X

$$\sigma_X^2 = \frac{2^{2h(X)}}{2\pi e}. \tag{A.23}$$

We can define $EP(X) \triangleq 2^{2h(X)}$ as the entropy power of X for any even variable X even when X is not Gaussian. Then $\frac{EP(X)}{2\pi e}$ can be interpreted as the variance of the corresponding Gaussian variable with the same entropy as X.

Since a Gaussian random variable has the largest entropy among all variables with the same variance, $\frac{EP(X)}{2\pi e} \leq Var(X)$ in general. Moreover, the equality holds only when X is Gaussian. Therefore, we can interpret the ratio $\frac{EP(X)}{(2\pi e)Var(X)}$ as some kind of measure of how "Gaussian" the variable X is.

So what is the entropy power inequality? It refers to

$$EP(X + Y) \geq EP(X) + EP(Y), \tag{A.24}$$

where X and Y are two arbitrary continuous random variables.

The proof of the entropy power inequality is rather involved [12], but the inequality itself is rather intuitive. It reflects the same spirit of the Central Limit Theorem that a sum of multiple independent continuous random variables will become more Gaussian than each original. Note that since $Var(\cdot)$ is a linear operator on independent variables, that is, $Var(X + Y) = Var(X) + Var(Y)$, to ensure that the sum is more Gaussian (ratio $\frac{EP(X+Y)}{(2\ pie)Var(X+Y)}$ is closer to 1), $EP(\cdot)$ should be *super*-linear, or $EP(X + Y) \geq EP(X) + EP(Y)$, and that is precisely (A.24).

We do have a conditional form of the entropy power inequality. If X and Y are conditionally independent given Z, or $X \leftrightarrow Z \leftrightarrow Y$, we have almost the same inequality as in (A.24) but with the entropies in the entropy powers replaced by the corresponding conditional entropies, namely,

$$2^{2h(X+Y|Z)} \geq 2^{2h(X|Z)} + 2^{2h(Y|Z)}. \tag{A.25}$$

Moreover, the following fact is often used along with the conditional entropy power inequality when we have two variables that are both independent to a third variable.

Fact A.2. *If random variables X and Y are both independent of Z, then we have $X \leftrightarrow Y \leftrightarrow Z$, that is, X and Z are conditionally independent given Y. By symmetry, we also have $Y \leftrightarrow X \leftrightarrow Z$.*

Proof. If both X and Y are independent of Z, we have $p(xy|z) = p(xy)$, which gives us $p(y|z)p(x|yz) = p(y)p(x|y)$. However, since $p(y|z) = p(y)$, we have $p(x|yz) = p(x|y)$, which is equivalent to $X \leftrightarrow Y \leftrightarrow Z$.

If we have vector variables X^n and Y^n instead, let us consider the corresponding Gaussian variables with the same entropies. In particular, let us assume that the Gaussian variables will have each component with the same variance and independent of each other. Then, from (A.22), we have $h(X^n) = \frac{n}{2}\log(2\pi e\sigma_X^2)$, where σ_X^2 is the variance of each component. Thus, $\sigma_X^2 \propto 2^{2h(X^n)/n} \triangleq EP(X^n)$. With the entropy power for vector variables as defined, we have similar entropy power and conditional entropy power inequalities for vector variables asnb follows,

$$2^{2h(X^n+Y^n)/n} \geq 2^{2h(X^n)/n} + 2^{2h(Y^n)/n} \tag{A.26}$$

for X^n and Y^n to be independent and

$$2^{2h(X^n+Y^n|Z)/n} \geq 2^{2h(X^n|Z)/n} + 2^{2h(Y^n|Z)/n} \tag{A.27}$$

for X^n and Y^n to be conditionally independent given Z.

A.6 Typicality

Consider a discrete independently and identically distributed (i.i.d.) source X defined by the probability mass function (pmf) $p(x)$. As usual, denote $\mathcal{X}$ as the alphabet of X. Let us draw the source n times to get a length-n sequence X^n, where X^n is shorthand notation for $X_1, X_2, \cdots, X_n$. We can analyze the characteristics of these sequences by introducing the notion of typicality. Just as the name suggests, typical sequences are expected to occur and nontypical sequences are unlikely to occur. Hence, we can focus on the former and neglect the latter.

There are many variations in defining typicality. We will adopt the notion of robust typicality following [21].

Definition A.10 *(Typical sequence).* A sequence x^n drawn from an i.i.d. source X specified by the pmf $p(x)$ is ϵ-typical or simply typical if $|\pi(x|x^n) - p(x)| \leq \epsilon p(x)$ for all $x \in \mathcal{X}$, where $\pi(x|x^n)$ is the empirical pmf of x^n, that is, $\pi(x|x^n)$ denotes the number of characters $x \in \mathcal{X}$ in the sequence x^n over n.

Definition A.11 *(Typical set).* A typical set is the set of all length-n typical sequences of X and is denoted by $\mathcal{T}_\epsilon^n(X)$.

The following lemma is useful in showing several properties of typical sequences.

Lemma A.16 *(Typical average lemma).* If $x^n \in \mathcal{T}_\epsilon^{(n)}(X)$, then for any non-negative function $g(\cdot)$,

$$(1 - \epsilon)E[g(X)] \leq \frac{1}{n} \sum_{i=1}^{n} g(x_i) \leq (1 + \epsilon)E[g(X)]. \tag{A.28}$$

Proof. From the definition of typicality,

$$|\pi(x|x^n) - p(x)| \leq \epsilon p(x)$$
$$\Rightarrow \pi(x|x^n) - p(x) \leq \epsilon p(x)$$
$$\Rightarrow \pi(x|x^n) \leq (1 + \epsilon)p(x)$$
$$\Rightarrow \frac{1}{n} \sum_{x \in \mathcal{X}} g(x)\pi(x|x^n) \leq \frac{1+\epsilon}{n} \sum_{x \in \mathcal{X}} g(x)p(x)$$
$$\Rightarrow \frac{1}{n} \sum_{i=1}^{n} g(x_i) \leq (1 + \epsilon)E[g(X)].$$

The other inequality can be shown in a similar manner. □

Lemma A.17 *(Probability of typical sequences).* If $x^n \in \mathcal{T}_\epsilon^{(n)}(X)$, then

$$2^{-n(H(X)+\delta(\epsilon))} \leq p(x^n) \leq 2^{-n(H(X)-\delta(c))},$$

where $\delta(\epsilon) \to 0$ as $\epsilon \to 0$.

Proof. Let $g(x_i) = -\log(p(x_i))$ and apply Lemma A.16. We have

$$(1 - \epsilon)E[-\log(p(g(X)))] \leq -\frac{1}{n} \sum_{i=1}^{n} \log(p(x_i))$$
$$\leq (1 + \epsilon)E[-\log(p(X))]$$
$$\Rightarrow -n(1 - \epsilon)H(X) \geq \log(p(x^n)) \geq -n(1 + \epsilon)H(X).$$

Let $\delta(\epsilon) = \epsilon H(X)$ and we obtain the lemma. □

Lemma A.18 *(Always typical).* As n goes to infinity, almost all sequences drawn from the i.i.d. source are typical. More precisely, for any $\epsilon > 0$, we can find n such that

$$Pr(X^n \in T_\epsilon^{(n)}(X)) \to 1.$$

Proof. Let $X^n \sim \prod_{i=1}^n p(x_i)$. The lemma follows immediately by the law of large numbers as[2]

$$\frac{\pi(x|X^n)}{n} = \frac{1}{n} \sum_{i=1}^n \delta(X_i = x) \to E[\delta(X = x)] = p(x),$$

when n goes to infinity. $\square$

Since almost all sequences drawn from the source are typical, from the definition of typicality, this in turn concludes that all sequences from the source have almost the same probability of occurrence. This property is sometimes referred to as the asymptotic equipartition property (AEP).

Lemma A.19 *(Size of typical set).* The size of the typical set of X, $|T_\epsilon^n(X)|$, is approximately equal to $2^{nH(X)}$. More precisely,

$$(1 - \epsilon)2^{n(H(X)-\delta(\epsilon))} \leq |T_\epsilon^n(X)| \leq 2^{n(H(X)+\delta(\epsilon))}. \tag{A.29}$$

Proof.

$$1 \geq Pr(X^n \in T_\epsilon^n(X))$$

$$= \sum_{x^n \in T_\epsilon^n(X)} p(x^n)$$

$$\overset{(a)}{\geq} \sum_{x^n \in T_\epsilon^n(X)} 2^{-n(H(X)+\delta(\epsilon))}$$

$$= |T_\epsilon^n(X)|2^{-n(H(X)+\delta(\epsilon))},$$

where (a) is from Lemma A.17.

For the other side for sufficiently large n, we have

$$1 - \epsilon \overset{(a)}{\leq} Pr(X^n \in T_\epsilon^n(X))$$

$$= \sum_{x^n \in T_\epsilon^n(X)} p(x^n)$$

---

2 Note that $\delta(\cdot)$ here is different from the $\delta(\epsilon)$ function used throughout this section and is referred to as the Kronecker-delta function, which is equal to 1 when the argument is satisfied and 0 otherwise.

$$\overset{(b)}{\leq} \sum_{x^n \in T_\epsilon^n(X)} 2^{-n(H(X)-\delta(\epsilon))}$$

$$= |T_\epsilon^n(X)| 2^{-n(H(X)-\delta(\epsilon))},$$

where (*a*) follows from Lemma A.18 and (*b*) is again from Lemma A.17.

A.6.1 Jointly Typical Sequences

For correlated sources X and Y we can define a jointly typical sequence similar to Definition A.10.

Definition A.12 *(Jointly typical sequence).* A pair of sequences (x^n, y^n) drawn from correlated sources X and Y is jointly ϵ-typical if

$$|\pi((x,y)|(x^n, y^n)) - p(x,y)| \leq \epsilon p(x,y),$$

where $\pi((x,y)|(x^n, y^n))$ is the number of (x,y) couples in (x^n, y^n). We will denote the set of all jointly typical sequences as $T_\epsilon^n(X,Y)$.

Note that if $(x^n, y^n) \in T_\epsilon^{(n)}(X,Y)$, then

$$|\pi(x|x^n) - p(x)| = \sum_y (\pi((x,y)|(x^n, y^n)) - p(x,y))|$$

$$\leq \sum_y |\pi((x,y)|(x^n, y^n)) - p(x,y)|$$

$$\leq \epsilon \sum_y p(x,y) = \epsilon p(x)$$

and thus $(x^n) \in T_\epsilon^{(n)}(X)$ and similarly $(y^n) \in T_\epsilon^{(n)}(Y)$.

Similar to the case of typical sequences, the probability of a jointly typical sequence can be shown to be approximately $2^{-nH(X,Y)}$, and the size of a typical set, $T_\epsilon^n(X,Y) \doteq 2^{nH(X,Y)}$. Moreover, almost all pairs of sequences drawn from the joint source are typical. The precise statements are described in the following lemmas without proof.

Lemma A.20 *(Probability of jointly typical sequences).* If $(x^n, y^n) \in T_\epsilon^{(n)}(X,Y)$, then

$$2^{-n(H(X,Y)+\delta(\epsilon))} \leq p(x^n, y^n) \leq 2^{-n(H(X,Y)-\delta(\epsilon))},$$

where $\delta(\epsilon) \to 0$ as $\epsilon \to 0$.

From the above lemma and Lemma A.17, immediately we have the following.

Lemma A.21 *(Conditional probability of jointly typical sequences).* If $(x^n, y^n) \in \mathcal{T}_\epsilon^{(n)}(X, Y)$, then

$$2^{-n(H(X|Y)+\delta(\epsilon))} \le p(x^n|y^n) \le 2^{-n(H(X|Y)-\delta(\epsilon))},$$

where $\delta(\epsilon) \to 0$ as $\epsilon \to 0$.

Lemma A.22 *(Always typical for joint sequences).* As the length of sequences, n, goes to infinity, almost all pair of sequences drawn from a joint source are jointly typical.

Lemma A.23 *(Size of typical set for joint sequences).* The size of the jointly typical set of X and Y, $|\mathcal{T}_\epsilon^n(X, Y)|$, is approximately equal to $2^{nH(X,Y)}$. More precisely,

$$(1 - \epsilon)2^{n(H(X,Y)-\delta(\epsilon))} \le |\mathcal{T}_\epsilon^n(X, Y)| \le 2^{n(H(X,Y)+\delta(\epsilon))}, \tag{A.30}$$

where $\delta(\epsilon) \to 0$ as $\epsilon \to 0$.

When x^n is typical and y^n is sampled from $p(y|x)$, one would expect that (x^n, y^n) will be jointly typical. The precise conditions are stated in the following lemma without proof.

Lemma A.24 *(Conditional typicality lemma).* If $x^n \in \mathcal{T}_{\epsilon'}^{(n)}(X)$ and $Y^n \sim p(y^n|x^n) = \prod_{i=1}^n p_{Y|X}(y_i|x_i)$, then for every $\epsilon > \epsilon'$,

$$Pr((x^n, Y^n) \in \mathcal{T}_\epsilon^{(n)}(X, Y)) \to 1$$

as $n \to \infty$.

Given three variables X, Y, and Z, the following lemma is useful in estimating the probability of their sequences being jointly typical.

Lemma A.25 *(Jointly typical lemma).* If $(x^n, y^n) \in \mathcal{T}_\epsilon^{(n)}(X, Y)$, and $Z^n \sim \prod_{i=1}^n p(z_i|x_i)$, then

$$(1 - \epsilon)2^{-n(I(X;Z|Y)+\delta(\epsilon))} \le Pr((x^n, y^n, Z^n) \in \mathcal{T}_\epsilon^{(n)}(X, Y, Z))$$
$$\le 2^{-n(I(X;Z|Y)-\delta(\epsilon))},$$

where $\delta(\epsilon) \to 0$ as $\epsilon \to 0$.

Proof.

$$Pr((x^n, y^n, Z^n) \in \mathcal{T}_\epsilon^{(n)}(X, Y, Z)) = \sum_{z^n \in \mathcal{T}_\epsilon^{(n)}(Z|x^n, y^n)} p(z^n|y^n)$$
$$\le |\mathcal{T}_\epsilon^{(n)}(Z|x^n, y^n)|2^{-n(H(Z|Y)-\epsilon H(Z|Y)}$$
$$\le 2^{n(H(Z|X,Y)+\epsilon H(Z|X,Y)}2^{-n(H(Z|Y)-\epsilon H(Z|Y)}$$
$$= 2^{-n(I(Y;Z|X)-\delta(\epsilon))}.$$

Similarly,

$$Pr((x^n, y^n, Z^n) \in \mathcal{T}_\epsilon^{(n)}(X, Y, Z)) = \sum_{z^n \in \mathcal{T}_\epsilon^{(n)}(Z|x^n, y^n)} p(z^n|y^n)$$

$$\geq |\mathcal{T}_\epsilon^{(n)}(Z|x^n, y^n)| 2^{-n(H(Z|Y) + \epsilon H(Z|Y))}$$

$$\geq (1 - \epsilon) 2^{n(H(Z|X,Y) - \epsilon H(Z|x,Y))}$$

$$2^{-n(H(Z|Y) + \epsilon H(Z|Y))}$$

$$= (1 - \epsilon) 2^{-n(I(Y;Z|X) + \delta(\epsilon))}.$$

□

Now, if X, Y, and Z form a Markov chain $X \leftrightarrow Y \leftrightarrow Z$, we have $I(X; Z|Y) = 0$ and this seems to suggest that the tuple (x^n, y^n, Z^n) will be jointly typical with probability one. This intuition is partially correct since there are some additional conditions involved as stated in the following Markov lemma. A proof of the lemma is omitted here but can be found in [21].

Lemma A.26 *(Markov lemma).* Suppose three variables X, Y, and Z form a Markov chain $X \leftrightarrow Y \leftrightarrow Z$. Then, $Pr((x^n, y^n, Z^n) \in \mathcal{T}_\epsilon^{(n)}(X, Y, Z)) \to 1$ if there exists some sufficiently small $\epsilon' < \epsilon$ such that $(x^n, y^n) \in \mathcal{T}_{\epsilon'}^{(n)}(X, Y)$ and $Z^n \sim p(z^n|y^n)$ along with

- $Pr((y^n, Z^n) \in \mathcal{T}_{\epsilon'}^{(n)}(Y, Z)) \to 1$
- For every $z^n \in \mathcal{T}_{\epsilon'}^{(n)}(Z|y^n)$,

$$2^{-n(H(Z|Y) + \epsilon')} \leq p(z^n|y^n) \leq 2^{-n(H(Z|Y) - \epsilon')}.$$

as $n \to \infty$.

A.7 Packing Lemmas and Covering Lemmas

When we randomly sample two sequences from a source, how likely is it that the two sequences will be jointly typical? Let X^n and Y^n be two such sequences, where $X^n \sim \prod_{i=1}^n p(x_i)$ and $Y^n \sim \prod_{i=1}^n p(y_i)$. Then by Lemma A.18,

$$Pr((X^n, Y^n) \in \mathcal{T}_\epsilon^{(n)}) = \sum_{\{(x^n, y^n)|(x^n, y^n) \in \mathcal{T}_\epsilon^{(n)}\}} p(x^n, y^n)$$

$$= \sum_{\{(x^n, y^n)|(x^n, y^n) \in \mathcal{T}_\epsilon^{(n)}\}} p(x^n)p(y^n)$$

$$\leq \sum_{\{(x^n, y^n)|(x^n, y^n) \in \mathcal{T}_\epsilon^{(n)}\}} 2^{-n(H(X) - \delta(\epsilon))} 2^{-n(H(Y) - \delta(\epsilon))}$$

$$\leq 2^{-n(I(X;Y) - 3\delta(\epsilon))}. \tag{A.31}$$

Similarly, we have

$$Pr((X^n, Y^n) \in \mathcal{T}_\epsilon^{(n)}) \geq (1 - \epsilon)2^{-n(I(X;Y)+3\delta(\epsilon))}. \tag{A.32}$$

Now, if we have several similar $X^n(m)$, $m = 1, \cdots, \lfloor 2^{nR} \rfloor$. As long as $R < I(X;Y) - 3\delta(\epsilon)$, we will have the probability of any of the $X^n(m)$ being jointly typical with Y^n going to 0 as n goes to infinity. This is known as the packing lemma and is stated more formally below.

Lemma A.27 (Packing lemma). For any $\epsilon > 0$, we can find n such that if $X^n(m)$, $m = 1, \cdots, \lfloor 2^{nR} \rfloor$ are drawn from $\prod_{i=1}^n p(x_i)$ and $Y^n \in \mathcal{T}_\epsilon^{(n)}(Y)$ independent of each $X^n(m)$, then

$$Pr((X^n(m), Y^n) \in \mathcal{T}_\epsilon^{(n)}(X, Y) \text{ for some } m) \rightarrow 0$$

if $R < I(X;Y) - \delta(\epsilon)$, where $\delta(\epsilon) \rightarrow 0$ as $\epsilon \rightarrow 0$.

The packing lemma is useful in proving the channel coding theorem, where we need to decide the maximum number of distinguishable codewords (and thus messages) that can be sent through a channel. Given n and the channel statistics, we can interpret this as the maximum number of codewords that can be *packed* into the system.

Instead of considering a sequence versus codewords in a codebook, we can also obtain the condition of when two codebooks can be *packed* together as in the following lemma.

Lemma A.28 (Mutual packing lemma). Let $U_1^n(l_1)$, $l_1 = 1, \cdots, \lfloor 2^{nR_1} \rfloor$ be drawn from $\prod_{i=1}^n p(u_{1,i})$, and $U_2^n(l_2)$, $l_2 = 1, \cdots, \lfloor 2^{nR_2} \rfloor$ be drawn from $\prod_{i=1}^n p(u_{2,i})$, then

$$Pr((U_1^n(l_1), U_2^n(l_2)) \in \mathcal{T}_\epsilon^{(n)}(U_1, U_2) \text{ for some } l_1, l_2) \rightarrow 0$$

if $R_1 + R_2 < I(U_1; U_2) - \delta(\epsilon)$, where $\delta(\epsilon) \rightarrow 0$ as $\epsilon \rightarrow 0$.

In contrast to the packing lemma, when $X^n(m)$ are independently drawn and $R > I(X;Y)$ we can also show that we can always find one $X^n(m)$ that is jointly typical with Y^n. This is known to be the covering lemma and is useful in studying the rate-distortion problems.

Lemma A.29 (Covering lemma). For any $\epsilon > 0$, we can find n such that if $X^n(m)$, $m = 1, \cdots, \lceil 2^{nR} \rceil$, are independently drawn from $\prod_{i=1}^n p(x_i)$ and $Y^n \in \mathcal{T}_{\epsilon'}^{(n)}(Y)$, $\epsilon' < \epsilon$, is independent of each $X^n(m)$, then

$$Pr((X^n(m), Y^n) \in \mathcal{T}_\epsilon^{(n)}(X, Y) \text{ for some } m) \rightarrow 1$$

if $R > I(X;Y) + \delta(\epsilon)$, where $\delta(\epsilon) \rightarrow 0$ as $\epsilon \rightarrow 0$.

Proof.

$$Pr((X^n(m), Y^n) \notin \mathcal{T}_\epsilon^{(n)}(X, Y) \text{ for all } m)$$

$$= \sum_{y^n} p(y^n) Pr((X^n(m), y^n) \notin \mathcal{T}_\epsilon^{(n)}(Y, X), \forall m | y^n)$$

$$= \sum_{y^n} p(y^n) \prod_{m=1}^{\lceil 2^{nR} \rceil} Pr((X^n(m), y^n) \notin \mathcal{T}_\epsilon^{(n)}(Y, X), \forall m | y^n)$$

$$\overset{(a)}{\leq} (1 - (1 - \epsilon)2^{-n(I(Y;X)+\delta(\epsilon))})^{\lceil 2^{nR} \rceil}$$

$$\leq \exp(-\lceil 2^{nR} \rceil (1 - \epsilon)2^{-n(I(Y;X)+\delta(\epsilon))})$$

$$\leq \exp(-(1 - \epsilon)2^{-n(I(Y;X)+\delta(\epsilon))}) \to 0 \text{ as } n \to \infty,$$

where (a) is due to A.32.

We can generalize the packing and covering lemmas by replacing the independence to the conditional independence conditions. Then we have the following lemmas.

Lemma A.30 *(Conditional packing lemma).* For any $\epsilon > 0$, we can find n such that if $X^n(m)$, $m = 1, \cdots, \lfloor 2^{nR} \rfloor$, are drawn from $\prod_{i=1}^n p(x_i|u_i)$ and $Y^n \in \mathcal{T}_\epsilon^{(n)}(Y)$ conditionally independent of each $X^n(m)$ given U^n, then

$$Pr((X^n(m), Y^n, U^n) \in \mathcal{T}_\epsilon^{(n)}(X, Y, U) \text{ for some } m) \to 0$$

if $R < I(X; Y|U) - \delta(\epsilon)$, where $\delta(\epsilon) \to 0$ as $\epsilon \to 0$.

Lemma A.31 *(Conditional covering lemma).* For any $\epsilon > 0$, we can find n such that if $X^n(m)$, $m = 1, \cdots, \lceil 2^{nR} \rceil$, are independently drawn from $\prod_{i=1}^n p(x_i|u_i)$ and $Y^n \in \mathcal{T}_{\epsilon'}^{(n)}(Y)$, $\epsilon' < \epsilon$, is conditionally independent of each $X^n(m)$ given U^n, then

$$Pr((X^n(m), Y^n, U^n) \in \mathcal{T}_\epsilon^{(n)}(X, Y, U) \text{ for some } m) \to 1$$

if $R > I(X; Y|U) + \delta(\epsilon)$, where $\delta(\epsilon) \to 0$ as $\epsilon \to 0$.

A.8 Shannon's Source Coding Theorem

Until now we have stated without proof that the entropy of a source quantifies the amount of information contained by the source. However, Shannon's Source Coding Theorem provides theoretical support for this statement.

In a nutshell, the theorem states that we need exactly $H(X)$ bits on average to store each sample from an i.i.d. source, no more and no less. This result intuitively follows from the typicality. Indeed, since the number of typical sequences is roughly $2^{nH(X)}$ and almost all sequences are typical, we can represent only typical sequences, for which we only need $H(X)$ bits per sample on average. We formally state the theorem next and give the formal proof.

Theorem A.2 *(Shannon's Source Coding Theorem).* For a discrete i.i.d. source X, denote R as the code rate for compressing X. As n goes to infinity, we can find a coding scheme to compress X losslessly at rate R (i.e., X can be reconstructed perfectly) if $R \geq H(X)$. On the other hand, if $R < H(X)$, there does not exist any coding scheme such that lossless compression is possible.

The above theorem is also called Shannon's First Coding Theorem or Shannon's Noiseless Coding Theorem.

Example A.2 *(Lossless compression of a binary source).* Consider a binary random source X with $Pr(X = 0) = 0.1$ and $Pr(X = 1) = 0.9$. Then, $H(X) = -Pr(X = 0) \log_2 Pr(X = 0) - Pr(X = 1) \log_2 Pr(X = 1) = 0.469.$ By Shannon's Source Coding Theorem we can compress 1000 samples of X losslessly with approximately 469 bits.

We will split the proof of the theorem into two parts: the forward part and the converse part. The achievability proof (forward part) is based on a simple counting argument and typicality, whereas we need Fano's inequality for the converse proof.

Achievability Proof of Shannon's Source Coding Theorem. Consider the coding scheme where the encoder takes n samples from the source and converts them into an index U, whereas decoding is done by simply mapping back the index U into the n samples. However, instead of dealing with individual samples, we will combine the n samples drawn from the source to form length-n sequences.

From Lemma A.19 there are at most $2^{n(H(X)+\epsilon)}$ typical sequences and we will index all of them. Then, the total number of bits required will be less than $n(H(X) + \epsilon)$ and the compression rate will be less than $H(X) + \epsilon$. Since ϵ can be made arbitrarily small as n goes to infinity, we can have the compression rate R arbitrarily close to $H(X)$.

When will such a scheme fail? It fails when the n samples drawn from the source do not form a typical sequence. However, from Lemma A.18 this probability and hence the error probability can be made arbitrarily small. Therefore, lossless compression is possible with this scheme (at the rate $H(X) + \epsilon$).

For the converse proof, we will consider the same general coding approach as the forward proof: mapping a length-n sequence X^n to U at the encoder and mapping it back at the decoder. The converse of the theorem basically states that no matter what mapping we use, we cannot compress the source losslessly if the rate, R, is less than $H(X)$. As mentioned earlier, the converse proof relies on Fano's inequality, which is stated in the following lemma.

Lemma A.32 *(Fano's inequality).* Let X be an i.i.d. random variable. X^n is compressed by mapping it to an index U, which is then transmitted to a decoder, which maps U back to an estimated sequence $\hat{X}^n$. Then,

$$\frac{1}{n}H(X^n|U) \le \frac{H(P_e)}{n} + P_e\log_2|\mathcal{X}|, \tag{A.33}$$

where $P_e = Pr(X^n \ne \hat{X}^n)$ is the probability of the error of reconstructing X^n and $|\mathcal{X}|$ is the size of the alphabet of X.

The detailed expression on the right-hand side of (A.33) is not very significant. The main point is that when P_e goes to zero, $\frac{1}{n}H(X^n|U)$ goes to zero as well. In other words, the average information content of X^n per symbol knowing U becomes insignificant. This is intuitive since when the probability of error goes to zero then $\hat{X}^n$, reconstructed from U, is essentially the same as X^n. We give the formal proof next.

Proof. Let E be the error event such that $E = 1$ if $X^n \ne \hat{X}^n$ and $E = 0$ otherwise. Hence, $Pr(E = 1) = P_e$. Then,

$$H(E, X^n|U) \overset{(a)}{=} H(X^n|U) + H(E|X^n, U) \tag{A.34}$$

$$\overset{(b)}{=} H(X^n|U), \tag{A.35}$$

where (a) comes from the chain rule for entropies and (b) is due to the fact that $\hat{X}^n$ is deterministic given U, and E is deterministic given X^n and $\hat{X}^n$ (and hence U). On the other hand,

$$H(E, X^n|U) \overset{(a)}{=} H(E|U) + H(X^n|U, E)$$

$$\overset{(b)}{\le} H(E) + H(X^n|U, E)$$

$$= H(E) + Pr(E = 0)H(X^n|U, E = 0)$$

$$\quad + Pr(E = 1)H(X^n|U, E = 1)$$

$$\overset{(c)}{=} H(E) + Pr(E = 1)H(X^n|U, E = 1)$$

$$\le H(E) + P_eH(X^n)$$

$$\overset{(d)}{=} H(E) + nP_eH(X)$$

$$\overset{(e)}{\le} H(E) + nP_e\log_2|\mathcal{X}|,$$

where (a) comes from the chain rule for entropies, (b) is due to Lemma A.1 (conditioning reduces entropy), (c) is from the fact that X^n is a deterministic given $\hat{X}^n$ (and thus given U) when there is no error, (d) is due to the fact that X is an i.i.d. source, and (e) is from Lemma A.1. Therefore, $H(X^n|U) \leq H(E) + nP_e\log_2|\mathcal{X}|$.

Now, we are ready for the converse proof of the source coding theorem. The converse proof below shows that if the probability of error goes to zero, the compression rate has to be larger than the entropy of the source, $H(X)$.

Converse Proof of Source Coding Theorem. If the probability of error $P_e = Pr(X^n \neq \hat{X}^n)$ goes to zero, then

$$nR \overset{(a)}{=} \log_2|U|$$

$$\overset{(b)}{\geq} H(U)$$

$$= H(U|X^n) + I(U;X^n)$$

$$\overset{(c)}{=} I(U;X^n)$$

$$= H(X^n) - H(X^n|U)$$

$$\overset{(d)}{=} H(X^n)$$

$$\overset{(e)}{=} nH(X),$$

where (a) is from the definition of coding rate, (b) is from Lemma A.1, (c) is from the fact that U is deterministic given X^n, (d) is due to Fano's inequality, and (e) is from the fact that X is an i.i.d. source.

A.9 Lossy Source Coding—Rate-distortion Theorem

Before ending this appendix we will give a brief description of lossy source coding. Unlike the the source coding problem we have just discussed, the reconstructed source at the decoder will not be exactly the same as the original source. To quantify the fidelity of the reconstructed source, we will introduce a distortion function $d : \hat{\mathcal{X}} \times \mathcal{X} \to \mathbb{R}^+ \cup \{0\}$. $d(\cdot, \cdot)$, which can be some common distortion measures such as square distance where $d(\hat{x}, x) = 0$ if and only if $\hat{x} = x$. The objective of the rate-distortion theory is to specify the minimum rate required to ensure that the average distortion of the reconstructed source $\hat{X}$ from the original is bounded by some predefined distortion D, namely, $E[d(X, \hat{X})] \leq D$.

The rate-distortion theory states that the required rate to guarantee a distortion less than or equal to D is given by

$$R^*(D) = \min_{p(\hat{x}|x)} I(X; \hat{X}), \tag{A.36}$$

where the optimization problem is minimizing over all conditional distribution $p(\hat{x}|x)$ such that $E[d(X, \hat{X})] = \sum_{x,\hat{x}} p(x) p(\hat{x}|x) d(x, \hat{x}) \leq D$.

We would like to note a couple properties of $R^*(D)$ before we proceed with the proof. First, $R^*(D)$ is apparently a monotonically decreasing function of D since the feasible set of $p(\hat{x}|x)$ in (A.36) grows with D. Second, one may easily show that $R^*(D)$ is a convex function of D [12]. This is due to the fact that $I(X; \hat{X})$ is a convex function with respect to $p(\hat{x}|x)$.

Achievability Proof of the Rate-distortion Theorem. We will give a sketch proof here. For a rigorous proof the reader should refer to [12]. Like the achievability proof of the Source Coding Theorem, we will approach this with the technique of typical sequences. Consider any conditional distribution $p(\hat{x}|x)$ with the corresponding $E[d(X, \hat{X})] \leq D$. Any joint typical sequence $(x^n, \hat{x}^n)$ that generated from $p(x) p(\hat{x}|x)$ will guarantee that $d(x^n, \hat{x}^n) \triangleq \frac{1}{n} \sum_{i=1}^{n} d(x_i, \hat{x}_i) \approx E[d(X, \hat{X})] \leq D$. Therefore, we just need to ensure that for an input x^n, the decoder can output an $\hat{x}^n$ such that the two sequences are jointly typical.

Let us consider the simplest possible scheme as follows. We first create a codebook composed of length-n sequence codewords generated from the marginal distribution $p(\hat{x}) = \sum_x p(x) p(\hat{x}|x)$. Given an input sequence x^n, the encoder simply searches a codeword $\hat{x}^n$ from the codebook such that x^n and $\hat{x}^n$ are jointly typical. If no such codeword exists, the encoder declares failure and gives up. Otherwise, the codeword index will be sent to the decoder and the decoder will simply output the codeword with the received index. For a given code rate R, the scheme can process 2^{nR} codewords. The key question here is how many codewords, and consequently how large R, are needed to *cover* all potential inputs such that a jointly typical sequence can always be found. From the covering lemma (Lemma A.29) we know that $R \geq I(X; \hat{X})$ would be sufficient and this concludes the proof.

Converse Proof of the Rate-distortion Theorem. For the converse proof, we need to show that for any code function pair (f, g) with $f : \mathcal{X}^n \to \{1, 2, \cdots, 2^{nR}\}$, $g : \{1, 2, \cdots, 2^{nR}\} \to \mathcal{X}^n$, and $\hat{x}^n = g(f(x^n))$, we have

$$R \leq R^*(D)$$

if $E[d(X^n, \hat{X}^n)] \triangleq E[\frac{1}{n} \sum_{i=1}^{n} d(X_i, \hat{X}_i)] \leq D$. This is because

$$nR \geq H(f(X^n)) \geq H(f(X^n)) - H(f(X^n)|X^n) = I(f(X^n); X^n)$$

$$\overset{(a)}{\geq} I(\hat{X}^n; X^n) = H(X^n) - H(X^n|\hat{X}^n)$$

$$= \sum_{i=1}^{n} H(X_i) - \sum_{i=1}^{n} H(X_i|\hat{X}^n, X^{i-1})$$

$$\geq \sum_{i=1}^{n} H(X_i) - \sum_{i=1}^{n} H(X - i|\hat{X}_i)$$

$$= \sum_{i=1}^{n} I(X_i; \hat{X}_i)$$

$$\geq \sum_{i=1}^{n} R^*(E[d(X_i, \hat{X}_i)]) = n \left(\frac{1}{n} \sum_{i=1}^{n} R^*(E[d(X_i; \hat{X}_i)]) \right)$$

$$\overset{(b)}{\geq} nR^* \left(\frac{1}{n} \sum_{i=1}^{n} E[d(X_i; \hat{X}_i)] \right)$$

$$= nR^* \left(E \left[\frac{1}{n} \sum_{i=1}^{n} d(X_i; \hat{X}_i) \right] \right)$$

$$= nR^*(E[d(X^n; \hat{X}^n)]) \geq nR^*(D),$$

where (a) is due to the data-processing inequality and (b) is due to the convexity of $R^*(D)$ [12], which in turn is due to the convexity of $I(X; \hat{X})$ with respect to $p(\hat{x}|x)$ (see Lemma A.15).

A.9.1 Rate-distortion Problem with Side Information

When some side information Y is given to both the encoder and decoder, one can imagine that the encoder and decoder can switch to a different coding scheme according to the current state information Y. So when $Y = y$, the rate-distortion function will be simply

$$R_{X|y}(D) = \min_{p(\hat{x}|x,y)} I(X; \hat{X}|y). \tag{A.37}$$

Averaging over all possible outcomes of Y we have

$$R_{X|Y}(D) = \min_{p(\hat{x}|x,y)} I(X; \hat{X}|Y). \tag{A.38}$$

B

Background on Channel Coding

The objective of channel codes or error-correcting codes is to protect a message going through a noisy channel. This is achieved by introducing redundancy to the message.

Let us consider a simple binary code where the encoder receives a length-k message and generates a codeword of length n. The codeword space, $\mathbb{Z}_2^n$, has in total 2^n code vectors. Of these code vectors, only 2^k will be chosen to be codewords to represent different messages. Therefore, the encoder simply maps the input (length-k) message into one of the codewords as a (length-n) coded message. When this coded message passes through a noisy channel, it may get corrupted and may no longer belong to the set of codewords. If this happens, the decoder will detect the transmission error and could try to correct it by searching for the codeword that "matches" best with the received code vector.

It is possible that the corrupted coded message also belongs to the codeword set, in which case the error will not be detected. However, this chance decreases as n increases or k decreases since there will be fewer allowed codewords relative to the entire code vector space.

This chapter is organized as follows. In the next section, we will start with linear block codes and describe the simplest decoding method, syndrome decoding. We will use Hamming codes as a concrete example for block coding. In Section B.2 we discuss convolutional codes and the Viterbi algorithm for decoding. Shannon's Channel Coding Theorem will be discussed in Section B.3. We will then digress from many coding textbooks in that we will skip algebraic channel coding but focus on the more efficient turbo and LDPC codes in Section B.4. We will start with the LDPC codes and describe belief propagation (BP) (the sum-product algorithm) in its most general form. We will then apply it for LDPC decoding and soft decoding for convolutional codes. We will briefly mention irregular repeat accumulate (IRA) codes, which can be viewed as a simple extension of LDPC codes. Last but not least, we will describe encoding and decoding of turbo codes.

Distributed Source Coding: Theory and Practice, First Edition. Shuang Wang, Yong Fang, and Samuel Cheng.
© 2017 John Wiley & Sons Ltd. Published 2017 by John Wiley & Sons Ltd.
Companion Website: www.wiley.com/go/cheng/dsctheoryandpractice

B.1   Linear Block Codes

One simple coding scheme is to confine codewords to linear combinations of some predefined bases. We call this a linear code. In the discussion we have had so far, the input message is processed one block at a time. This is known as a block code. Another main type of code, convolutional codes, is described in Section B.2.

Encoding can be precisely defined by the generator matrix of the codes, as defined below.

Definition B.1 *(Generator matrix).* For a linear code, given a length-k message $\mathbf{m}$, the codeword output by the encoder is $G^T\mathbf{m}$, where G, a $k \times n$ matrix, is known to be the *generator matrix*.

Apparently, the set of all codewords forms a *linear* vector space (i.e., the column space of G) inside the n-dimensional codeword space. Hence, there exists a matrix H such that the right null space of H is equivalent to the set of all codewords. In other words, for any codeword $\mathbf{x}$, we have $H\mathbf{x} = \mathbf{0}$. Such a matrix H is commonly known as the parity check matrix of the code.

Definition B.2 *(Parity check matrix).* A code vector $\mathbf{x}$ is a codeword if and only if $H\mathbf{x} = \mathbf{0}$.

Linearity is obvious since the set of codewords is nothing but a linear vector space. In particular, the sum of two codewords has to be a codeword as well. This is true because $H(\mathbf{x} + \mathbf{y}) = H\mathbf{x} + H\mathbf{y} = \mathbf{0}$ if both $\mathbf{x}$ and $\mathbf{y}$ are codewords.

Note that G has to be full rank since otherwise the codeword set will have fewer than 2^k vectors and will not be able to accommodate all messages. Since the right nullspace of H is the codeword set, the minimum row dimension of H is $n - k$. Without loss of generality, we will let H have dimension $(n - k) \times n$ and thus H will be full rank as well. Moreover, for any message $\mathbf{m}$, we have $G^T\mathbf{m}$ being a codeword. Thus, $HG^T\mathbf{m} = \mathbf{0}, \forall\mathbf{m}$, therefore $HG^T = \mathbf{0}$.

While G apparently defines a unique block code, H does not. For example, we can take an invertible $k \times k$ matrix T and form a new generator matrix $G' = TG$ and still have the same parity check matrix as before since $HG'^T = HG^TT^T = 0$. Basically, we simply shuffle the input message before encoding. On the other hand, the distribution of the codewords are completely determined by H. Therefore, when messages are symmetrical in the sense that we do not need to provide better protection for some messages than others (which is true for most cases), it is sufficient to design a code based on H alone.

Generally, the k-bit message may or may not appear in the actual n-bit codeword. If the message does appear in the first k bits of the codeword, G will have the form $\begin{bmatrix} I \\ P \end{bmatrix}$ and essentially we just add $n - k$ redundant bits to the message to generate the coded message. These redundant bits are known as *parity* bits. Moreover, such a coding scheme is known to be *systematic*. Systematic coding is convenient since after correcting the channel error at the decoder, we can read out the message directly from the decoded codeword. For *nonsystematic* coding we have an extra step to translate the corrected codeword to the actual message.

B.1.1   Syndrome Decoding of Block Codes

As mentioned earlier, the principle of channel decoding is very simple: to find the best matched codeword with the received decoder input. Typically, the noise from the channel is relatively small and the received code vector is usually very similar to the original one. Therefore, this reduces to finding the codeword that is "closest" to the received code vector. For binary codes, the distance between two code vectors can be measured by the Hamming distance.

Definition B.3   *(Hamming distance and Hamming weight).*   The *Hamming distance* between two equal-length binary sequences is the number of bits by which they differ. The Hamming distance between a binary vector and the all-zero vector of the same length is called the *Hamming weight*.

Finding the closest codeword among 2^k codewords every time can be computationally infeasible in practice. A simple solution is to leverage space with computation time and precompute and store the closest codewords for all possible 2^n input vectors. It turns out that we can shrink this table significantly. Denote $\mathbf{y} = \mathbf{x} + \mathbf{e}$ as the received code vector, where $\mathbf{x}$ is the coded message and $\mathbf{e}$ is the error vector introduced by the noisy channel. At the decoder, we can compute the *syndrome* of $\mathbf{y}$ as $\mathbf{s} = H\mathbf{y}$. Note that

$$\mathbf{s} = H\mathbf{y} = H\mathbf{x} + H\mathbf{e} = H\mathbf{e}. \tag{B.1}$$

Thus, instead of building a table that stores the closest $\mathbf{x}$ for all $\mathbf{y}$, we can construct a table that only stores the error vector $\mathbf{e}$ with smallest Hamming weight (thus most likely to happen) for all syndromes s. With the error vector $\mathbf{e}$ known, we can get back $\mathbf{x}$ as $y - e$. Note that there is significant space saving since there are only 2^{n-k} syndromes in contrast to 2^n total code vectors.

B.1.2   Hamming Codes, Packing Bound, and Perfect Codes

Hamming codes are one of the simplest classes of block codes. For any integer $m \geq 3$, we can construct a $(2^m - 1, 2^m - 1 - m)$ Hamming code as follows.

Definition B.4   *(Hamming code).*   The parity check matrix H of a Hamming code includes all length-m vectors except the all-zero vector as column vectors.

The exact order of column vectors is not very important and can vary. For example, for $m = 3$, the parity check matrix is

$$H = \begin{bmatrix} 1010101 \\ 0110011 \\ 0001111 \end{bmatrix} \tag{B.2}$$

and for $m = 4$, the parity check matrix is

$$H = \begin{bmatrix} 101010101010101 \\ 011001100110011 \\ 000111100001111 \\ 000000011111111 \end{bmatrix}. \tag{B.3}$$

In general, H has width $n = 2^m - 1$ and height m. Thus the resulting Hamming code is a $(2^m - 1, 2^m - 1 - m)$ code as mentioned.

By inspection, we observe that the linear combination of fewer than three column vectors of H is nonzero. Therefore, a Hamming code satisfies the following lemma.

Lemma B.1 *(The weight of a Hamming codeword).* The Hamming weight of any codeword of a Hamming code is larger than 2.

Proof. Assume that this is not true and say there exists a weight-2 vector x such that $\mathbf{x} = [\,\overbrace{0 \cdots 0}^{i-1} 1 0 \cdots \underbrace{0 1 0 \cdots 0}_{j-1}]$ and $H\mathbf{x} = \mathbf{0}$. This implies that the sum of the ith and jth columns of H is equal to the zero vector and contradicts our earlier observation. Similarly, a Hamming code cannot have any weight-1 codeword.

Lemma B.1 implies that any two codewords of a Hamming code have the Hamming distance at least 3 bits away. Therefore, if a codeword corrupted by noise is flipped by 1 bit, the corrupted codeword will still be closer to the original codeword than any other codewords (they are at least 3 bits away). Thus, the Hamming code can inherently correct any 1-bit error.

Consider any codeword $\mathbf{x}$ that passes through a noisy channel. If the channel output $\mathbf{y}$ is no more than 1 bit away from $\mathbf{x}$ and the channel output $\mathbf{y}$ is not too far away from $\mathbf{x}$ we will be able to correct the error. Now consider the set of correctable code vectors for each codeword. These sets cannot overlap, as shown in Figure B.1. Therefore, we have the *packing bound* below.

Lemma B.2 *(Packing bound).*

Number of codewords × average cardinality of sets of "correctable"

code vectors ≤ number of length-n code vectors. (B.4)

Figure B.1 Packing of a channel code. Dark circles and light circles represent codeword and noncodeword vectors, respectively. Each cell includes code vectors that can be corrected to the codeword in the cell.

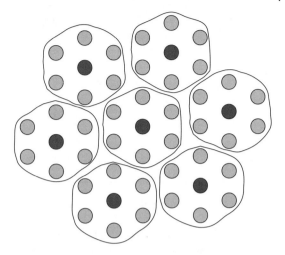

Moreover, if a code satisfies the packing bound with equality, we refer to the code as a *perfect* code.

For a Hamming code, since it can correct all 1-bit errors, the size of the set of correctable code vectors for any codeword is $n + 1 = 2^m$. It can then be verified that both the left-hand side and right-hand side of (B.4) are equal to 2^n. Therefore, a Hamming code satisfies the packing bound with equality and is a perfect code.

B.2   Convolutional Codes

Compared to block codes, convolutional codes are very different in that a message is not coded block by block. Instead, the encoder processes message bitstreams continuously. The name convolutional codes come from the fact that the output bitstreams are essentially the convolution of the input bitstream and the code coefficients.

Consider a very simple convolutional code, as shown in Figure B.2. Note that this code has rate $1/2$ since there are 2 bits out for every bit coming in. The D block in Figure B.2(a) represents a delay buffer. Therefore, the current output x_1 is equal to the sum of the current input bit of m and the bit of m two time steps before. Mathematically, we can write $x_1^{(n)} = m^{(n)} + m^{(n-2)}$, where we use $a^{(n)}$ to represent the bit of a bitstream a at time step (n). Alternatively, we have $\mathbf{x}_1 = \mathbf{m} * [1, 0, 1]$, where $*$ is the convolution operation. Similarly, $x_2^{(n)} = m^{(n)} + m^{(n-1)} + m^{(n-2)}$ and $\mathbf{x}_2 = \mathbf{m} * [1, 1, 1]$. The code is also sometimes known as a $[5, 7]$ rate-$1/2$ convolutional code, since 5 and 7 are the octal representations of the code coefficients 101 and 111, respectively.

While the representation shown in Figure B.2(a) is most commonly used for convolutional codes, the trellis representation shown in Figure B.2(b) is more

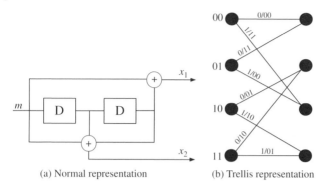

(a) Normal representation   (b) Trellis representation

Figure B.2 Different representations of a convolutional code.

useful in describing the encoding and decoding procedures. On the left-hand side, each trellis node corresponds to the current internal state of the encoder, which is essentially the values of the delay buffers. On the right-hand side, the trellis node corresponds to the internal state after an input bit is processed. The three bits shown along each edge are correspondingly the input bit (before the backslash) and the two output bits (after the backslash).

For example, assuming that the current state is 01 (the values in the first and second delay buffers are respectively 0 and 1) and the input bit is 0, the encoder should make a transition through the upper branch (0/11), as shown in Figure B.2(b). Thus, the output bits will be 11. Moreover, since the branch is connected to state 00 to the right, the 00 will be the state after encoding 0. Similarly, if the next incoming bit is 1, the encoder will transit from state 00 to state 10 and 11 will be output. Encoding will continue until all input bits are exhausted. To improve decoding accuracy, it is beneficial to have the encoder stop at state 00. This can be done easily by adding extra zeros to the end of the input bitstream.

Example B.1 *(Encoding of the convolutional code).* Encoding of bit-stream 0, 1, 1, 1, 0, 0 using the [5, 7] rate-1/2 convolutional code is shown in Figure B.3. As bitstream is read, and the state of the encoder transits from one state to another. We can read from the figure that the states are $00, 00, 10, 11, 11, 01, 00$ as time changes. Moreover, the output bits are 00, 11, 10, 01, 10, 11, respectively.

B.2.1 Viterbi Decoding Algorithm

The decoder will be able to recover the precise message if it can find the path that the encoder has passed through in the trellis diagram. Due to channel noise, this may not always be possible. Instead, the best we can do is to find the most probable path given the received coded bits. This corresponds to the

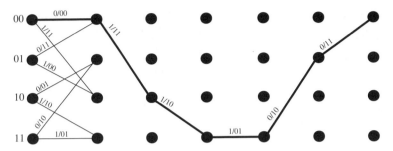

Figure B.3 Encoding of convolutional code. The input is [0, 1, 1, 1, 0, 0] and the encoded message is [00, 11, 10, 01, 10, 11].

maximum *a posteriori* (MAP) decoder. In practice, if the input is uniformly distributed and thus all paths have the same prior probability, the MAP decoder is equivalent to maximizing the probability of the received coded bits given a particular path (i.e., the likelihood of the path) over all allowed paths. This is known to be the maximum likelihood decoder.

At time step n, let $x_1^{(n)}$ and $x_2^{(n)}$ be the coded bits and $y_1^{(n)}$ and $y_2^{(n)}$ be the received bits. The decoder tries to find the path that maximizes the likelihood, where the likelihood of a path is:

$$Likelihood(path)$$
$$=p(y_1^{(1)},y_2^{(1)},y_1^{(2)},y_2^{(2)},\cdots,y_1^{(n)},y_2^{(n)},\cdots$$
$$|x_1^{(1)},x_2^{(1)},x_1^{(2)},x_2^{(2)},\cdots,x_1^{(n)},x_2^{(n)},\cdots)$$
$$\overset{(a)}{=}\prod_n p(y_1^{(n)},y_2^{(n)}|x_1^{(1)},x_2^{(1)},x_1^{(2)},x_2^{(2)},\cdots,x_1^{(n)},x_2^{(n)},\cdots)$$
$$\overset{(b)}{=}\prod_n p(y_1^{(n)},y_2^{(n)}|x_1^{(n)},x_2^{(n)}),$$

where (a) is due to the Markov chain $(y_1^{(n_1)},y_2^{(n_1)}) \leftrightarrow (x_1^{(1)},x_2^{(1)},x_1^{(2)},x_2^{(2)},\cdots,x_1^{(n)},x_2^{(n)},\cdots) \leftrightarrow (y_1^{(n_2)},y_2^{(n_2)})$ for any time instances n_1, n_2, and (b) is due to the Markov chain $(y_1^{(n)},y_2^{(n)}) \leftrightarrow (x_1^{(n)},x_2^{(n)}) \leftrightarrow (x_1^{(1)},x_2^{(1)},x_1^{(2)},x_2^{(2)},\cdots,x_1^{(n-1)},x_2^{(n-1)},x_1^{(n+1)},x_2^{(n+1)},\cdots)$. Taking the log of the likelihood function, we have

$$\log Likelihood(path) = \sum_n \log p(y_1^{(n)},y_2^{(n)}|x_1^{(n)},x_2^{(n)}).$$

Therefore, maximizing the overall likelihood is equivalent to minimizing the sum of negative log-likelihoods, $-\log p(y_1^{(n)},y_2^{(n)}|x_1^{(n)},x_2^{(n)})$, over all time steps. Thus, if we assign each edge in the trellis diagram with a "length" $\log p(y_1^{(n)},y_2^{(n)}|x_1^{(n)},x_2^{(n)})$, the maximum likelihood path is simply the shortest distance path from the 00 state at beginning time step to the 00 state at the ending time step.

Example B.2 *(Likelihood assignment for the binary symmetric channel).* Let us assume that the coded bits have passed through a binary symmetric channel with crossover probability p. For a particular time step, let y_1 and y_2 be the received bits, and x_1 and x_2 be the anticipated output bits if the encoder makes a transition along an edge. The likelihood is given by

$$p(y_1, y_2 | x_1, x_2) = \begin{cases} (1-p)^2, & \text{if } y_1 = x_1 \text{and } y_2 = x_2, \\ p^2, & \text{if } y_1 \neq x_1 \text{and } y_2 \neq x_2, \\ p(1-p), & \text{otherwise.} \end{cases}$$

Moreover, the "length" of the corresponding edge should be set to $-\log p(y_1, y_2 | x_1, x_2)$.

If the input bitstream has length 1000, the total number of possible paths will be 8^{1000} for [5, 7] convolutional code. Such a huge number makes an exhaustive search of the shortest path impossible. The Viterbi algorithm, which is a special case of dynamic programming, allows us to find the shortest path very efficiently. The main idea is based on a very simple observation, as shown in Figure B.4: if the shortest path passes through an intermediate state s, the sub-path from the starting node (00 at time 0) to state s has also to be the shortest path. Therefore, at each time step we only need to store the shortest path to the states at the current time step. For the case of [5, 7] convolutional code, the decoder only needs to store the four shortest paths, each of which connects one of the four states. After all the received bits have been processed, the shortest path and thus the message can be found by backtracking.

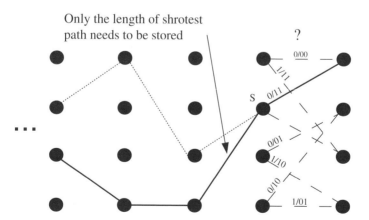

Figure B.4 The main idea of Viterbi decoding. The thick black line is the shortest path, which passes through an intermediate state s. The sub-path from state s to the starting node also needs to be a shortest path. Therefore, only the shortest path to the current state s needs to be stored and the rest can be safely discarded.

At time step n (after n tuples of output bits are processed), let $L^{(n)}(s)$ be the length of the shortest path to state s and $l^{(n)}(s_1, s_2)$ be the negative log-likelihood of receiving the output bits given the edge transition from s_1 to s_2 if such a transition is valid, otherwise set $l^{(n)}(s_1, s_2) = \infty$, that is,

$$l^{(n)}(s_1, s_2) = \begin{cases} p(\mathbf{y}|\mathbf{x}(s_1, s_2)), & \text{if transition } s_1 \text{to } s_2 \text{ is valid,} \\ \infty, & \text{otherwise.} \end{cases} \tag{B.5}$$

The Viterbi algorithm is summarized in Algorithm B.1. Starting from the first time step, the algorithm iteratively finds the shortest path to each state of every time step using dynamic programming. The result is then extracted through backtracking and stored in SHORTEST_PATH_STATE. Given the shortest path, the original message can be readily read out, as in Figure B.3.

Algorithm B.1: Viterbi decoding algorithm

- **Initialize**: Set $n = 0$ and

$$L^{(0)}(s) = \begin{cases} 0, & s = 00, \\ \infty, & \text{otherwise.} \end{cases}$$

- **Shortest path update**: Increment n, set

$$L^{(n)}(s) = \min_{s'} L^{(n-1)}(s')l^{(n)}(s', s)$$

and

$$\text{BACKTRACKING}^{(n)}(s) = \arg\min_{s'} L^{(n-1)}(s')l^{(n)}(s', s),$$

where $l^{(n)}(s', s)$ is computed according to (B.5).
Repeat this step until all input bits are exhausted.
- **Backtracking**: Trace back the shortest path using the variable BACKTRACKING as follows.
  a) Set SHORTEST_PATH_STATE(n) = 00 and $s = 00$.
  b) Set SHORTEST_PATH_STATE(n) = BACKTRACKING$^{(n)}(s)$;
      set s = SHORTEST_PATH_STATE(n); decrement n.
  c) Repeat (2) until n reaches 0.

B.3 Shannon's Channel Coding Theorem

Before Shannon's seminal paper [147], most people thought that it was impossible to transmit information over a noisy channel without error. It was thought that the rate of error would increase as the noise of the channel increased.

It turns out that as long as the transmission rate is below the capacity of the channel, error-free transmission is possible through the use of channel code.

In this chapter we will confine ourselves to the so-called discrete memoryless channel (DMC), as defined below.

Definition B.5 *(Discrete memoryless channel).* A channel with discrete input X and a discrete random output Y is described by conditional distribution $p(y|x)$. The channel is memoryless, meaning there is consecutive use of the channel with inputs $x_1, x_2, x_3, \cdots, x_n$. The probability of obtaining output $y_1, y_2, y_3, \cdots, y_n = p(y_1, \cdots, y_n | x_1, \cdots, x_n) = p(y_1|x_1)p(y_2|x_2)\cdots p(y_n|x_n)$, that is, consecutive outputs are conditionally independent given the inputs.

Although we already mentioned channel code and code rate (or transmission rate) previously, let us define them more precisely as follows.

Definition B.6 *(Channel code and code rate).* A *channel code* can be described by a mapping function from a message to a codeword for transmission. We will always assume codewords with equal length here, that is, for a random message $M \in \{1, 2, \cdots, \mathcal{M}\}$, a realization m will be coded to the length-N codeword $c^N(m) = c_1(m)c_2(m)\cdots c_N(m)$ (note that $c_i(m) \in \mathcal{X}$) for transmission.

Since $H(M)$ bits of information are transmitted per N channel use, the *code rate* (or *transmission rate*) is $\frac{1}{N}H(M)$. When M is uniform (it is usually the case), the code rate is equal to $\frac{1}{N}\log_2 \mathcal{M}$.

Theorem B.1 *(Channel Coding Theorem).* For every channel, we have a value C known to be the capacity $(C = \max_{p(x)} I(X; Y))$, where we can on average send up to C bits of information losslessly through the channel per each channel use.

More precisely, the Channel Coding Theorem states that for any code transmitting at a rate higher than the capacity, the transmission error probability will be strictly larger than 0. On the other hand, for any rate less than the capacity, there exists a code at the requested rate and with no transmission error.[1]

Example B.3 *(Binary symmetric channel).* Consider a binary symmetric channel with input X and output Y. Let the crossover probability be p (i.e., $p_{Y|X}(1|0) = p_{Y|X}(0|1) = p$ and $p_{Y|X}(0|0) = p_{Y|X}(1|1) = 1 - p$). Then,

$$\text{Capacity} = \max_{p(x)} I(X; Y) \tag{B.6}$$

---

1 Note that if we fix the statistics of the input to the channel to $p(x)$, the maximum rate of information that can pass through the channel is simply $I(X; Y)$ (the mutual information between X and Y).

$$= \max_{p(x)} H(Y) - H(Y|X) \tag{B.7}$$

$$= \max_{p(x)} H(Y) - H(p) = 1 - H(p), \tag{B.8}$$

where $H(Y)$ is maximized when $Pr(Y = 0) = Pr(Y = 1) = 0.5$. This happens if the input is also uniform ($Pr(X = 0) = Pr(X = 1) = 0.5$).

When p is 0, the capacity is 1 bit because we can pass 1 bit losslessly per channel use. Note that the capacity is 1 bit also when the transmission error probability is 1. This is reasonable since the decoder can simply flip all incoming bit from 1 to 0 and vice versa.

When $p = 0.5$, the input and output are actually independent. No information can pass through the channel.

B.3.1 Achievability Proof of the Channel Coding Theorem

If the code rate $r < C = \max_{p(x)} I(X; Y)$, according to the Channel Coding Theorem, we should be able to find a code with encoding mapping $c : m \in \{1, 2, \cdots, 2^{nr}\} \rightarrow \{0, 1\}^n$ and the error probability of transmitting any message $m \in \{1, 2, \cdots, 2^{nr}\}$, $p_e(m)$, is arbitrarily small.

The main tool of the proof is random coding. Let $p^*(x)$ be the distribution that maximizes $I(X; Y)$ that is,

$$p^*(x) = \arg \max_{p(x)} I(X; Y) \tag{B.9}$$

Generate codewords from the discrete memoryless source (DMS) $p^*(x)$ by sampling 2^n length-n sequences from the source. This denotes the resulting codewords as

$$\mathbf{c}(1) = (x_1(1), x_2(1), \cdots, x_n(1))$$
$$\mathbf{c}(2) = (x_1(2), x_2(2), \cdots, x_n(2))$$
$$\cdots$$
$$\mathbf{c}(2^n) = (x_1(2^n), x_2(2^n), \cdots, x_n(2^n))$$

The encoding and decoding procedures will be as follows.

Encoding: For input message m, output $\mathbf{c}(m) = (x_1(m), x_2(m), \cdots, x_n(m))$.
Decoding: Upon receiving sequence $\mathbf{y} = (y_1, y_2, \cdots, y_n)$, pick the sequence $\mathbf{c}(m)$ from $\{\mathbf{c}(1), \cdots, \mathbf{c}(2^{nr})\}$ such that $(\mathbf{c}(m), \mathbf{y})$ are jointly typical, that is $p_{X^n, Y^n}(\mathbf{c}(m), \mathbf{y}) \sim 2^{-nH(X,Y)}$. If no such $\mathbf{c}(m)$ exists or more than one such sequence exists, announce an error. Otherwise output the decoded message as m.

Instead of studying the performance of a particular code instance, let us look into the average performance over all codes randomly generated from $p^*(x)$.

Without loss of generality, let us assume $m = 1.A$. A decoding error occurs when (1) $P_1 = (\mathbf{c}(1), \mathbf{y}) \notin A_\epsilon^n(X, Y)$ and (2) $P_2 : \exists m' \neq 1$ and $(\mathbf{c}(m'), \mathbf{y}) \in A_\epsilon^n(X, Y)$. Thus $P(error) = P(error|m = 1) \leq P_1 + P_2$.

a) Since $(\mathbf{c}(1), \mathbf{y})$ is coming out of the joint source X, Y, the probability that they are not jointly typical goes to 0 as $n \to \infty$ for any finite n. $P_1 < \delta(n)$ and $\delta(n) \to 0$ as $n \to \infty$.

b) Note that $\mathbf{c}(m')$ and $\mathbf{y}$ are independent and for any other $m' \neq 1$,

$$p((\mathbf{c}(m'), \mathbf{y}) \in A_\epsilon^n(X, Y)) = \sum_{\substack{(x^n, y^n) \in A_\epsilon^n(X,Y); \\ x^n \text{ and } y^n \text{ are independent}}} p(x^n, y^n) \tag{B.10}$$

$$= \sum_{(x^n, y^n) \in A_\epsilon^n(X,Y)} p(x^n)p(y^n) \tag{B.11}$$

$$\leq 2^{n(H(X,Y)+\epsilon)} 2^{-n(H(X)-\epsilon)} 2^{-n(H(Y)-\epsilon)} \tag{B.12}$$

$$\leq 2^{-n(I(X;Y)+3\epsilon)}. \tag{B.13}$$

Since we have $2^{nr} - 1$ different m', $P_2 \leq 2^{nr} 2^{n(I(X;Y)-3\epsilon)} = 2^{n(r-I(X;Y)+3\epsilon)}$. So as long as $r < I(X; Y), P(error) \leq P_1 + P_2 \leq \delta + \gamma$, where $\gamma = 2^{-n(I(X;Y)-3\epsilon-r)} \to 0$ as $n \to \infty$.

Note that this analysis only addresses average performance over all codes. More precisely, we should have written,

$$\frac{1}{|C|} \sum_{c' \in C} P(error|\mathbf{c}') \leq \delta + \gamma \to 0 \text{ as } n \to \infty, \tag{B.14}$$

where C is the set of all random codes generated out of $p^*(x)$.

Note that there must be one code with performance better or equal to the average (otherwise, the average would not happen!), that is, $\exists \mathbf{c}^* \in C$ s.t.

$$Pr(error|\mathbf{c}^*) \leq \frac{1}{|C|} \sum_{c' \in C} Pr(error|\mathbf{c}') \leq \delta + \gamma. \tag{B.15}$$

Here we only claimed that the average error is bounded, but we want to say something even stronger: for any message transmitted, $p(error)$ can be arbitrarily small, t. i.e., That is,

$$Pr(error|\mathbf{c}^*, m) \to 0 \text{ as } n \to \infty \text{ for any } m.$$

Note that $Pr(error|\mathbf{c}^*) = \frac{1}{2^{nr}} \sum_m Pr(error|\mathbf{c}^*, m)$ (assuming that messages are equally likely) and if we discard the worse half of the messages with higher error probabilities, the remaining messages should have error probability at most double the average, that is, for $m \in$ set of the better half of the messages, $Pr(error|\mathbf{c}^*, m) \leq 2Pr(error|\mathbf{c}^*) \leq 2(\gamma + \delta) \to 0$ as $n \to \infty$.

Note that the rate now reduces from r to $r - \frac{1}{n}$ (number of messages from $2^{nr} \to 2^{nr-1}$), but we can still make the final rate arbitrarily close to the capacity as $n \to \infty$.

B.3.2 Converse Proof of Channel Coding Theorem

For the converse, we want to show that if the transmission rate is smaller than the capacity $(\max_{p(x)} I(X; Y))$, then the error probability is bounded away from 0.

To finish the proof, we need to use a result from Fano.

Lemma B.3 *(Fano's inequality).* Let M be an input message, and X^n, Y^n, $\hat{M}$ are respectively the corresponding codeword, the received codeword at the decoder, and the decoded message. Let $P_e = \Pr(M \neq \hat{M})$, then $H(M|Y^n) \leq 1 + P_e H(M)$. Intuitively, if $P_e \to 0$, on average we know M for certain given y and thus $\frac{1}{n}H(M|Y^n) \to 0$.

Proof. Let E be an indicator variable such that

$$E = \begin{cases} 0, & \text{if } M = \hat{M}, \\ 1, & \text{otherwise.} \end{cases} \tag{B.16}$$

Then $H(M, E|Y^n) = H(M|Y^n) + H(E|Y^n, M) = H(E|Y^n) + H(M|Y^n, E)$. However, $H(E|Y^n, M) = 0$ since E is deterministic given both input and output. Therefore,

$$H(M|Y^n) = H(E|Y^n) + H(M|Y^n, E) \tag{B.17}$$

$$\overset{(a)}{\leq} H(E) + H(M|Y^n, E) \tag{B.18}$$

$$\overset{(b)}{\leq} 1 + P(E = 0)H(M|Y^n, E = 0) + P(E = 1)H(M|Y^n, E = 1) \tag{B.19}$$

$$\overset{(c)}{\leq} 1 + 0 + P_E H(M|Y^n, E = 1) \overset{(d)}{\leq} 1 + P_e H(M), \tag{B.20}$$

where (a) is because conditioning reduces entropy, (b) is from the fact that E is binary, (c) is because M is deterministic given Y^n and when there is no error $(E = 0)$, and (d) is again due to the fact that conditioning reduces entropy.

Now we are ready to prove the converse,

$$R = \frac{H(M)}{n} = \frac{1}{n}[I(M; Y^n) + H(M|Y^n)] \tag{B.21}$$

$$\overset{(a)}{\leq} \frac{1}{n}[I(X^n; Y^n) + H(M|Y^n)] \tag{B.22}$$

$$= \frac{1}{n}[H(Y^n) - H(Y^n|X^n) + H(M|Y^n)] \tag{B.23}$$

$$= \frac{1}{n}\left[H(Y^n) - \sum_i H(Y_i|X^n, Y^{i-1}) + H(M|Y^n)\right] \tag{B.24}$$

$$\overset{(b)}{=} \frac{1}{n}\left[H(Y^n) - \sum_i H(Y_i|X_i) + H(M|Y^n)\right] \tag{B.25}$$

$$\leq \frac{1}{n}\left[\sum_i H(Y_i) - \sum_i H(Y_i|X_i) + H(M|Y^n)\right] \quad (B.26)$$

$$= \frac{1}{n}\left[\sum_i I(X_i;Y_i) + H(M|Y^n)\right] \quad (B.27)$$

$$= I(X;Y) + \frac{H(M|Y^n)}{n}. \quad (B.28)$$

Now, by Fano's inequality,

$$R \leq I(X;Y) + \frac{1}{n}[1 + P_e H(M)] \quad (B.29)$$

$$\leq I(X;Y) + \frac{1}{n} + P_e R. \quad (B.30)$$

Thus, as $n \to \infty$, $R \leq I(X;Y) \leq C$ if $P_e \to 0$.

B.4 Low-density Parity-check Codes

While one can construct channel codes rather arbitrarily by specifying generator and parity check matrices, whether it is possible to decode the code is another story. Before 1990s, the strategy for channel code was always to look for codes that could be decoded optimally. This led to a wide range of so-called algebraic codes. It turns out the "optimally-decodable" codes are usually poor codes. Until then, researchers had basically agreed that the Shannon capacity was restricted to theoretical interest and could not be reached in practice.

The introduction of turbo codes was a huge shock to the research community. The community were so dubious about the amazing performance of turbo codes that they did not accept the finding initially until independent researchers had verified the results. Low-density parity-check (LDPC) codes were later rediscovered and both LDPC codes and turbo codes are based on the same philosophy, which differs from that of codes in the past. Instead of designing and using codes that can be decoded "optimally", let us just pick some *random* codes and perform decoding "sub-optimally". As the sub-optimal decoding performs reasonably well, we expect that the coding scheme will also probably work quite well since the proof of the Channel Coding Theorem suggests that random codes that tend to achieve the Shannon capacity are very good.

B.4.1 A Quick Summary of LDPC Codes

As the name suggests, LDPC codes refer to codes that with sparse (low-density) parity check matrices. In other words, there are only a few in a parity check

Figure B.5 The Tanner graph of the (7,4) Hamming code described in (B.2).

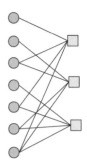

matrix and the rest are all zeros. The necessity of sparsity is apparent as we discuss the decoding procedure in the following subsections.

We learn from the proof of the Channel Coding Theorem that a random code is asymptotically optimum. This suggests that if we just use a code with a randomly generated parity-check matrix and increase the code length it is likely that we will get a very good code. However, the problem is how do we perform decoding? Note that syndrome coding is not usually feasible in practice. Consider rate-$1/2$ LDPC codes with code length $10,000$, which is not really a very long code, but there will still be $2^{5000} \approx 10^{1667}$ different syndromes! Moreover, due to the lack of structure in a random code, tricks that enable fast decoding for structured algebraic codes that were widely used before the 1990s are unrealizable here.

The solution is (loopy) BP or the sum-product algorithm, an inference technique to compute a marginal probability of joint random variables that was used mainly within the artificial intelligence community. We will describe BP in a general setting in the next subsection. Before that, we would like to introduce a graph formulation of a block code that facilitates our discussion later on.

A *Tanner graph* of a block code with a parity-check matrix $H_{m \times n}$ is a bipartite graph containing a set of n variable nodes and a set of m check nodes. The ith variable node is connected to the jth check node if and only if $H_{j,i} = 1$. The Tanner graph of the $(7, 4)$ Hamming code described by (B.2) is shown in Figure B.5.

The number of edges connecting a node is known to be the degree of a node, for example the degree of the first variable node is one and that of the first check node is three. If the degrees of all variable nodes are the same and so are the degrees of all check nodes, we call the code a *regular* code. Otherwise, the code is known as *irregular*.

B.4.2   Belief Propagation Algorithm

The goal of BP (or sum-product algorithm) is to efficiently compute the marginal distribution out of the joint distribution of multiple variables. This is essential for inferring the outcome of a particular variable with insufficient information.

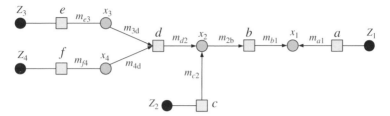

Figure B.6 The factor graph of the inference example and the direction of messages passed by the BP algorithm.

The BP algorithm is usually applied to problems modeled by a undirected graph or a factor graph. The factor graph model is generally more flexible and is preferred when there is less structure in the graph, therefore we will only confine ourselves to the formulation based on the factor graph. Rather than giving a rigorous proof of the algorithm, we will provide a simple example to illustrate the basic idea.

A factor graph is a bipartite graph describing the correlation among several random variables. It generally contains two different types of nodes: variable nodes and factor nodes. A variable node is usually shown as a circle and corresponds to a random variable. A factor node is usually shown as a square and connects variable nodes whose corresponding variables are immediately related.

A factor graph example is shown in Figure B.6. We have eight *discrete* random variables, x_1^4 and z_1^4, depicted by eight variable nodes. Among the variable nodes, random variables x_1^4 (indicated by light circles) are unknown and variables z_1^4 (indicated by dark circles) are observed with known outcomes $\tilde{z}_1^4$. The relationships among variables are captured entirely by the figure. For example, given x_1^4, z_1, z_2, z_3, and z_4 are conditionally independent of each other. Moreover, (x_3, x_4) are conditionally independent of x_1 given x_2. Thus, the joint probability $p(x^4, z^4)$ of all variables can be decomposed into factor functions with subsets of all variables as arguments in the following:

$$p(x^4, z^4) = p(x^4)p(z_1|x_1)p(z_2|x_2)p(z_3|x_3)p(z_4|x_4) \qquad (B.31)$$

$$= \underbrace{p(x_1, x_2)}_{f_b(x_1,x_2)} \underbrace{p(x_3, x_4|x_2)}_{f_d(x_2,x_3,x_4)} \underbrace{p(z_3|x_3)}_{f_e(x_3,z_3)} \underbrace{p(z_1|x_1)}_{f_a(x_1,z_1)} \underbrace{p(z_4|x_4)}_{f_f(x_4,z_4)} \underbrace{p(z_2|x_2)}_{f_c(x_2,z_2)} \qquad (B.32)$$

$$= f_b(x_1, x_2)f_d(x_2, x_3, x_4)f_e(x_3, z_3)f_a(x_1, z_1)f_f(x_4, z_4)f_c(x_2, z_2). \qquad (B.33)$$

Note that each factor function corresponds to a factor node in the factor graph. Moreover, the arguments of the factor function correspond to the variable nodes that the factor node connects to.

One common problem in probability inference is to estimate the value of a variable given incomplete information. For example, we may want to estimate x_1 given z^4 as $\tilde{z}^4$. The optimum estimate $\hat{x}_1$ will satisfy

$$\hat{x}_1 = \arg\max_{x_1} p(x_1|\tilde{z}^4) \tag{B.34}$$

$$= \arg\max_{x_1} \frac{p(x_1,\tilde{z}^4)}{p(\tilde{z}^4)} \tag{B.35}$$

$$= \arg\max_{x_1} p(x_1,\tilde{z}^4). \tag{B.36}$$

This requires us to compute the marginal distribution $p(x_1,\tilde{z}^4)$ out of the joint probability $p(x^4,\tilde{z}^4)$. Note that

$$p(x_1,\tilde{z}^4) = \sum_{x_2^4} p(x^4,\tilde{z}^4) \tag{B.37}$$

$$= \sum_{x_2^4} f_a(x_1,\tilde{z}_1)f_b(x_1,x_2)f_c(x_2,\tilde{z}_2)f_d(x_2,x_3,x_4)f_e(x_3,\tilde{z}_3)f_f(x_4,\tilde{z}_4) \tag{B.38}$$

$$= \underbrace{f_a(x_1,\tilde{z}_1)}_{m_{a1}} \underbrace{\sum_{x_2} f_b(x_1,x_2)}_{m_{2b}} \underbrace{f_c(x_2,\tilde{z}_2)}_{m_{c2}} \underbrace{\sum_{x_3,x_4} f_d(x_2,x_3,x_4)}_{m_{d2}} \underbrace{f_e(x_3,\tilde{z}_3)}_{m_{3d}} \underbrace{f_f(x_4,\tilde{z}_4)}_{m_{4d}} \tag{B.39}$$

We can see from (B.39) that the joint probability can be computed by combining a sequence of messages passing from a variable node i to a factor node a (m_{ia}) and vice versa (m_{ai}).[2] More precisely, we can write

$$m_{a1}(x_1) \leftarrow f_a(x_1,\tilde{z}_1) = \sum_{z_1} f_a(x_1,z_1)\, \underbrace{p(z_1)}_{m_{1a}}\,, \tag{B.40}$$

$$m_{c2}(x_2) \leftarrow f_c(x_2,\tilde{z}_2) = \sum_{z_2} f_c(x_2,z_2)\, \underbrace{p(z_2)}_{m_{2c}}\,, \tag{B.41}$$

$$m_{e3}(x_3) \leftarrow f_e(x_3,\tilde{z}_3) = \sum_{z_3} f_e(x_3,z_3)\, \underbrace{p(z_3)}_{m_{3e}}\,, \tag{B.42}$$

$$m_{f4}(x_4) \leftarrow f_f(x_4,\tilde{z}_4) = \sum_{z_4} f_f(x_4,z_4)\, \underbrace{p(z_4)}_{m_{4f}}\,, \tag{B.43}$$

2 We use a shorthand notation m_{ia} to represent a message from node i to node a. When potential conflict in notation may occur, we will write the message more explicitly as $m_{x_i \to a}(x_i)$.

$$m_{3d}(x_3) \leftarrow m_{e3}(x_3), \tag{B.44}$$

$$m_{4d}(x_4) \leftarrow m_{f4}(x_4), \tag{B.45}$$

$$m_{d2}(x_2) \leftarrow \sum_{x_3, x_4} f_d(x_2, x_3, x_4) m_{3d}(x_3) m_{4d}(x_4), \tag{B.46}$$

$$m_{2b}(x_2) \leftarrow m_{c2}(x_2) m_{d2}(x_2), \tag{B.47}$$

$$m_{b1}(x_1) \leftarrow \sum_{x_2} f_b(x_1, x_2) m_{2b}(x_2), \tag{B.48}$$

$$p(x_1, \tilde{z}^4) \leftarrow m_{a1}(x_1) m_{b1}(x_1), \tag{B.49}$$

where we intentionally use arrow signs instead of equality signs because the operations are indeed assignments and should be computed sequentially.

By inspection, we can see that there are very regular update patterns in generating the outgoing messages. Moreover, the update rules are different for variable and factor nodes. From (B.47), we see that the message going out of a variable node is equal to the product of the incoming messages from nodes, excluding the output node. More precisely, for the message sending out from a variable node i to a factor node a,

$$m_{ia}(x_i) \leftarrow \prod_{b \in N(i) \backslash a} m_{bi}(x_i), \tag{B.50}$$

where $N(i)$ denotes the set of neighboring nodes of i. Note that this update equation is also verified in (B.44) and (B.45) as there is only one incoming message for each of the variable nodes and thus the outgoing message is equivalent to the incoming one.

The factor node update rule is a little bit more complicated. We can see from (B.46) and (B.48) that the outgoing message from a factor node is the sum of products of the factor function and the incoming messages from nodes excluding the output node. This is why the BP algorithm is also known as the sum-product algorithm. Mathematically, a message sending from a factor node a to a variable node i is given by

$$m_{ai}(x_i) \leftarrow \sum_{\mathbf{x}_a \backslash x_i} f_a(\mathbf{x}_a) \prod_{j \in N(a) \backslash i} m_{ja}(x_j), \tag{B.51}$$

where $\mathbf{x}_a$ denotes the set of the arguments of the factor node a. This updating rule can also be validated from (B.40) if we interpret $f_a(x_1, \tilde{z}_1) = \sum_{z_1} f_c(x_1, z_1) m_{1a}$, where m_{1a} is the message that is initialized as the prior probability $p(z_1) = \delta(z_1 - \tilde{z}_1)$. Similar observations can be made for (B.41)–(B.43).

(B.40)–(B.43) also suggest the rule in initializing the messages. For a variable node i that we have some known prior probability for regarding its outcome, we can set any initial outgoing message from node i equal to this prior probability. Finally, (B.49) suggests that the final belief of a variable is just equal to the product of all its incoming messages. Thus, for a variable node i, its final belief is given by

$$\beta_i(x_i) \leftarrow \prod_{a \in N(i)} m_{ai}(x_i) \tag{B.52}$$

Throughout our discussion of "deriving" the BP algorithm, we do not assume the precise physical meanings of the factor functions themselves. The only assumption we make is that the joint probability can be decomposed into the factor functions and apparently this decomposition is not unique.

Algorithm B.2: Summary of the BP algorithm

• **Initialization**: For any variable node i, if the prior probability of x_i is known and equal to $p(x_i)$, for $a \in N(i)$,

$$m_{ia}(x_i) \leftarrow p(x_i), \tag{B.53}$$

otherwise

$$m_{ia}(x_i) \leftarrow 1/|\mathcal{X}_i|, \tag{B.54}$$

where $|\mathcal{X}_i|$ is the cardinality of alphabet of x_i.
• **Messages update**:
  – Factor to variable node update:

$$m_{ai}(x_i) \leftarrow \sum_{\mathbf{x}_a} f_a(\mathbf{x}_a) \prod_{j \in N(a)\backslash i} m_{ja}(x_j) \tag{B.55}$$

  – Variable to factor node update:

$$m_{ia}(x_i) \leftarrow \prod_{b \in N(i)\backslash a} m_{bi}(x_i), \tag{B.56}$$

• **Belief update**: Denote the net belief in terms of log-likelihood ratio as $\beta_i(x_i)$ for variable node i, then

$$\beta_i(x_i) \leftarrow \prod_{a \in N(i)} m_{ai}(x_i) \tag{B.57}$$

• **Stopping criteria**: Repeat message update and/or belief update until the algorithm stops when the maximum number of iterations is reached or some other conditions are satisfied.

The BP algorithm as shown above is exact only because the corresponding graph is a tree and has no loop. If a loop exists, the algorithm is not exact and generally the final belief may not even converge. In this general case, factor and variable node updates are supposed to be performed repeatedly and alternatively for all nodes. An exception is that the message from a factor node to an observation node should not be computed since we assume the observation is deterministic in the above formulation. For example, $m_{e \to z_3}(z_3)$ is meaningless and should never be used to overwrite the initial belief of z_3 that is considered to be the ground truth. As a consequence, we can see that messages $m_{a1}(x_1)$, $m_{c2}(x_2)$, $m_{e3}(x_3)$, and $m_{f4}(x_4)$ will never be modified. As a result, the observation variable nodes (z_1^4 in our example) are usually omitted and absorbed into the factor nodes (a, c, e, f) instead. Finally, while the result is no longer exact, applying the BP algorithm for general graphs (sometimes referred to as loopy BP) works well in many applications, such as LDPC decoding. The overall algorithm is summarized in Algorithm B.2.

B.4.3 LDPC Decoding using BP

In this section we will apply the BP algorithm directly to LDPC decoding. Assume that x_1^n is the codeword sent by the encoder and y_1^n is the noisy version received at the decoder after passing though a channel. Thus, each x_i is directly related to y_i through the noisy channel. Since x^n is a codeword, it needs to satisfy the condition that all its parity checks are zero. Thus, a x_i and x_j will be immediately related if they are involved in the same check equation (they are both connected to the same check node in the Tanner graph). The above correlations can be represented by the factor nodes attached to the corresponding variable nodes of x^n and y^n, as shown in Figure B.7. We will call them the channel factor nodes (left) and the factor nodes which describe parity checks as check factor nodes (right). To avoid confusion, let us call variable nodes corresponding to x_i *unknown variable nodes* and those corresponding to y_i *observation variable nodes*.

The channel factor node connecting x_i and y_i will have its factor function defined as

$$f_i(x, y) \triangleq p(y|x). \qquad (B.58)$$

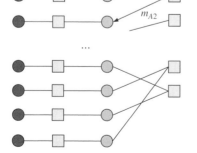

Figure B.7 LDPC decoding. The decoder estimates initial probabilities $p(x_i|y_i)$ for each bit.

On the other hand, since a check node a will constrain the sum of the values of all its neighboring nodes to 0, the corresponding factor function can be written as

$$f_a(\mathbf{x}) = \begin{cases} 0, & \mathbf{x} \text{ contains even number of 1,} \\ 1, & \mathbf{x} \text{ contains odd number of 1.} \end{cases} \tag{B.59}$$

One can easily verify that

$$p(x^n, y^n) = K \prod_i f_i(x_i, y_i) \prod_a f_a(\mathbf{x}_a), \tag{B.60}$$

for some constant K. Note that the constant K will not affect the result of BP as long as we always scale the sum of messages to one, that is, for any variable node i and factor node a, $\sum_{x_i} m_{ia}(x_i) = 1$ and $\sum_{x_i} m_{ai}(x_i) = 1$.

Since we do not have any prior information from any x_i, the initial message to a neighboring node a should bear no information, that is, $m_{ia}(1) = m_{ia}(0) = 0.5$. On the other hand, a message from the node of y_j, m_{ja}, should satisfy

$$m_{ja}(y) = \delta(y - y_j). \tag{B.61}$$

After assigning the initial factor to variable node messages, we can update the factor to variable and variable to factor node messages alternatively as described previously. As discussed in the previous section, messages from observation variable nodes to channel factor nodes and thus messages from channel factor nodes to unknown variable nodes are never updated. In particular, the message from a channel factor node i to a unknown variable node i is $m_{ii}(x) = p(y_i|x)$. Therefore, the actual message updates are confined to those between unknown variable nodes and check factor nodes.

Since the unknown variables are binary, it is more convenient to represent the messages using likelihood or log-likelihood ratios. We can define

$$l_{ai} \triangleq \frac{m_{ai}(0)}{m_{ai}(1)}, \quad L_{ai} \triangleq \log l_{ai} \tag{B.62}$$

and

$$l_{ia} \triangleq \frac{m_{ia}(0)}{m_{ia}(1)}, \quad L_{ia} \triangleq \log l_{ia} \tag{B.63}$$

for any variable node i and factor node a.

Therefore, the variable to check node update can be written as

$$L_{ia} \leftarrow \sum_{b \in N(i) \setminus i} L_{ia}. \tag{B.64}$$

Assuming that we have three variable nodes 1, 2, and 3 connecting to the check node a, then the check to variable node updates become

$$m_{a1}(1) \leftarrow m_{2a}(1)m_{3a}(0) + m_{2a}(0)m_{3a}(1) \tag{B.65}$$

$$m_{a1}(0) \leftarrow m_{2a}(0)m_{3a}(0) + m_{2a}(1)m_{3a}(1) \tag{B.66}$$

Substituting in the likelihood ratios and log-likelihood ratios, we have

$$l_{a1} \triangleq \frac{m_{a1}(0)}{m_{a1}(1)} \leftarrow \frac{1 + l_{2a}l_{3a}}{l_{2a} + l_{3a}} \tag{B.67}$$

and

$$e^{L_{a1}} = l_{a1} \leftarrow \frac{1 + e^{L_{2a}}e^{L_{3a}}}{e^{L_{2a}} + e^{L_{3a}}}. \tag{B.68}$$

Note that

$$\tanh\left(\frac{L_{a1}}{2}\right) = \frac{e^{\frac{L_{a1}}{2}} - e^{-\frac{L_{a1}}{2}}}{e^{\frac{L_{a1}}{2}} + e^{-\frac{L_{a1}}{2}}} = \frac{e^{L_{a1}} - 1}{e^{L_{a1}} + 1} \tag{B.69}$$

$$\leftarrow \frac{1 + e^{L_{2a}}e^{L_{3a}} - e^{L_{2a}} - e^{L_{3a}}}{1 + e^{L_{2a}}e^{L_{3a}} + e^{L_{2a}} + e^{L_{3a}}} \tag{B.70}$$

$$= \frac{(e^{L_{2a}} - 1)(e^{L_{3a}} - 1)}{(e^{L_{2a}} + 1)(e^{L_{3a}} + 1)} \tag{B.71}$$

$$= \tanh\left(\frac{L_{2a}}{2}\right)\tanh\left(\frac{L_{3a}}{2}\right). \tag{B.72}$$

When we have more than three variable nodes connecting to the check node a, it is easy to show using induction that

$$\tanh\left(\frac{L_{ai}}{2}\right) \leftarrow \prod_{j \in N(a) \backslash i} \tanh\left(\frac{L_{ja}}{2}\right). \tag{B.73}$$

Hard thresholding can be applied to the final belief to estimate the value of each bit, that is,

$$\hat{x}_i = \begin{cases} 0, & \text{if } \beta_i \geq 0, \\ 1, & \text{otherwise,} \end{cases} \tag{B.74}$$

where $\beta_i = \sum_{a \in N(i)} L_{ai}$. The algorithms may terminate after some predefined number of iterations. It is also common to terminate the algorithm whenever the estimate variables satisfy all checks. The overall algorithm is summarized in Algorithm B.3.

B.4.4 IRA Codes

LDPC codes are generally not systematic codes (see Section B.1), therefore we need an extra step to further translate the corrected codeword at the decoder to the actual message. Moreover, the generator matrix is generally not sparse. Thus, the encoding complexity can be rather high (of order $n \times m$). The irregular repeat accumulate (IRA) codes are a simple extension of LDPC codes to address this issue.

The Tanner graph for a typical IRA code is shown in Figure B.8. The left-hand column of variable nodes in Figure B.8 corresponds to the input message **x**

Algorithm B.3: Summary of LDPC decoding using BP

- **Initialization:**

$$L_{ii} = \log\frac{p(0|y_i)}{p(1|y_i)} \qquad (B.75)$$

- **Messages update:**
  - Check factor to variable node update:

$$L_{ai} \leftarrow 2\tanh^{-1}\left(\prod_{j\in N(a)\backslash i}\tanh\left(\frac{L_{ja}}{2}\right)\right) \qquad (B.76)$$

  - Variable to check factor node update:

$$L_{ia} \leftarrow \sum_{b\in N(i)\backslash a} L_{ia}, \qquad (B.77)$$

- **Belief update:** Denote the net belief in terms of log-likelihood ratio as β_i for variable node i, then

$$\beta_i \leftarrow \sum_{a\in N(i)} L_{ai} \qquad (B.78)$$

- **Stopping criteria:** Repeat message update and belief update steps until the maximum number of iterations is reached or all checks are satisfied. Output the hard threshold values of net beliefs as the estimated coded bits.

---

whereas the right-hand column of variable nodes corresponds to the parity bits **y**. A variable node corresponding to a parity bit is added for each check bit. Apart from the first check bit connecting to only the first parity bit, each check bit is connected to the last and current parity bits. Let the connectivity between the message and check nodes be captured by the matrix H'. Since the checks have all to be satisfied, we have

$$H'\mathbf{x} + U\mathbf{y} = \mathbf{0}, \qquad (B.79)$$

where $U = \begin{bmatrix} 1 & 0 & \cdots & \\ 1 & 1 & 0 & \cdots \\ 0 & 1 & 1 & 0 \\ 0 & 0 & 1 & 1 & \cdots \\ & & \cdots & \\ & & \cdots & 1 & 1 \end{bmatrix}$. Thus

$$\mathbf{y} = U^{-1}H'\mathbf{x}, \qquad (B.80)$$

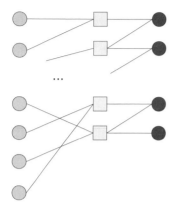

Figure B.8 Tanner graph for an IRA code.

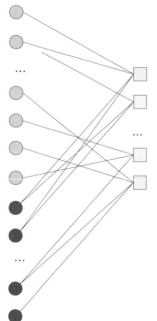

Figure B.9 IRA decoding. Exactly the same BP decoding can be applied. Parity bits are treated the same as systematic bits.

and it is easy to verify that $U^{-1} = \begin{bmatrix} 1 & 0 & \cdots & & \\ 1 & 1 & 0 & \cdots & \\ 1 & 1 & 1 & 0 & \\ 1 & 1 & 1 & 1 & \cdots \\ & & \cdots & & \\ 1 & 1 & \cdots & 1 & 1 \end{bmatrix}$. Thus, U^{-1} in effect sums

all the previous checks at each output parity. This is what "repeat accumulate" refers to in the name IRA. "Irregular" comes from the fact that generally the code degree for each node is different.

As shown in Figure B.9, we can simply move the parity bits along with the message bits and treat them indifferently. Then, the decoding of IRA codes is essentially the same as for LDPC codes.

C

Approximate Inference

As shown in the previous chapter, the sum-product algorithm is a powerful probabilistic inference technique for efficiently computing posterior probabilities over discrete variables of small alphabet sizes or continuous variables of linear Gaussian distributions. However, it cannot handle a discrete variable with a medium alphabet size as the computational complexities increase exponentially with the alphabet size. Moreover, for continuous variables with nonlinear Gaussian distribution, the sum-product is also infeasible as the integration may not have a closed-form solution. To tackle these difficulties, two common workarounds in approximate inference are either to discretize the variable through sampling techniques or to parametrize the variables through variational inference, where the sampling and the variational methods are also known as stochastic and deterministic approximation schemes. In this chapter we will focus on both the stochastic approximation and the deterministic approximation.

C.1 Stochastic Approximation

Stochastic approximation is an approximate technique based on numerical sampling, which is also known as the Monte Carlo technique. In many applications, posterior distributions are employed to make predictions by evaluating expectations. For example, considering a function $f(\mathbf{x})$ with continuous variable $\mathbf{x}$, the expectation of function $f(\mathbf{x})$ with respect to probability distribution $p(\mathbf{x})$ can be written as

$$E[f] = \int f(\mathbf{x})p(\mathbf{x})d\mathbf{x}, \tag{C.1}$$

where the integral can be replaced by summation for discrete variables. However, for arbitrary functions and probability distributions, the exact evaluation over the integral may not always be tractable. One workaround is to use the stochastic approximation method. The general idea of this sampling method is

Distributed Source Coding: Theory and Practice, First Edition. Shuang Wang, Yong Fang, and Samuel Cheng.
© 2017 John Wiley & Sons Ltd. Published 2017 by John Wiley & Sons Ltd.
Companion Website: www.wiley.com/go/cheng/dsctheoryandpractice

to independently draw a set of samples $\mathbf{x}^l$, $l = 1, 2, \cdots, L$ from the distribution $p(\mathbf{x})$, and then (C.1) can be approximated by a finite sum

$$\hat{f} = \frac{1}{L} \sum_{l=1}^{L} f(\mathbf{x}^l), \tag{C.2}$$

where the estimator $\hat{f}$ has the correct mean $E[\hat{f}] = E[f]$, and the variance $\mathrm{var}[\hat{f}] = \frac{1}{L}E[(f - E[f])^2]$. Since the estimator does not depend on the dimensionality of x, usually a relatively small number (e.g., 10 or 20) of independent samples may achieve a sufficient accuracy in principle. However, the problem is that samples might not always be independent in practice, thus the effective sample size might be much smaller than the actual sample size. Therefore, to achieve sufficient accuracy, a relatively large sample size is usually required in the stochastic approximation technique. This is why stochastic approximation methods are highly computationally demanding. In the rest of this section, we will concisely review some sampling strategies (i.e., importance sampling and MCMC).

C.1.1 Importance Sampling Methods

The general idea of importance sampling is to draw samples from an easily sampled proposal distribution $q(\mathbf{x})$ instead of the distribution $p(\mathbf{x})$. Then the expectation form based on a finite sum in (C.2) can be rewritten as

$$
\begin{aligned}
E[f] &= \int f(\mathbf{x})p(\mathbf{x})dx \\
&= \int f(\mathbf{x})\frac{p(\mathbf{x})}{q(\mathbf{x})}q(\mathbf{x})dx \\
&\simeq \frac{1}{L} \sum_{l=1}^{L} \frac{p(\mathbf{x}^l)}{q(\mathbf{x}^l)}f(\mathbf{x}^l) \\
&= \frac{1}{L} \sum_{l=1}^{L} r_l f(\mathbf{x}^l)
\end{aligned}
\tag{C.3}
$$

where $r_l = p(\mathbf{x}^l)/q(\mathbf{x}^l)$ is an important weight.

Considering the evaluation of $p(\mathbf{x})$ and $q(\mathbf{x})$ in terms of normalization constant, (C.3) can be written as

$$
\begin{aligned}
E[f] &= \int f(\mathbf{x})p(\mathbf{x})dx \\
&= \frac{Z_q}{Z_p} \int f(\mathbf{x})\frac{\tilde{p}(\mathbf{x})}{\tilde{q}(\mathbf{x})}q(\mathbf{x})dx \\
&\simeq \frac{Z_q}{Z_p}\frac{1}{L} \sum_{l=1}^{L} \tilde{r}_l f(\mathbf{x}^l),
\end{aligned}
\tag{C.4}
$$

where $\tilde{r}_l = \tilde{p}(\mathbf{x})/\tilde{q}(\mathbf{x})$, and the ratio Z_q/Z_p can also be evaluated through the same samples

$$\frac{Z_p}{Z_q} = \frac{1}{Z_q} \int \tilde{p}(\mathbf{x})d\mathbf{x} = \int \frac{\tilde{p}(\mathbf{x})}{\tilde{q}(\mathbf{x})} q(\mathbf{x})d\mathbf{x} \simeq \frac{1}{L} \sum_{l-1}^{L} \tilde{r}_l. \tag{C.5}$$

Thus,

$$E[f] \simeq \sum_{l=1}^{L} w_l f(\mathbf{x}^l) \tag{C.6}$$

with

$$w_l = \frac{\tilde{r}_l}{\sum_m \tilde{r}_m} = \frac{\tilde{p}(\mathbf{x}^l)/\tilde{q}(\mathbf{x}^l)}{\sum_m \tilde{p}(\mathbf{x}^m)/\tilde{q}(\mathbf{x}^m)}. \tag{C.7}$$

Since importance sampling technology introduces a proposal distribution $q(\mathbf{x})$ from which it is much easier to draw samples, the expectation accuracy depends on the selection of the proposal distribution. In the case where $p(\mathbf{x})$ has much of its mass concentrated on a small region of $\mathbf{x}$ space, while the proposal distribution is small or zero in the regions where $p(\mathbf{x})$ has large weight, the expectation accuracy will be very wrong.

C.1.2 Markov Chain Monte Carlo

MCMC is a more general and powerful sampling technique that overcomes the limitation within importance sampling and works for a large class of distributions. Before we discuss the MCMC framework, we will first review some general properties of Markov chains.

Markov Chains

Let us denote by $\mathbf{x}^1, \cdots, \mathbf{x}^t$ a series of random values. Then we say that the random variables are treated as a first-order Markov chain as long as the following conditional independence property holds:

$$p(\mathbf{x}^{t+1}|\mathbf{x}^1, \cdots, \mathbf{x}^t) = p(\mathbf{x}^{t+1}|\mathbf{x}^t). \tag{C.8}$$

The transition probability $p(\mathbf{x}^{t+1}|\mathbf{x}^t)$ indicates that the future state of a random variable only depends on the current state. Moreover, if the transition probabilities are the same for all the states, the Markov chain is called homogeneous.

Markov Chain Monte Carlo

Similar to importance sampling methods, the Markov Chain Monte Carlo (MCMC) method also uses a proposal distribution, with which one can draw samples much more easily. However, in MCMC the proposal distribution $q(\mathbf{x}) = q(\mathbf{x}|\mathbf{x}^t)$ depends on the current state $\mathbf{x}^t$, which satisfies the first-order Markov chain. In addition, the candidate sample drawn from the proposal distribution is accepted with a probability. For example, in the Metropolis algorithm, where the proposal distribution is symmetric, the probability of accepting a candidate sample can be expressed as

$$a(\mathbf{x}^*, \mathbf{x}^t) = \min\left(1, \frac{\tilde{p}(\mathbf{x}^*)}{\tilde{p}(\mathbf{x}^t)}\right). \tag{C.9}$$

If a candidate sample is accepted, then the sample at the next state will take the candidate sample value $\mathbf{x}^{t+1} = \mathbf{x}^*$, otherwise the sample at the next state will take the current state sample value $\mathbf{x}^{t+1} = \mathbf{x}^t$. Then the next candidate sample can be drawn from the updated distribution $q(\mathbf{x}|\mathbf{x}^{t+1})$.

C.2 Deterministic Approximation

The existing deterministic approximation methods include Laplace approximation (LA), variational Bayes (VB), expectation propagation (EP) and so on. In this section we will focus on the discussion of EP, since EP usually shows a higher accuracy compared with other deterministic approximation methods if the selected model matches the problem. This section contains three parts. First, we will review some preliminaries for EP. Then, the EP algorithm will be described. Finally, the relationship between EP and other variational methods will be discussed.

C.2.1 Preliminaries

Exponential Family
The exponential family of distributions over $\mathbf{x}$ is a set of distributions with the form

$$p(\mathbf{x}; \theta) = h(\mathbf{x})g(\theta)\exp\left(\theta^T \mathbf{u}(\mathbf{x})\right), \tag{C.10}$$

where measurement $\mathbf{x}$ may be scalar or vector, discrete or continuous, θ are parameters of the distribution, $h(\mathbf{x})$ and $\mathbf{u}(\mathbf{x})$ are some functions of $\mathbf{x}$, and the function $g(\theta)$ is a normalization factor as

$$g(\theta) \int h(\mathbf{x})\exp\left(\theta^T \mathbf{u}(\mathbf{x})\right) d\mathbf{x} = 1. \tag{C.11}$$

In addition, if the variables are discrete, the integration can simply be replaced with summation.

The exponential family has many properties, which may simplify computations. For example, if a likelihood function is one of the members in the exponential family, the posterior can be expressed in a closed-form expression by choosing a conjugate prior within the exponential family. Moreover, the exponential family has a wide range of members such as Gaussian, Bernoulli, discrete multinomial, Poisson and so on, thus it is applicable to many different inference models.

Kullback–Leibler Divergence

Kullback–Leibler (KL) divergence (148) (or relative entropy) is a measure to quantify the difference between a probabilistic distribution $p(\mathbf{x})$ and an approximate distribution $q(\mathbf{x})$. For the distributions $p(\mathbf{x})$ and $q(\mathbf{x})$ over continuous variables, KL divergence is defined as

$$D_{KL}(p(\mathbf{x}) \parallel q(\mathbf{x})) = \int p(\mathbf{x}) \log \frac{p(\mathbf{x})}{q(\mathbf{x})} d\mathbf{x}, \qquad (\text{C.12})$$

where for discrete variables integration is replaced with summation. Moreover, KL divergence is nonsymmetric, which means $D_{KL}(p(\mathbf{x})\|q(\mathbf{x})) \neq D_{KL}(q(\mathbf{x})\|p(\mathbf{x}))$. To give readers an intuitive view about the difference between these two forms of KL divergence, we assume that the true distribution $p(\mathbf{x})$ is multimodal and the candidate distribution $q(\mathbf{x})$ is unimodal. By minimizing $D_{KL}(q(\mathbf{x})\|p(\mathbf{x}))$, the approximate distribution $q(\mathbf{x})$ will pick one of the modes in $p(\mathbf{x})$, which is usually used in variational Bayes method. However, the best approximate distribution $q(\mathbf{x})$ obtained by minimizing $D_{KL}(p(\mathbf{x})\|q(\mathbf{x}))$ will be the average of all modes. The latter case is used in the approximate inference procedure of EP. Since this section focuses on a review of the EP algorithm, we will study the property of minimizing $D_{KL}(p(\mathbf{x})\|q(\mathbf{x}))$ first. The difference between minimizing $D_{KL}(p(\mathbf{x})\|q(\mathbf{x}))$ and $D_{KL}(q(\mathbf{x})\|p(\mathbf{x}))$ will be discussed later in this chapter.

To ensure a tractable solution for minimizing KL divergence $D_{KL}(p(\mathbf{x})\|q(\mathbf{x}))$, the approximate distribution $q(\mathbf{x})$ is usually restricted to a member of the exponential family. Thus, according to (C.10), $q(\mathbf{x})$ can be written as

$$q(\mathbf{x}; \theta) = h(\mathbf{x})g(\theta) \exp\left(\theta^T \mathbf{u}(\mathbf{x})\right), \qquad (\text{C.13})$$

where θ are the parameters of the given distribution.

By substituting $q(\mathbf{x}; \theta)$ into the KL divergence $D_{KL}(p(\mathbf{x})\|q(\mathbf{x}))$, we get

$$D_{KL}(p(\mathbf{x})\|q(\mathbf{x})) = -\ln g(\theta) - \theta^T \mathbb{E}_{p(\mathbf{x})}[\mathbf{u}(\mathbf{x})] + \text{const}, \qquad (\text{C.14})$$

where the const represents all the terms that are independent of parameters θ. To minimize KL divergence, taking the gradient of $D_{KL}(p(\mathbf{x})\|q(\mathbf{x}))$ with respect to θ to zero, we get

$$-\nabla \ln g(\theta) = \mathbb{E}_{p(\mathbf{x})}[\mathbf{u}(\mathbf{x})]. \qquad (\text{C.15})$$

Moreover, for (C.11), taking the gradient of both sides with respect to θ, we get

$$\nabla g(\theta) \int h(\mathbf{x}) \exp\left\{\theta^T \mathbf{u}(\mathbf{x})\right\} d\mathbf{x} + g(\theta) \int h(\mathbf{x}) \exp\{\theta^T \mathbf{u}(\mathbf{x})\}\mathbf{u}(\mathbf{x}) d\mathbf{x} = 0. \qquad (\text{C.16})$$

Then by rearranging and reusing (C.11), we get

$$-\nabla \ln g(\theta) = \mathbb{E}_{q(\mathbf{x})}[\mathbf{u}(\mathbf{x})]. \qquad (\text{C.17})$$

By comparing (C.15) and (C.17), we obtain

$$\mathbb{E}_{p(\mathbf{x})}[\mathbf{u}(\mathbf{x})] = \mathbb{E}_{q(\mathbf{x})}[\mathbf{u}(\mathbf{x})]. \tag{C.18}$$

Thus, from (C.18) we see that the minimization of KL divergence is equivalent to matching the expected sufficient statistics. For example, for minimizing KL divergence with a Gaussian distribution $q(\mathbf{x}; \theta)$, we only need to find the mean and covariance of $q(\mathbf{x}; \theta)$ that are equal to the mean and covariance of $p(\mathbf{x}; \theta)$, respectively.

Assumed-density Filtering

Assumed-density filtering (ADF) is a technique to construct a tractable approximation to complex probability distribution. EP can be viewed as an extension on ADF. Thus, we first provide a concise review of ADF and then extend it to the EP algorithm.

Let us consider the same problem of computing the factorization of the Bayes' rule

$$p(\mathbf{x}|\mathbf{y}) = \frac{p(\mathbf{x})p(\mathbf{y}|\mathbf{x})}{p(\mathbf{y})}$$

$$= \frac{1}{Z} p_0(\mathbf{x}) \prod_{i=1}^{N} p(y_i|\mathbf{x}),$$

$$= \frac{1}{Z} \prod_{i=0}^{N} p_i(\mathbf{x}), \tag{C.19}$$

where Z is a normalization constant and $p_i(\mathbf{x})$ is a simplified notation of each corresponding factor in (C.19), where $p_0(\mathbf{x}) = p_0(\mathbf{x})$ and $p_i(\mathbf{x}) = p_i(y_i|\mathbf{x})$ for $i > 0$. If we assume that the likelihood function $p(y_i|\mathbf{x})$ has a complex form, the direct evaluation of the posterior distribution would be infeasible. For example, if each likelihood function is a mixture of two Gaussian distributions and there is a total of $N = 100$ observed data then to get the posterior distribution we need to evaluate a mixture of 2^{100} Gaussians.

To solve this problem, approximate inference methods try to seek an approximate posterior distribution that can be very close to the true posterior distribution $p(\mathbf{x}|\mathbf{y})$. Usually the approximate distributions are chosen within the exponential family to ensure the computational feasibility. Then the best approximate distribution can be found by minimizing KL divergence:

$$\theta^* = \arg \min_{\theta} D_{KL}(p(\mathbf{x})\|q(\mathbf{x}; \theta)). \tag{C.20}$$

However, we can see that it is difficult to solve (C.20) directly. ADF solves this problem by iteratively including each factor function in the true posterior distribution. Thus, at first ADF chooses $q(\mathbf{x}; \theta^0)$ to best approximate the factor function $p_0(\mathbf{x})$ through

$$\theta^0 = \arg\min_\theta D_{KL}(p_0(\mathbf{x})\|q(\mathbf{x};\theta)). \tag{C.21}$$

Then ADF will update the approximation by incorporating the next factor function $p_i(y_i|\mathbf{x})$ until all the factor functions have been involved, which gives us the following update rule

$$\theta^i = \arg\min_\theta D_{KL}(p_i(\mathbf{x})q(\mathbf{x};\theta^{i-1})\|q(\mathbf{x};\theta)). \tag{C.22}$$

As shown in Section C.2.1, if $q(\mathbf{x};\theta)$ is chosen from the exponential family, the optimal solution of (C.22) is to match the expected sufficient statistics between the approximate distribution $q(\mathbf{x};\theta^i)$ and the target distribution $p_i(\mathbf{x})q(\mathbf{x};\theta^{i-1})$. Moreover, according to (C.22) we can see that the current best approximation is based on the previous best approximation. For this reason, the estimation performance of ADF may be sensitive to the process order of factor functions, which may produce an extremely poor approximation in some cases. In the next section we will provide another perspective of the ADF update rule, which results in the EP algorithm and provides a way of avoiding the drawback associated with the ADF algorithm.

C.2.2 Expectation Propagation

By taking another perspective, ADF can be seen as sequentially approximating the factor function $p_i(\mathbf{x})$ by the approximate factor function $\tilde{p}_i(\mathbf{x})$, which is restricted within the exponential family, and then exactly updating the approximate distribution $q(\mathbf{x};\theta)$ by multiplying these approximate factor functions. This alternative view of ADF can be described as

$$\tilde{p}_i(\mathbf{x}) \propto \frac{q(\mathbf{x};\theta^i)}{q(\mathbf{x};\theta^{i-1})}, \tag{C.23}$$

which also produces the EP algorithm. The EP algorithm initializes each factor function $p_i(\mathbf{x})$ by a corresponding approximate factor function $\tilde{p}_i(\mathbf{x})$. Then, at later iterations, EP revisits each approximated factor function $\tilde{p}_i(\mathbf{x})$ and refines it by multiplying all the current best estimates by one true factor function $p_i(\mathbf{x})$ of the revisiting term. After multiple iterations, the approximation is obtained according to (C.24):

$$q(\mathbf{x};\theta^*) \propto \prod_i \tilde{p}_i(\mathbf{x}). \tag{C.24}$$

Since this procedure does not depend on the process order of the factor function, EP provides a more accurate approximation than ADF.

The general process of EP is given as follows:

a) Initialize the term approximation $\tilde{p}_i(\mathbf{x})$, which can be chosen from one of the members in the exponential family based on the problem.

b) Compute the approximate distribution

$$q(\mathbf{x};\theta) = \frac{1}{Z} \prod_i \tilde{p}_i(\mathbf{x}), \tag{C.25}$$

where $Z = \int \prod_i \tilde{p}_i(\mathbf{x})d\mathbf{x}$.

c) Until all $\tilde{p}_i(\mathbf{x})$ converge:

  a) Choose $\tilde{p}_i(\mathbf{x})$ to refine the approximate.

  b) Remove $\tilde{p}_i(\mathbf{x})$ from the current approximated distribution $q(\mathbf{x};\theta)$ with a normalization factor:

$$q(\mathbf{x};\theta^{\backslash i}) \propto \frac{q(\mathbf{x};\theta)}{\tilde{p}_i(\mathbf{x})}. \tag{C.26}$$

  c) Update $q(\mathbf{x};\theta)$, where we first combine $q(\mathbf{x};\theta^{\backslash i})$, current $p_i(\mathbf{x})$ and a normalizer Z_i, and then minimize the KL divergence through moment matching projection (C.18) (i.e. the **Proj**$(\cdot)$ operator):

$$q(\mathbf{x};\theta) = \mathbf{Proj}\left(\frac{1}{Z_i}q(\mathbf{x};\theta^{\backslash i})p_i(\mathbf{x})\right). \tag{C.27}$$

  d) Update $\tilde{p}_i(\mathbf{x})$ as

$$\tilde{p}_i(\mathbf{x}) = Z_i \frac{q(\mathbf{x};\theta)}{q(\mathbf{x};\theta^{\backslash i})}. \tag{C.28}$$

d) Get the final approximate distribution through

$$p(\mathbf{x}) \approx q(\mathbf{x};\theta^*) \propto \prod_i \tilde{p}_i(\mathbf{x}). \tag{C.29}$$

Relationship with BP

This section shows that the BP algorithm is a special case of EP, where EP provides an improved approximation for models in which BP is generally intractable.

Let us first take a quick review of the BP algorithm discussed in Appendix B. The procedure of the BP algorithm is to iteratively update all variable nodes, then update all factor nodes through sending messages in parallel, and finally update the belief of each variable after each iteration. By taking another viewpoint, BP can be viewed as updating the belief over a variable x_i by incorporating one factor node at each time. Under this perspective, the belief of variable x_i will be updated as

$$b(x_i) = \frac{m_{X_i \to f_s}(x_i)m_{f_s \to X_i}(x_i)}{Z_i}, \tag{C.30}$$

where $Z_i = \int m_{X_i \to f_s}(x_i)m_{f_s \to X_i}(x_i)dx_i$ is the normalization factor. Moreover, we can loosely interpret $m_{X_i \to f_s}(x_i)$ and $m_{f_s \to X_i}(x_i)$ as the prior and likelihood messages, respectively.

Let us suppose that each likelihood message $m_{f_s \to X_i}(x_i)$ has a complex form, for example a mixture of multiple Gaussian distributions. Then the computational complexity to evaluate the exact beliefs over all variables would not be feasible. Instead of propagating exact likelihood message $m_{f_s \to X_i}(x_i)$, EP passes an approximate message $\tilde{m}_{f_s \to X_i}(x_i)$, where $\tilde{m}_{f_s \to X_i}(x_i)$ is obtained by using the projection operation, as shown in the general process of EP. Moreover, $\tilde{m}_{f_s \to X_i}(x_i)$ is usually chosen from the exponential family to make the problem tractable. Thus, the approximate belief in EP has the following form

$$b(x_i) \approx q(x_i) \propto \prod_{s \in N(X_i)} \tilde{m}_{f_s \to X_i}(x_i). \tag{C.31}$$

To show BP as a special case of EP, we further define the partial belief of a variable node as

$$b(x_i)^{\backslash f_s} = \frac{b(x_i)}{\tilde{m}_{f_s \to X_i}(x_i)} \propto \prod_{s' \in N(X_i) \backslash s} \tilde{m}_{f_{s'} \to X_i}(x_i) \tag{C.32}$$

and the partial belief of a factor node as

$$b(f_s)^{\backslash X_i} = \frac{b(f_s)}{\tilde{m}_{X_i \to f_s}(x_i)}, \tag{C.33}$$

where $b(f_s) = \prod_{j \in N(f_s)} \tilde{m}_{X_j \to f_s}(x_j)$ is defined as the belief of the factor node f_s. By comparing to (C.27) and (C.28), the factor node message updating rule in EP can be written as

$$\begin{aligned}
\tilde{m}_{f_s \to X_i}(x_i) &= \frac{\mathbf{Proj}(b(x_i)^{\backslash f_s} m_{f_s \to X_i}(x_i))}{b(x_i)^{\backslash f_s}} \\
&= \frac{\mathbf{Proj}\left(b(x_i)^{\backslash f_s} \int_{\mathbf{x}_s \backslash x_i} f_s(\mathbf{x}_s) b(f_s)^{\backslash X_i} \right)}{b(x_i)^{\backslash f_s}}
\end{aligned} \tag{C.34}$$

where the integral works over continuous variables. For discrete variables, one can simply replace the integral with summation. Further, the new belief $b(x_i)$ will be approximated as

$$b(x_i) \approx q_i(x_i) = \frac{b(x_i)^{\backslash f_s} \tilde{m}_{f_s \to X_i}(x_i)}{Z_i}, \tag{C.35}$$

where $Z_i = \int_{x_i} b(x_i)^{\backslash f_s} \tilde{m}_{f_s \to X_i}(x_i)$.

Now, if the integral in (C.34) is tractable (e.g., a linear Gaussian model) even without using the projection to approximate $m_{f_s \to X_i}$ then $b(x_i)^{\backslash f_s}$ in (C.34) can be canceled. Finally, the factor node message update rule in EP reduces to the standard BP case.

C.2.3 Relationship with Other Variational Inference Methods

In this section we will describe the relationship between EP and other variational inference algorithms, for example VB. The problem considered here is the same as that in previous sections, where $\mathbf{y}$ are the observed data and $\mathbf{x}$ are some hidden variables. The Bayesian probabilistic model specifies the joint distribution $p(\mathbf{x}, \mathbf{y})$ where all the hidden variables in $\mathbf{x}$ are given prior distributions. The goal is to find the best approximation for the posterior distribution $p(\mathbf{x}|\mathbf{y})$. Let us take a look at the decomposition of the log joint distribution as follows:

$$\log p(\mathbf{x}, \mathbf{y}) = \log p(\mathbf{x}|\mathbf{y}) + \log p(\mathbf{y}). \tag{C.36}$$

By rearranging (C.36) and taking the integral of both sides of the rearranged equation with respect to a given distribution $q(\mathbf{x})$, we get the log model evidence

$$\begin{aligned}
\log p(\mathbf{y}) &= \int q(\mathbf{x}) \log(p(\mathbf{y})) d\mathbf{x} \\
&= \int q(\mathbf{x}) \log(p(\mathbf{x}, \mathbf{y})) - \int q(\mathbf{x}) \log(p(\mathbf{x}|\mathbf{y})) d\mathbf{x},
\end{aligned} \tag{C.37}$$

where $\int q(\mathbf{x})d\mathbf{x} = 1$. Then, by reformatting (C.37), we get

$$\log p(\mathbf{y}) = \mathcal{L}(q(\mathbf{x})) + D_{KL}(q(\mathbf{x}) \| p(\mathbf{x})), \tag{C.38}$$

where we define

$$\mathcal{L}(q(\mathbf{x})) = \int q(\mathbf{x}) \log(\frac{p(\mathbf{x}, \mathbf{y})}{q(\mathbf{x})}) d\mathbf{x}, \tag{C.39}$$

$$D_{KL}(q(\mathbf{x}) \| p(\mathbf{x})) = \int q(\mathbf{x}) \log(\frac{q(\mathbf{x})}{p(\mathbf{x}|\mathbf{y})}) d\mathbf{x}. \tag{C.40}$$

Since $D_{KL}(q(\mathbf{x}) \| p(\mathbf{x}))$ is a non-negative functional, $\mathcal{L}(q(x))$ gives the lower bound of $\log p(\mathbf{y})$. Then the maximization of the lower bound $\mathcal{L}(q(x))$ with respect to the distribution $q(\mathbf{x})$ is equivalent to minimizing $D_{KL}(q(\mathbf{x}) \| p(\mathbf{x}))$, which happens when $q(\mathbf{x}) = p(\mathbf{x}|\mathbf{y})$. However, working with the true posterior distribution $p(\mathbf{x}|\mathbf{y})$ may be intractable. Thus, we assume that the elements of $\mathbf{x}$ can be partitioned into M disjoint groups $\mathbf{x}_i$, $i = 1, 2, \cdots, M$. We then further assume that the factorization of the approximate distribution $q(\mathbf{x})$ with respect to these groups has the form

$$q(\mathbf{x}) = \prod_{i}^{M} q_i(\mathbf{x}_i). \tag{C.41}$$

Note that the factorized approximation corresponds to the mean filed theory, which was developed in physics. Given the aforementioned assumptions, we now try to find any possible distribution $q(\mathbf{x})$ over which the lower bound $\mathcal{L}(q(\mathbf{x}))$ is largest. Since the direct maximization of (C.39) with respect to $q(\mathbf{x})$

is difficult, we instead optimize (C.39) with respect to each of the factors in (C.41). By substituting (C.41) into (C.39) we get

$$\mathcal{L}(q(\mathbf{x})) = \int q_j(\mathbf{x}_j) \left(\int \log(p(\mathbf{x}, \mathbf{y})) \prod_{i \neq j} q_i(\mathbf{x}_i) d\mathbf{x}_i \right) d\mathbf{x}_j$$

$$- \int q_j(\mathbf{x}_j) \log(q_j(\mathbf{x}_j)) d\mathbf{x}_j + \text{const}$$

$$= - \int q_j(\mathbf{x}_j) \log(\frac{q_j(\mathbf{x}_j)}{\tilde{p}(\mathbf{x}_j, \mathbf{y})}) d\mathbf{x}_j + \text{const}$$

$$= -D_{KL}(q_j(\mathbf{x}_j) \| \tilde{p}(\mathbf{x}_j, \mathbf{y})) + \text{const}, \tag{C.42}$$

where we define $\tilde{p}(\mathbf{x}_j, \mathbf{y})$ as

$$\tilde{p}(\mathbf{x}_j, \mathbf{y}) = \exp \left(\int \log(p(\mathbf{x}, \mathbf{y})) \prod_{i \neq j} q_i(\mathbf{x}_i) d\mathbf{x}_i \right)$$

$$= \exp(\mathbb{E}_{i \neq j}[\log p(\mathbf{x}, \mathbf{y})]). \tag{C.43}$$

Thus, if we keep all the factors $q_i(\mathbf{x}_i)$ for $i \neq j$ fixed, then the maximization of (C.42) with respect to $q_j(\mathbf{x}_j)$ is equivalent to the minimization of $D_{KL}(q_j(\mathbf{x}_j) \| \tilde{p}(\mathbf{x}_j, \mathbf{y}))$. In practice we need to initialize all the factors $q_i(\mathbf{x}_i)$ first, and then iteratively update each of the factors $q_j(\mathbf{x}_j)$ by minimizing the $D_{KL}(q_j(\mathbf{x}_j) \| \tilde{p}(\mathbf{x}_j, \mathbf{y}))$ until the algorithm convergences.

Now we can see that the key difference between EP and VB is the way the KL divergence is minimized. The advantage of VB is that it provides a lower bound during each optimizing step, thus the convergence is guaranteed. However, VB may cause under-estimation of variance. In EP, minimizing $D_{KL}(p(\mathbf{x}) \| q(\mathbf{x}))$ is equivalent to the "moment matching", but convergence is not guaranteed. However, EP has a fix point and if it does converge, the approximation performance of EP usually outperforms VB.

D

Multivariate Gaussian Distribution

D.1 Introduction

Gaussian or normal distribution is the most important and widely used distribution in engineering. In this chapter we present the basic tools to manipulate the multivariate Gaussian distribution.

A quick note on convention is helpful here. Vectors are in bold. Random variables are in upper case and realizations of random variables are in lower case. A vector random variable is therefore in bold upper case.

D.2 Probability Density Function

The probability density function (pdf) of a multivariate Gaussian random variable $\mathbf{X}$ with mean $\boldsymbol{\mu}$ and covariance matrix Σ is given by

$$p_{\mathbf{X}}(\mathbf{x}) = \frac{1}{\sqrt{\det(2\pi\Sigma)}} \exp\left(-\frac{1}{2}(\mathbf{x} - \boldsymbol{\mu})^T \Sigma^{-1}(\mathbf{x} - \boldsymbol{\mu})\right). \tag{D.1}$$

For convenience, we will also denote the multivariate Gaussian pdf with $\mathcal{N}(\mathbf{x}; \boldsymbol{\mu}, \Sigma)$. Note that $\mathbf{x}$ and $\boldsymbol{\mu}$ are symmetric in $\mathcal{N}(\mathbf{x}; \boldsymbol{\mu}, \Sigma)$. We have $\mathcal{N}(\mathbf{x}; \boldsymbol{\mu}, \Sigma) = \mathcal{N}(\boldsymbol{\mu}; \mathbf{x}, \Sigma) = \mathcal{N}(\boldsymbol{\mu} - \mathbf{x}; 0, \Sigma) = \mathcal{N}(0; \boldsymbol{\mu} - \mathbf{x}, \Sigma)$. These equations are trivial but are also very handy at times.

To avoid confusion in notation, a common convention is to use $p(\mathbf{x})$ to denote $p_{\mathbf{X}}(\mathbf{x})$[1]. Of course, for the above definition to be well defined we need to make sure that Σ^{-1} exists. Since Σ is symmetric, the eigenvalues are real and the eigenvectors can be made orthogonal. If all eigenvalues are strictly larger than 0, then Σ^{-1} exists. On the contrary, if there is a zero eigenvalue, it means that there is no variation along the direction of the corresponding eigenvector, that is, if we

---

1 Note that some books may also use $f_{\mathbf{X}}(\mathbf{x})$ instead of $p_{\mathbf{X}}(\mathbf{x})$ for pdfs.

Distributed Source Coding: Theory and Practice, First Edition. Shuang Wang, Yong Fang, and Samuel Cheng.
© 2017 John Wiley & Sons Ltd. Published 2017 by John Wiley & Sons Ltd.
Companion Website: www.wiley.com/go/cheng/dsctheoryandpractice

project the variable along that eigenvector, the projected value is just a constant (instead of stochastic). So if we ignore these degenerated cases, we can safely assume that Σ^{-1} exists and the pdf is well defined.

D.3 Marginalization

Consider $\mathbf{Z} \sim \mathcal{N}(\boldsymbol{\mu}_{\mathbf{Z}}, \Sigma_{\mathbf{Z}})$ and let us say $\mathbf{X}$ is a segment of $\mathbf{Z}$, that is, $\mathbf{Z} = \begin{pmatrix} \mathbf{X} \\ \mathbf{Y} \end{pmatrix}$ for some $\mathbf{Y}$. Then how should $\mathbf{X}$ behave?

We can find the pdf of $\mathbf{X}$ by just marginalizing that of $\mathbf{Z}$, that is

$$p(\mathbf{x}) = \int p(\mathbf{x}, \mathbf{y}) d\mathbf{y} \tag{D.2}$$

$$= \frac{1}{\sqrt{\det(2\pi\Sigma)}} \int \exp\left(-\frac{1}{2}\begin{pmatrix} \mathbf{x} - \boldsymbol{\mu}_{\mathbf{X}} \\ \mathbf{y} - \boldsymbol{\mu}_{\mathbf{Y}} \end{pmatrix}^T \Sigma^{-1} \begin{pmatrix} \mathbf{x} - \boldsymbol{\mu}_{\mathbf{X}} \\ \mathbf{y} - \boldsymbol{\mu}_{\mathbf{Y}} \end{pmatrix}\right) d\mathbf{y}. \tag{D.3}$$

Let us denote Σ^{-1} as Λ, which is usually known to be the precision matrix. Partition both Σ and Λ into $\Sigma = \begin{pmatrix} \Sigma_{\mathbf{XX}} & \Sigma_{\mathbf{XY}} \\ \Sigma_{\mathbf{YX}} & \Sigma_{\mathbf{YY}} \end{pmatrix}$ and $\Lambda = \begin{pmatrix} \Lambda_{\mathbf{XX}} & \Lambda_{\mathbf{XY}} \\ \Lambda_{\mathbf{YX}} & \Lambda_{\mathbf{YY}} \end{pmatrix}$. Then we have

$$p(\mathbf{x}) = \frac{1}{\sqrt{\det(2\pi\Sigma)}} \int \exp\left(-\frac{1}{2}[(\mathbf{x} - \boldsymbol{\mu}_{\mathbf{X}})^T \Lambda_{\mathbf{XX}}(\mathbf{x} - \boldsymbol{\mu}_{\mathbf{X}})\right. \tag{D.4}$$

$$+ (\mathbf{y} - \boldsymbol{\mu}_{\mathbf{Y}})^T \Lambda_{\mathbf{YX}}(\mathbf{x} - \boldsymbol{\mu}_{\mathbf{X}}) + (\mathbf{x} - \boldsymbol{\mu}_{\mathbf{X}})^T \Lambda_{\mathbf{XY}}(\mathbf{y} - \boldsymbol{\mu}_{\mathbf{Y}}) \tag{D.5}$$

$$+ (\mathbf{y} - \boldsymbol{\mu}_{\mathbf{Y}})^T \Lambda_{\mathbf{YY}}(\mathbf{y} - \boldsymbol{\mu}_{\mathbf{Y}})])d\mathbf{y} \tag{D.6}$$

$$= \frac{e^{-\frac{(\mathbf{x} - \boldsymbol{\mu}_{\mathbf{X}})^T \Lambda_{\mathbf{XX}}(\mathbf{x} - \boldsymbol{\mu}_{\mathbf{X}})}{2}}}{\sqrt{\det(2\pi\Sigma)}} \int \exp\left(-\frac{1}{2}[(\mathbf{y} - \boldsymbol{\mu}_{\mathbf{Y}})^T \Lambda_{\mathbf{YX}}(\mathbf{x} - \boldsymbol{\mu}_{\mathbf{X}})\right. \tag{D.7}$$

$$+ (\mathbf{x} - \boldsymbol{\mu}_{\mathbf{X}})^T \Lambda_{\mathbf{XY}}(\mathbf{y} - \boldsymbol{\mu}_{\mathbf{Y}}) + (\mathbf{y} - \boldsymbol{\mu}_{\mathbf{Y}})^T \Lambda_{\mathbf{YY}}(\mathbf{y} - \boldsymbol{\mu}_{\mathbf{Y}})])d\mathbf{y} \tag{D.8}$$

To proceed, we use the "completing square" trick that was probably learnt in high school. Basically, a quadratic expression $ax^2 + 2bx + c$ may be rewritten as $a(x^2 + 2\frac{b}{a}x) + c = a(x + \frac{b}{a})^2 + c - \frac{b^2}{a}$. By doing this, we immediately can see that the minimum (assuming a is positive) of $ax^2 + 2bx + c$ is $c - \frac{b^2}{a}$ and occurs when $x = -\frac{b}{a}$.

Now let us apply the completing square trick on $(\mathbf{y} - \boldsymbol{\mu}_{\mathbf{Y}})^T \Lambda_{\mathbf{YX}}(\mathbf{x} - \boldsymbol{\mu}_{\mathbf{X}}) + (\mathbf{x} - \boldsymbol{\mu}_{\mathbf{X}})^T \Lambda_{\mathbf{XY}}(\mathbf{y} - \boldsymbol{\mu}_{\mathbf{Y}}) + (\mathbf{y} - \boldsymbol{\mu}_{\mathbf{Y}})^T \Lambda_{\mathbf{YY}}(\mathbf{y} - \boldsymbol{\mu}_{\mathbf{Y}})$. For the ease of exposition, let us denote $\tilde{\mathbf{x}}$ as $\mathbf{x} - \boldsymbol{\mu}_{\mathbf{X}}$ and $\tilde{\mathbf{y}}$ as $\mathbf{y} - \boldsymbol{\mu}_{\mathbf{Y}}$. We have

$$\tilde{\mathbf{y}}^T \Lambda_{\mathbf{YX}}\tilde{\mathbf{x}} + \tilde{\mathbf{x}}^T \Lambda_{\mathbf{XY}}\tilde{\mathbf{y}} + \tilde{\mathbf{y}}^T \Lambda_{\mathbf{YY}}\tilde{\mathbf{y}} \tag{D.9}$$

$$= (\tilde{\mathbf{y}} + \Lambda_{\mathbf{YY}}^{-1}\Lambda_{\mathbf{YX}}\tilde{\mathbf{x}})^T \Lambda_{\mathbf{YY}}(\tilde{\mathbf{y}} + \Lambda_{\mathbf{YY}}^{-1}\Lambda_{\mathbf{YX}}\tilde{\mathbf{x}}) - \tilde{\mathbf{x}}^T \Lambda_{\mathbf{XY}}\Lambda_{\mathbf{YY}}^{-1}\Lambda_{\mathbf{YX}}\tilde{\mathbf{x}}, \tag{D.10}$$

where we use the fact that $\Lambda = \Sigma^{-1}$ is symmetric and so $\Lambda_{XY} = \Lambda_{YX}$. Therefore, we have

$$p(\mathbf{x}) = \frac{e^{-\frac{\tilde{\mathbf{x}}^T(\Lambda_{XX} - \Lambda_{XY}\Lambda_{YY}^{-1}\Lambda_{YX})\tilde{\mathbf{x}}}{2}}}{\sqrt{\det(2\pi\Sigma)}} \int e^{-\frac{(\tilde{\mathbf{y}} + \Lambda_{YY}^{-1}\Lambda_{YX}\tilde{\mathbf{x}})^T\Lambda_{YY}(\tilde{\mathbf{y}} + \Lambda_{YY}^{-1}\Lambda_{YX}\tilde{\mathbf{x}})}{2}} d\mathbf{y} \tag{D.11}$$

$$= \frac{\sqrt{\det(2\pi\Lambda_{YY}^{-1})}}{\sqrt{\det(2\pi\Sigma)}} \exp\left(-\frac{\tilde{\mathbf{x}}^T(\Lambda_{XX} - \Lambda_{XY}\Lambda_{YY}^{-1}\Lambda_{YX})\tilde{\mathbf{x}}}{2}\right) \tag{D.12}$$

$$\overset{(a)}{=} \frac{\sqrt{\det(2\pi\Lambda_{YY}^{-1})}}{\sqrt{\det(2\pi\Sigma)}} \exp\left(-\frac{\tilde{\mathbf{x}}^T\Sigma_{XX}^{-1}\tilde{\mathbf{x}}}{2}\right) \tag{D.13}$$

$$\overset{(b)}{=} \frac{1}{\sqrt{\det(2\pi\Sigma_{XX})}} \exp\left(-\frac{\tilde{\mathbf{x}}^T\Sigma_{XX}^{-1}\tilde{\mathbf{x}}}{2}\right) \tag{D.14}$$

$$= \frac{1}{\sqrt{\det(2\pi\Sigma_{XX})}} \exp\left(-\frac{(\mathbf{x} - \boldsymbol{\mu}_X)^T\Sigma_{XX}^{-1}(\mathbf{x} - \boldsymbol{\mu}_X)}{2}\right), \tag{D.15}$$

where (a) is due to Lemma D.1 and (b) is due to Corollary D.2. In conclusion, $X \sim \mathcal{N}(\boldsymbol{\mu}_X, \Sigma_{XX})$, which is probably what one may expect from the beginning. Note that for illustrative purposes we kept track of the normalization factor $\frac{1}{\sqrt{\det(2\pi\Sigma_{XX})}}$ in the above derivation but this was really not necessary because we know $p(\mathbf{x})$ should still be a density function and thus will be normalized to one. In future sections we will mostly just keep track of the exponent.

D.4   Conditioning

Consider the same $Z \sim \mathcal{N}(\boldsymbol{\mu}_Z, \Sigma_Z)$ and $\mathbf{Z} = \begin{pmatrix} \mathbf{X} \\ \mathbf{Y} \end{pmatrix}$. What will $\mathbf{X}$ be like if $\mathbf{Y}$ is observed to be $\mathbf{y}$? Basically, we want to find $p(\mathbf{x}|\mathbf{y}) = p(\mathbf{x}, \mathbf{y})/p(\mathbf{y})$. From the previous section, we have $p(\mathbf{y}) = \mathcal{N}(\mathbf{y}; \boldsymbol{\mu}_Y, \Sigma_{YY})$. Therefore,

$$p(\mathbf{x}|\mathbf{y}) \propto \exp\left(-\frac{1}{2}\left[\begin{pmatrix} \tilde{\mathbf{x}} \\ \tilde{\mathbf{y}} \end{pmatrix}^T \Sigma^{-1} \begin{pmatrix} \tilde{\mathbf{x}} \\ \tilde{\mathbf{y}} \end{pmatrix} - \tilde{\mathbf{y}}^T\Sigma_{YY}^{-1}\tilde{\mathbf{y}}\right]\right) \tag{D.16}$$

$$\propto \exp\left(-\frac{1}{2}[\tilde{\mathbf{x}}^T\Lambda_{XX}\tilde{\mathbf{x}} + \tilde{\mathbf{x}}^T\Lambda_{XY}\tilde{\mathbf{y}} + \tilde{\mathbf{y}}^T\Lambda_{YX}\tilde{\mathbf{x}}]\right), \tag{D.17}$$

where we use $\tilde{\mathbf{x}}$ and $\tilde{\mathbf{y}}$ as shorthands of $\mathbf{x} - \boldsymbol{\mu}_X$ and $\mathbf{y} - \boldsymbol{\mu}_Y$ as before. Completing the square for $\tilde{\mathbf{x}}$, we have

$$p(\mathbf{x}|\mathbf{y}) \propto \exp\left(-\frac{1}{2}(\tilde{\mathbf{x}} + \Lambda_{XX}^{-1}\Lambda_{XY}\tilde{\mathbf{y}})^T\Lambda_{XX}(\tilde{\mathbf{x}} + \Lambda_{XX}^{-1}\Lambda_{XY}\tilde{\mathbf{y}})\right) \tag{D.18}$$

$$= \exp\left(-\frac{1}{2}(\mathbf{x} - \boldsymbol{\mu}_X + \Lambda_{XX}^{-1}\Lambda_{XY}(\mathbf{y} - \boldsymbol{\mu}_Y))^T \right.$$
$$\left. \Lambda_{XX}(\mathbf{x} - \boldsymbol{\mu}_X + \Lambda_{XX}^{-1}\Lambda_{XY}(\mathbf{y} - \boldsymbol{\mu}_Y))\right) \tag{D.19}$$

Therefore $\mathbf{X}|\mathbf{y}$ is Gaussian distributed with mean $\boldsymbol{\mu}_{\mathbf{X}} - \Lambda_{\mathbf{XX}}^{-1}\Lambda_{\mathbf{XY}}(\mathbf{y} - \boldsymbol{\mu}_{\mathbf{Y}})$ and covariance $\Lambda_{\mathbf{XX}}^{-1}$. Note that since $\Lambda_{\mathbf{XX}}\Sigma_{\mathbf{XY}} + \Lambda_{\mathbf{XY}}\Sigma_{\mathbf{YY}} = 0$, $\Lambda_{\mathbf{XX}}^{-1}\Lambda_{\mathbf{XY}} = -\Sigma_{\mathbf{XY}}\Sigma_{\mathbf{YY}}^{-1}$ and from Lemma D.1 we have

$$\mathbf{X}|\mathbf{y} \sim \mathcal{N}(\boldsymbol{\mu}_{\mathbf{X}} + \Sigma_{\mathbf{XY}}\Sigma_{\mathbf{YY}}^{-1}(\mathbf{y} - \boldsymbol{\mu}_{\mathbf{Y}}), \Sigma_{\mathbf{XX}} - \Sigma_{\mathbf{XY}}\Sigma_{\mathbf{YY}}^{-1}\Sigma_{\mathbf{YX}}). \tag{D.20}$$

One can make some intuitive interpretation for the conditioning result above. Let us say both $\mathbf{X}$ and $\mathbf{Y}$ are scalar.[2] When the observation of $\mathbf{Y}$ is exactly the mean, the conditioned mean does not change. Otherwise, it needs to be modified and the size of the adjustment decreases with $\Sigma_{\mathbf{YY}}$, the variance of $\mathbf{Y}$ for the 1D case. This is reasonable as the observation is less reliable with the increase in $\Sigma_{\mathbf{YY}}$. The adjustment is finally scaled by $\Sigma_{\mathbf{XY}}$, which translates the variation of $\mathbf{Y}$ to the variation of $\mathbf{X}$. In particular, if $\mathbf{X}$ and $\mathbf{Y}$ are negatively correlated, the direction of the adjustment will be shifted. As for the variance of the conditioned variable, it always decreases and the decrease is larger if $\Sigma_{\mathbf{YY}}$ is smaller and $\Sigma_{\mathbf{XY}}$ is larger ($\mathbf{X}$ and $\mathbf{Y}$ are more correlated).

Corollary D.1 Given multivariate Gaussian variables $X, Y,$ and Z, we have X and Y conditionally independent given Z if $\rho_{XZ}\rho_{YZ} = \rho_{XY}$, where $\rho_{XZ} = \frac{E[(X-E(X))(Z-E(Z))]}{\sqrt{E[(X-E(X))^2]E[(Z-E(Z))^2]}}$ is the correlation coefficient between X and Z. Similarly, ρ_{YZ} and ρ_{XY} are the correlation coefficients between Y and Z, and X and Y, respectively.

Proof. Without loss of generality, we can assume the variables are all zero-mean with unit variance. Thus, $\begin{pmatrix} X \\ Y \\ Z \end{pmatrix} \sim \mathcal{N}(\mathbf{0}, \Sigma)$, where $\Sigma = \begin{pmatrix} 1 & \rho_{XY} & \rho_{XZ} \\ \rho_{XY} & 1 & \rho_{YZ} \\ \rho_{XZ} & \rho_{YZ} & 1 \end{pmatrix}$. Then from (D.20), we have

$$\Sigma_{\begin{pmatrix} X \\ Y \end{pmatrix}|Z} = \begin{pmatrix} 1 & \rho_{XY} \\ \rho_{XY} & 1 \end{pmatrix} - \begin{pmatrix} \rho_{XZ} & \rho_{YZ} \end{pmatrix} \sigma_{YY}^{-1} \begin{pmatrix} \rho_{XZ} \\ \rho_{YZ} \end{pmatrix}$$

$$= \begin{pmatrix} 1 - \rho_{XZ}^2 & \rho_{XY} - \rho_{XZ}\rho_{YZ} \\ \rho_{XY} - \rho_{XZ}\rho_{YZ} & 1 - \rho_{YZ}^2 \end{pmatrix}$$

Therefore, X and Y are uncorrelated given Z when $\sigma_{XY|Z} = \rho_{XY} - \rho_{XZ}\rho_{YZ} = 0$ or $\rho_{XY} = \rho_{XZ}\rho_{YZ}$ since for Gaussian variables uncorrelatedness implies independence. This concludes the proof. □

D.5   Product of Gaussian pdfs

Assume that we try to recover some vector parameter $\mathbf{x}$, which is subject to multivariate Gaussian noise. Say we made two measurements $\mathbf{y}_1$ and

---

[2] For consistency, we will stick with the vector notation for the rest of this section.

$\mathbf{y}_2$, where $\mathbf{Y}_1 \sim \mathcal{N}(\mathbf{x}, \Sigma_{\mathbf{Y}_1})$ and $\mathbf{Y}_2 \sim \mathcal{N}(\mathbf{x}, \Sigma_{\mathbf{Y}_2})$. Note that even though both measurements have mean $\mathbf{x}$, they have different covariance. This variation, for instance, can be due to environment change between the two measurements. Now, if we want to compute the overall likelihood, $p(\mathbf{y}_1, \mathbf{y}_2 | \mathbf{x})$. Assuming that $\mathbf{Y}_1$ and $\mathbf{Y}_2$ are conditionally independent given $\mathbf{X}$, we have

$$p(\mathbf{y}_1, \mathbf{y}_2 | \mathbf{x}) = p(\mathbf{y}_1 | \mathbf{x}) p(\mathbf{y}_2 | \mathbf{x}) \tag{D.21}$$

$$= \mathcal{N}(\mathbf{y}_1; \mathbf{x}, \Sigma_{\mathbf{Y}_1}) \mathcal{N}(\mathbf{y}_2; \mathbf{x}, \Sigma_{\mathbf{Y}_2}). \tag{D.22}$$

Essentially, we just need to compute the product of two Gaussian pdfs. Such computation is very useful and it occurs often when one needs to perform inference.

As in previous sections, the product turns out to be "Gaussian" also. However, unlike the previous case, the product is not a pdf and so it does not normalize to 1. So we have to compute both the scaling factor and the exponent explicitly. Let us start with the exponent.

$$\mathcal{N}(\mathbf{y}_1; \mathbf{x}, \Sigma_{\mathbf{Y}_1}) \mathcal{N}(\mathbf{y}_2; \mathbf{x}, \Sigma_{\mathbf{Y}_2}) \tag{D.23}$$

$$\propto \exp\left(-\frac{1}{2}[(\mathbf{x} - \mathbf{y}_1)^T \Lambda_{\mathbf{Y}_1}(\mathbf{x} - \mathbf{y}_1) + (\mathbf{x} - \mathbf{y}_2)^T \Lambda_{\mathbf{Y}_2}(\mathbf{x} - \mathbf{y}_2)]\right) \tag{D.24}$$

$$\propto \exp\left(-\frac{1}{2}[\mathbf{x}^T(\Lambda_{\mathbf{Y}_1} + \Lambda_{\mathbf{Y}_2})\mathbf{x} - (\mathbf{y}_2^T \Lambda_{\mathbf{Y}_2} + \mathbf{y}_1^T \Lambda_{\mathbf{Y}_1})\mathbf{x} - \mathbf{x}^T(\Lambda_{\mathbf{Y}_2}\mathbf{y}_2 + \Lambda_{\mathbf{Y}_1}\mathbf{y}_1)]\right) \tag{D.25}$$

$$\propto e^{-\frac{1}{2}[(\mathbf{x} - (\Lambda_{\mathbf{Y}_1} + \Lambda_{\mathbf{Y}_2})^{-1}(\Lambda_{\mathbf{Y}_2}\mathbf{y}_2 + \Lambda_{\mathbf{Y}_1}\mathbf{y}_1))^T(\Lambda_{\mathbf{Y}_1} + \Lambda_{\mathbf{Y}_2})(\mathbf{x} - (\Lambda_{\mathbf{Y}_1} + \Lambda_{\mathbf{Y}_2})^{-1}(\Lambda_{\mathbf{Y}_2}\mathbf{y}_2 + \Lambda_{\mathbf{Y}_1}\mathbf{y}_1))]} \tag{D.26}$$

$$\propto \mathcal{N}(\mathbf{x}; (\Lambda_{\mathbf{Y}_1} + \Lambda_{\mathbf{Y}_2})^{-1}(\Lambda_{\mathbf{Y}_2}\mathbf{y}_2 + \Lambda_{\mathbf{Y}_1}\mathbf{y}_1), (\Lambda_{\mathbf{Y}_2} + \Lambda_{\mathbf{Y}_1})^{-1}). \tag{D.27}$$

Therefore,

$$\mathcal{N}(\mathbf{y}_1; \mathbf{x}, \Sigma_{\mathbf{Y}_1}) \mathcal{N}(\mathbf{y}_2; \mathbf{x}, \Sigma_{\mathbf{Y}_2})$$
$$= K(\mathbf{y}_1, \mathbf{y}_2, \Sigma_{\mathbf{Y}_1}, \Sigma_{\mathbf{Y}_2}) \mathcal{N}(\mathbf{x}; (\Lambda_{\mathbf{Y}_1} + \Lambda_{\mathbf{Y}_2})^{-1}(\Lambda_{\mathbf{Y}_2}\mathbf{y}_2 + \Lambda_{\mathbf{Y}_1}\mathbf{y}_1), (\Lambda_{\mathbf{Y}_2} + \Lambda_{\mathbf{Y}_1})^{-1}) \tag{D.28}$$

for some scaling factor $K(\mathbf{y}_1, \mathbf{y}_2, \Sigma_{\mathbf{Y}_1}, \Sigma_{\mathbf{Y}_2})$ independent of $\mathbf{x}$. Note that we used $\Lambda_{\mathbf{Y}_1} = \Sigma_{\mathbf{Y}_1}^{-1}$ and $\Lambda_{\mathbf{Y}_2} = \Sigma_{\mathbf{Y}_2}^{-1}$ to denote the precision matrices of $\mathbf{Y}_1$ and $\mathbf{Y}_2$ above.

Of course, one can compute the scaling factor $K(\mathbf{y}_1, \mathbf{y}_2, \Sigma_{\mathbf{Y}_1}, \Sigma_{\mathbf{Y}_2})$ directly. However, it is much easier to realize that

$$\mathcal{N}(\mathbf{y}_1; \mathbf{x}, \Sigma_{\mathbf{Y}_1}) \mathcal{N}(\mathbf{y}_2; \mathbf{x}, \Sigma_{\mathbf{Y}_2}) = \mathcal{N}(\mathbf{y}_1; \mathbf{x}, \Sigma_{\mathbf{Y}_1}) \mathcal{N}(\mathbf{x}; \mathbf{y}_2, \Sigma_{\mathbf{Y}_2}) = p(\mathbf{y}_1, \mathbf{x} | \mathbf{y}_2) \tag{D.29}$$

for $\mathbf{X}$ and $\mathbf{Y}_1$ to be conditionally independent given $\mathbf{Y}_2$, with the setup as shown in Figure D.1.

Then, marginalizing $\mathbf{x}$ out from $p(\mathbf{y}_1, \mathbf{x} | \mathbf{y}_2)$, we have $p(\mathbf{y}_1 | \mathbf{y}_2) = \int p(\mathbf{y}_1, \mathbf{x} | \mathbf{y}_2) d\mathbf{x}$. However, from Figure D.1,

$$\int p(\mathbf{y}_1, \mathbf{x} | \mathbf{y}_2) d\mathbf{x} = p(\mathbf{y}_1 | \mathbf{y}_2) = \mathcal{N}(\mathbf{y}_1; \mathbf{y}_2, \Sigma_{\mathbf{Y}_2} + \Sigma_{\mathbf{Y}_1}) \tag{D.30}$$

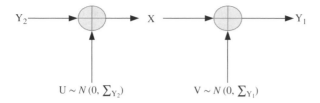

Figure D.1 The conditional pdf $p(\mathbf{y}_1, \mathbf{x}|\mathbf{y}_2) = p(\mathbf{y}_1|\mathbf{x})p(\mathbf{x}|\mathbf{y}_2) = \mathcal{N}(\mathbf{y}_1; \mathbf{x}, \Sigma_{\mathbf{Y}_1})\mathcal{N}(\mathbf{x}; \mathbf{y}_2, \Sigma_{\mathbf{Y}_2})$ if $\mathbf{X} = \mathbf{U} + \mathbf{Y}_2$ and $\mathbf{Y}_1 = \mathbf{V} + \mathbf{X}$, where $\mathbf{U} \sim (0, \Sigma_{\mathbf{Y}_2})$ is independent of $\mathbf{Y}_2$ and $\mathbf{V} \sim (0, \Sigma_{\mathbf{Y}_1})$ is independent of $\mathbf{X}$.

but from D.28,

$$\int p(\mathbf{y}_1, \mathbf{x}|\mathbf{y}_2)d\mathbf{x} = \int \mathcal{N}(\mathbf{y}_1; \mathbf{x}, \Sigma_{\mathbf{Y}_1})\mathcal{N}(\mathbf{y}_2; \mathbf{x}, \Sigma_{\mathbf{Y}_2})d\mathbf{x} \quad\quad\text{(D.31)}$$

$$= \int K(\mathbf{y}_1, \mathbf{y}_2, \Sigma_{\mathbf{Y}_1}, \Sigma_{\mathbf{Y}_2})\mathcal{N}(\mathbf{x}; \quad\quad\quad\quad\text{(D.32)}$$

$$(\Lambda_{\mathbf{Y}_1} + \Lambda_{\mathbf{Y}_2})^{-1}(\Lambda_{\mathbf{Y}_2}\mathbf{y}_2 + \Lambda_{\mathbf{Y}_1}y), (\Lambda_{\mathbf{Y}_2} + \Lambda_{\mathbf{Y}_1})^{-1})d\mathbf{x}$$

$$= K(\mathbf{y}_1, \mathbf{y}_2, \Sigma_{\mathbf{Y}_1}, \Sigma_{\mathbf{Y}_2}). \quad\quad\quad\quad\text{(D.33)}$$

In summary,

$$\mathcal{N}(\mathbf{y}_1; \mathbf{x}, \Sigma_{\mathbf{Y}_1})\mathcal{N}(\mathbf{y}_2; \mathbf{x}, \Sigma_{\mathbf{Y}_2})$$

$$=\mathcal{N}(\mathbf{y}_1; \mathbf{y}_2, \Sigma_{\mathbf{Y}_2} + \Sigma_{\mathbf{Y}_1})\mathcal{N}(\mathbf{x}; (\Lambda_{\mathbf{Y}_1} + \Lambda_{\mathbf{Y}_2})^{-1}(\Lambda_{\mathbf{Y}_2}\mathbf{y}_2 + \Lambda_{\mathbf{Y}_1}y), (\Lambda_{\mathbf{Y}_2} + \Lambda_{\mathbf{Y}_1})^{-1}).$$
$$\text{(D.34)}$$

Let us try to interpret the product as the overall likelihood after making two observations. For simplicity, let us also assume that $\mathbf{X}$, $\mathbf{Y}_1$ and $\mathbf{Y}_2$ are all scaler.[3] The mean considering both observations, $(\Lambda_{\mathbf{Y}_1} + \Lambda_{\mathbf{Y}_2})^{-1}(\Lambda_{\mathbf{Y}_2}\mathbf{y}_2 + \Lambda_{\mathbf{Y}_1}y)$, is essentially a weighted average of observations $\mathbf{y}_2$ and $\mathbf{y}_1$. The weight is higher when the precision $\Lambda_{\mathbf{Y}_2}$ or $\Lambda_{\mathbf{Y}_1}$ is larger, and the overall variance $(\Lambda_{\mathbf{Y}_2} + \Lambda_{\mathbf{Y}_1})^{-1}$ is always smaller than the individual variance $\Sigma_{\mathbf{Y}_2}$ and $\Sigma_{\mathbf{Y}_1}$. This can be understood since we are more certain with $\mathbf{x}$ after considering both $\mathbf{y}_1$ and $\mathbf{y}_2$. Finally, the scaling factor, $\mathcal{N}(\mathbf{y}_1; \mathbf{y}_2, \Sigma_{\mathbf{Y}_2} + \Sigma_{\mathbf{Y}_1})$, can be interpreted as how much one can believe in the overall likelihood. The value is reasonable since when the two observations are far away with respect to the overall variance $\Sigma_{\mathbf{Y}_2} + \Sigma_{\mathbf{Y}_1}$, the likelihood will become less reliable. The scaling factor is especially important when we deal with a mixture of Gaussian in Section D.7.

---

3 Again, for consistency, we will keep using the vector convention for the rest of this section.

D.6   Division of Gaussian pdfs

To compute $\frac{\mathcal{N}(\mathbf{x};\mu_1,\Sigma_1)}{\mathcal{N}(\mathbf{x};\mu_2,\Sigma_2)}$, note that from the product formula D.34

$$
\mathcal{N}(\mathbf{x};\mu_2,\Sigma_2)\mathcal{N}(\mathbf{x};(\Lambda_1-\Lambda_2)^{-1}(\Lambda_1\mu_1-\Lambda_2\mu_2),(\Lambda_1-\Lambda_2)^{-1})
$$
$$
=\mathcal{N}(\mu_1;(\Lambda_1-\Lambda_2)^{-1}(\Lambda_1\mu_1-\Lambda_2\mu_2),\Lambda_2^{-1}+(\Lambda_1-\Lambda_2)^{-1})\mathcal{N}(\mathbf{x};\mu_1,\Sigma_1).
$$
$$
\text{(D.35)}
$$

Therefore,

$$
\frac{\mathcal{N}(\mathbf{x};\mu_1,\Sigma_1)}{\mathcal{N}(\mathbf{x};\mu_2,\Sigma_2)} = \frac{\mathcal{N}(\mathbf{x};(\Lambda_1-\Lambda_2)^{-1}(\Lambda_1\mu_1-\Lambda_2\mu_2),(\Lambda_1-\Lambda_2)^{-1})}{\mathcal{N}(\mu_1,(\Lambda_1-\Lambda_2)^{-1}(\Lambda_1\mu_1-\Lambda_2\mu_2);\Lambda_2^{-1}+(\Lambda_1-\Lambda_2)^{-1})}
$$
$$
\text{(D.36)}
$$

$$
= \frac{\mathcal{N}(\mathbf{x};\mu,(\Lambda_1-\Lambda_2)^{-1})}{\mathcal{N}(\mu_1;\mu,\Lambda_2^{-1}+(\Lambda_1-\Lambda_2)^{-1})},
$$
$$
\text{(D.37)}
$$

where $\mu = (\Lambda_1-\Lambda_2)^{-1}(\Lambda_1\mu_1-\Lambda_2\mu_2)$. Note that the final pdf will be Gaussian-like if $\Lambda_1 \succeq \Lambda_2$. Otherwise, one can still write out the pdf using the precision matrix, but the covariance matrix will not be defined.

D.7   Mixture of Gaussians

Consider an electrical system that outputs signals of different statistics when it is on and off. When the system is on, the output signal S behaves like $\mathcal{N}(5, 1)$. When the system is off, S behaves like $\mathcal{N}(0, 1)$. If someone measuring the signal does not know the status of the system but only knows that the system is on 40% of the time then to the observer the signal S behaves like a mixture of Gaussians. The pdf of S will be $0.4\mathcal{N}(s; 5, 1) + 0.6\mathcal{N}(s; 0, 1)$, as shown in Figure D.2.

The main limitation of Gaussian distribution is that it is unimodal. By mixing Gaussian pdfs of different means, mixtures of Gaussian pdfs are multimodal and can virtually model any pdfs. However, there is a computational cost for this extra power. Let us illustrate this with the following example.

Consider two mixtures of Gaussian likelihood x given two observations y_1 and y_2 as follows:

$$
p(y_1|x) = 0.6\mathcal{N}(x; 0, 1) + 0.4\mathcal{N}(x; 5, 1); \tag{D.38}
$$
$$
p(y_2|x) = 0.5\mathcal{N}(x; -2, 1) + 0.5\mathcal{N}(x; 4, 1). \tag{D.39}
$$

What is the overall likelihood, $p(y_1, y_2|x)$?

As usual, it is reasonable to assume the observations to be conditionally independent given x. Then, the overall likelihood $p(y_1, y_2|x)$ just equal to the product

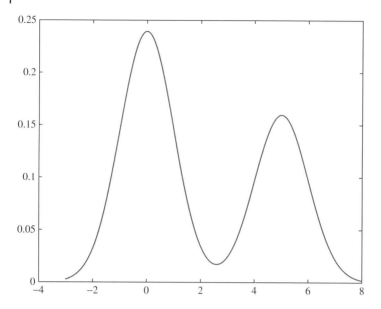

Figure D.2 The pdf of a mixture of Gaussians $(0.4\mathcal{N}(5,1) + 0.6\mathcal{N}(0,1))$.

of likelihoods $p(y_1|x)p(y_2|x)$, that is,

$$p(y_1, y_2|x) = (0.6\mathcal{N}(x; 0, 1) + 0.4\mathcal{N}(x; 5, 1))(0.5\mathcal{N}(x; -2, 1) \qquad \text{(D.40)}$$
$$+ 0.5\mathcal{N}(x; 4, 1))$$
$$= 0.3\mathcal{N}(x; 0, 1)\mathcal{N}(x; -2, 1) + 0.2\mathcal{N}(x; 5, 1)\mathcal{N}(x; -2, 1)$$
$$+ 0.3\mathcal{N}(x; 0, 1)\mathcal{N}(x; 4, 1) + 0.2\mathcal{N}(x; 5, 1)\mathcal{N}(x; 4, 1).$$
$$\text{(D.41)}$$

This involves computing products of Gaussians but we have learned this in previous sections. Using (D.34),

$$p(y_1, y_2|x) = 0.3\mathcal{N}(-2; 0, 2)\mathcal{N}(x; -1, 0.5) + 0.2\mathcal{N}(-2; 5, 2)$$
$$\mathcal{N}(x; 1.5, 0.5) + 0.3\mathcal{N}(4; 0, 2)\mathcal{N}(x; 2, 0.5)$$
$$+ 0.2\mathcal{N}(4; 5, 2)\mathcal{N}(x; 4.5, 0.5). \qquad \text{(D.42)}$$

So the overall likelihood is a mixture of four Gaussians.

D.7.1 Reduce the Number of Components in Gaussian Mixtures

Let us say we have a likelihood of x given that an observation is a mixture of two Gaussians, as just discussed, and we have n such similar observations. The overall likelihood will be a mixture of 2^n Gassians, therefore the computation will quickly become intractable as the number of observations increases. Fortunately, in reality, some of the Gaussians in the mixture tend to have a very

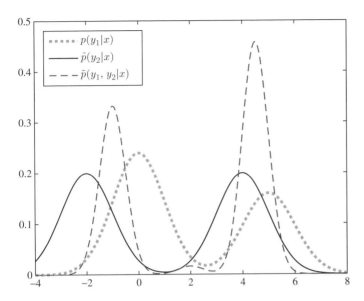

Figure D.3 Likelihood functions: $p(y_1|x) = 0.6\mathcal{N}(x; 0, 1) + 0.4\mathcal{N}(x; 5, 1)$, $p(y_2|x) = 0.5\mathcal{N}(x; -2, 1) + 0.5\mathcal{N}(x; 4, 1)$, $p(y_1, y_2|x) = p(y_1|x)p(y_2|x) = 0.4163\mathcal{N}(x; -1, 0.5) + 3.5234 \times 10^{-6}\mathcal{N}(x; 1.5, 0.5) + 0.0202\mathcal{N}(x; 2, 0.5) + 0.5734\mathcal{N}(x; 4.5, 0.5)$.

small weight. For instance, in our previous numerical example, if we continue our numerical computation in (D.42), we have

$$p(y_1, y_2|x) = 0.4163\mathcal{N}(x; -1, 0.5) + 3.5234 \times 10^{-6}\mathcal{N}(x; 1.5, 0.5)$$
$$+ 0.0202\mathcal{N}(x; 2, 0.5) + 0.5734\mathcal{N}(x; 4.5, 0.5). \tag{D.43}$$

We can see that the weight for the component at mean 1.5 is very small and the component at mean 2 also has a rather small weight. Even with the four Gaussian components, the overall likelihood is essentially just a bimodal distribution, as shown in Figure D.3. Therefore, we may approximate $p(y_1, y_2|x)$ with only two of its original component as $0.4163/(0.4163 + 0.5734)$ $\mathcal{N}(x; -1, 0.5) + 0.5734/(0.4163 + 0.5734)\mathcal{N}(x; 4.5, 0.5) = 0.4206\mathcal{N}(x; -1, 0.5)$ $+ 0.5794\mathcal{N}(x; 4.5, 0.5)$.

However, it is not always a good approximation strategy just to discard the small components in a Gaussian mixture. For example, consider

$$p(x) = 0.1\mathcal{N}(x; -0.2, 1) + 0.1\mathcal{N}(x; -0.1, 1) + 0.1\mathcal{N}(x; 0, 1)$$
$$+ 0.1\mathcal{N}(x; 0.1, 1) + 0.1\mathcal{N}(x; 0.2, 1) + 0.5\mathcal{N}(x; 5, 1). \tag{D.44}$$

Let's say we want to reduce $p(x)$ to only a mixture of two Gaussians. It is tempting to just discard the four smallest ones and renormalized the weight. For

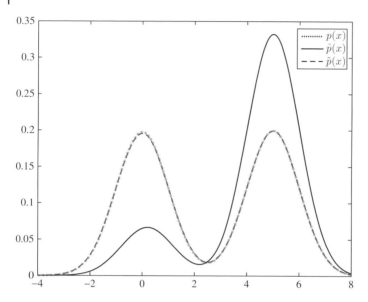

Figure D.4 Approximate $p(x) = 0.1\mathcal{N}(x; -0.2, 1) + 0.1\mathcal{N}(x; -0.1, 1) + 0.1\mathcal{N}(x; 0, 1) + 0.1\mathcal{N}(x; 0.1, 1) + 0.1\mathcal{N}(x; 0.2, 1) + 0.5\mathcal{N}(x; 5, 1)$ by discarding the smallest weight components ($\hat{p}(x) = 1/6\mathcal{N}(x; 0.2, 1) + 5/6\mathcal{N}(x; 5, 1)$) and merging similar components ($\tilde{p}(x) = 0.5\mathcal{N}(x; 0, 1.02) + 0.5\mathcal{N}(x; 5, 1)$). The latter approximation does so well that $p(x)$ and $\tilde{p}(x)$ essentially overlap each other.

example, if we choose to remove the first four components, we have

$$\hat{p}(x) = 1/6\mathcal{N}(x; 0.2, 1) + 5/6\mathcal{N}(x; 5, 1), \tag{D.45}$$

which is significantly different from $p(x)$, as shown in Figure D.4.

The problem is that while the first five components are all relatively small compared to the last one, they are all quite similar and their combined contribution is comparable to the latter. Actually, as one can see from Figure D.4, the first five components are so similar that their combined contribution can be accurately modeled as one Gaussian. So one can get a much more accurate approximation by merging these components rather than discarding them. Such an approximation $\tilde{p}(x)$ is also illustrated in Figure D.4. However, to successfully obtain this approximation $\tilde{p}(x)$ we have to answer two questions: which components should be merged and how do we merge them? We will address these questions in the following section [149].

Which Components to Merge?

It is reasonable to pick similar components to merge. The question is how to gauge the similarity between two components. Consider two pdfs $p(\mathbf{x})$ and $q(\mathbf{x})$.

Note that we can define an inner product of $p(\mathbf{x})$ and $q(\mathbf{x})$ by

$$\langle p(\mathbf{x}), q(\mathbf{x}) \rangle = \int p(\mathbf{x})q(\mathbf{x})d\mathbf{x}. \tag{D.46}$$

Such an inner product is well defined, in particular $\langle p(\mathbf{x}), p(\mathbf{x}) \rangle \geq 0$. Therefore, by the Cauchy–Schwartz inequality,

$$\frac{\langle p(\mathbf{x}), q(\mathbf{x}) \rangle}{\sqrt{\langle p(\mathbf{x}), p(\mathbf{x}) \rangle \langle q(\mathbf{x}), q(\mathbf{x}) \rangle}} = \frac{\int p(\mathbf{x})q(\mathbf{x})d\mathbf{x}}{\sqrt{\int p(\mathbf{x})^2 d\mathbf{x} \int q(\mathbf{x})^2 d\mathbf{x}}} \leq 1. \tag{D.47}$$

Moreover, the latter equality holds only when $p(\mathbf{x}) = q(\mathbf{x})$. This suggests a very reasonable similarity measure between the two pdfs. Let us define

$$Sim(p(\mathbf{x}), q(\mathbf{x})) \triangleq \frac{\int p(\mathbf{x})q(\mathbf{x})d\mathbf{x}}{\sqrt{\int p(\mathbf{x})^2 d\mathbf{x} \int q(\mathbf{x})^2 d\mathbf{x}}}. \tag{D.48}$$

In particular, if $p(\mathbf{x}) = \mathcal{N}(\mathbf{x}; \boldsymbol{\mu}_p, \Sigma_p)$ and $q(\mathbf{x}) = \mathcal{N}(\mathbf{x}; \boldsymbol{\mu}_q, \Sigma_q)$, we have

$$Sim(\mathcal{N}(\boldsymbol{\mu}_p, \Sigma_p), \mathcal{N}(\boldsymbol{\mu}_q, \Sigma_q)) = \frac{\mathcal{N}(\boldsymbol{\mu}_p; \boldsymbol{\mu}_q, \Sigma_p + \Sigma_q)}{\sqrt{\mathcal{N}(0; 0, 2\Sigma_p)\mathcal{N}(0; 0, 2\Sigma_q)}}, \tag{D.49}$$

which can be computed very easily and is equal to one only when means and covariances are the same.

How to Merge Components?
Say we have n components $\mathcal{N}(\boldsymbol{\mu}_1, \Sigma_1), \mathcal{N}(\boldsymbol{\mu}_2, \Sigma_2), \cdots, \mathcal{N}(\boldsymbol{\mu}_n, \Sigma_n)$ with weights $w_1, w_2, \cdots, w_n$. What should the combined component be like? First of all, the combined component obviously will have weight equal to the combined weight $\sum_{i=1}^n w_i$ and its mean will simply be $\sum_{i=1}^n \hat{w}_i \boldsymbol{\mu}_i$, where $\hat{w}_i = \frac{w_i}{\sum_{i=1}^n w_i}$.

It is tempting to write the combined covariance as $\sum_{i=1}^n \hat{w}_i \Sigma_i$. However, the covariance is more than that. Because the sum only counts the contribution of variation among each component, it does not take into account the variation due to different means across components. Instead, let us denote $\mathbf{X}$ as the variable sampled from the mixture, that is, $\mathbf{X} \sim \mathcal{N}(\boldsymbol{\mu}_i, \Sigma_i)$ with probability $\hat{w}_i$. Then we have the combined covariance Σ given by

$$\Sigma = E[\mathbf{X}\mathbf{X}^T] - E[\mathbf{X}]E[\mathbf{X}]^T \tag{D.50}$$

$$= \sum_{i=1}^n \hat{w}_i(\Sigma_i + \boldsymbol{\mu}_i\boldsymbol{\mu}_i^T) - \sum_{i=1}^n \sum_{j=1}^n \hat{w}_i\hat{w}_j\boldsymbol{\mu}_i\boldsymbol{\mu}_j^T. \tag{D.51}$$

Now, go back to our previous numerical example. Recall that $p(x) = 0.1\mathcal{N}(x; -0.2, 1) + 0.1\mathcal{N}(x; -0.1, 1) + 0.1\mathcal{N}(x; 0, 1) + 0.1\mathcal{N}(x; 0.1, 1) + 0.1\mathcal{N}(x; 0.2, 1) + 0.5\mathcal{N}(x; 5, 1)$. If we merge the five smallest components (one can easily check that they are also more similar to each other than to the last component), we have $\tilde{p}(x) = 0.5\mathcal{N}(x; 0, 1.02) + 0.5\mathcal{N}(x; 5, 1)$, as shown in Figure D.4, where the approximate pdf is virtually indistinguishable from the original.

D.8 Summary

Below we assume $Y_2 \sim \mathcal{N}(\mu, \Sigma)$, $Y_2 = \begin{pmatrix} X \\ Y \end{pmatrix}$, $\mu = \begin{pmatrix} \mu_X \\ \mu_Y \end{pmatrix}$, $\Sigma = \begin{pmatrix} \Sigma_{XX} & \Sigma_{XY} \\ \Sigma_{YX} & \Sigma_{YY} \end{pmatrix}$,

$\Sigma^{-1} = \Lambda = \begin{pmatrix} \Lambda_{XX} & \Lambda_{XY} \\ \Lambda_{YX} & \Lambda_{YY} \end{pmatrix}$.

Marginal pdf of X:

$$\boxed{X \sim \mathcal{N}(\mu_X, \Sigma_{XX})} \tag{D.52}$$

Conditional pdf of X given observation y:

$$\boxed{X|y \sim \mathcal{N}(\mu_X + \Sigma_{XY}\Sigma_{YY}^{-1}(y - \mu_Y), \Sigma_{XX} - \Sigma_{XY}\Sigma_{YY}^{-1}\Sigma_{YX})} \tag{D.53}$$

Product of Gaussian pdfs:

$$\boxed{\begin{aligned} &\mathcal{N}(x; y_1, \Sigma_{Y_1})\mathcal{N}(x; y_2, \Sigma_{Y_2}) \\ &= \mathcal{N}(y_1; y_2, \Sigma_{Y_2} + \Sigma_{Y_1})\mathcal{N}(x; (\Lambda_{Y_1} + \Lambda_{Y_2})^{-1}(\Lambda_{Y_2}y_2 + \Lambda_{Y_1}y_1), \\ &(\Lambda_{Y_2} + \Lambda_{Y_1})^{-1}) \end{aligned}} \tag{D.54}$$

Division of Gaussian pdfs:

$$\boxed{\frac{\mathcal{N}(x; \mu_1, \Sigma_1)}{\mathcal{N}(x; \mu_2, \Sigma_2)} = \frac{\mathcal{N}(x; (\Lambda_1 - \Lambda_2)^{-1}(\Lambda_1\mu_1 - \Lambda_2\mu_2), (\Lambda_1 - \Lambda_2)^{-1})}{\mathcal{N}(\mu_1, (\Lambda_1 - \Lambda_2)^{-1}(\Lambda_1\mu_1 - \Lambda_2\mu_2); \Lambda_2^{-1} + (\Lambda_1 - \Lambda_2)^{-1})}}$$

$$\tag{D.55}$$

Measure Similarity between Two Gaussian pdfs:

$$\boxed{Sim(\mathcal{N}(\mu_p, \Sigma_p), \mathcal{N}(\mu_q, \Sigma_q)) = \frac{\mathcal{N}(\mu_p; \mu_q, \Sigma_p + \Sigma_q)}{\sqrt{\mathcal{N}(0; 0, 2\Sigma_p)\mathcal{N}(0; 0, 2\Sigma_q)}}} \tag{D.56}$$

Merging n Gaussian Components in a Mixture:

Merging n components $\mathcal{N}(\mu_1, \Sigma_1)$, $\mathcal{N}(\mu_2, \Sigma_2)$, $\cdots$, $\mathcal{N}(\mu_n, \Sigma_n)$ with weights $w_1, w_2, \cdots, w_n$. Let the combined weight and combined component be w and

$\mathcal{N}(\boldsymbol{\mu}, \Sigma)$. Then, w, $\boldsymbol{\mu}$, and Σ are given by the following:

$$w = \sum_{i=1}^{n} w_i \tag{D.57}$$

$$\boldsymbol{\mu} = \sum_{i=1}^{n} \frac{w_i}{w} \boldsymbol{\mu}_i \tag{D.58}$$

$$\Sigma = \frac{1}{w} \sum_{i=1}^{n} w_i (\Sigma_i + \boldsymbol{\mu}_i \boldsymbol{\mu}_i^T) - \frac{1}{w^2} \sum_{i=1}^{n} \sum_{j=1}^{n} w_i w_j \boldsymbol{\mu}_i \boldsymbol{\mu}_j^T \tag{D.59}$$

Appendix: Matrix Equations

Throughout this section, we assume $\Sigma^{-1} = \Lambda$, $\Sigma = \begin{pmatrix} \Sigma_{XX} & \Sigma_{XY} \\ \Sigma_{YX} & \Sigma_{YY} \end{pmatrix}$, and $\Lambda = \begin{pmatrix} \Lambda_{XX} & \Lambda_{XY} \\ \Lambda_{YX} & \Lambda_{YY} \end{pmatrix}$. Note that many of the equations have "symmetry" and other similar forms hold as well. For example, apparently we also have $\Sigma_{YY} = \Lambda_{YY} - \Lambda_{YX} \Lambda_{XX}^{-1} \Lambda_{XY}$ from Lemma D.1. Without loss of generality, we simply pick one arbitrary form for each lemma in this section.

Lemma D.1 $\Sigma_{XX}^{-1} = \Lambda_{XX} - \Lambda_{XY} \Lambda_{YY}^{-1} \Lambda_{YX}$

Proof. Since $\Lambda = \Sigma^{-1}$, we have $\Sigma_{XX} \Lambda_{XY} + \Sigma_{XY} \Lambda_{YY} = 0$ and $\Sigma_{XX} \Lambda_{XX} + \Sigma_{XY} \Lambda_{YX} = I$. Inserting an identity into the latter equation, we have $\Sigma_{XX} \Lambda_{XX} + \Sigma_{XY} (\Lambda_{YY} \Lambda_{YY}^{-1}) \Lambda_{YX} = \Sigma_{XX} \Lambda_{XX} - (\Sigma_{XX} \Lambda_{XY}) \Lambda_{YY}^{-1} \Lambda_{YX} = \Sigma_{XX} (\Lambda_{XX} - \Lambda_{XY} \Lambda_{YY}^{-1} \Lambda_{YX}) = I$. □

Lemma D.2 $\det(\Sigma) = \det(\Sigma_{YY}) \det(\Lambda_{XX}^{-1})$

Proof.

$$\det(\Sigma) = \det \begin{pmatrix} \Sigma_{XX} & \Sigma_{XY} \\ \Sigma_{YX} & \Sigma_{YY} \end{pmatrix} \tag{D.60}$$

$$= \det \left(\begin{pmatrix} I & 0 \\ 0 & \Sigma_{YY} \end{pmatrix} \begin{pmatrix} \Sigma_{XX} & \Sigma_{XY} \\ \Sigma_{YY}^{-1} \Sigma_{YX} & I \end{pmatrix} \right) \tag{D.61}$$

$$= \det \left(\begin{pmatrix} I & 0 \\ 0 & \Sigma_{YY} \end{pmatrix} \begin{pmatrix} I & \Sigma_{XY} \\ 0 & I \end{pmatrix} \begin{pmatrix} \Sigma_{XX} - \Sigma_{XY} \Sigma_{YY}^{-1} \Sigma_{YX} & 0 \\ \Sigma_{YY}^{-1} \Sigma_{YX} & I \end{pmatrix} \right) \tag{D.62}$$

$$= \det \begin{pmatrix} I & 0 \\ 0 & \Sigma_{YY} \end{pmatrix} \det \begin{pmatrix} I & \Sigma_{XY} \\ 0 & I \end{pmatrix} \det \begin{pmatrix} \Sigma_{XX} - \Sigma_{XY} \Sigma_{YY}^{-1} \Sigma_{YX} & 0 \\ \Sigma_{YY}^{-1} \Sigma_{YX} & I \end{pmatrix} \tag{D.63}$$

$$= \det \Sigma_{YY} \det(\Sigma_{XX} - \Sigma_{XY}\Sigma_{YY}^{-1}\Sigma_{YX}) \qquad (D.64)$$

$$= \det \Sigma_{YY} \det \Lambda_{XX}^{-1}, \qquad (D.65)$$

where the last equality is from Lemma D.1. $\qquad \square$

Note that since the width (height) of Σ is equal to the sum of the widths of Σ_{XX} and Σ_{YY} the equation below follows immediately.

Corollary D.2 $\det(a\Sigma) = \det(a\Sigma_{YY}) \det(a\Lambda_{XX}^{-1})$ for any constant a.

Bibliography

1 D. Slepian and J. Wolf. Noiseless coding of correlated information sources. *IEEE Trans. Inform. Theory*, **19**:471–480, 1973.

2 A. Wyner and J. Ziv. The rate-distortion function for source coding with side information at the decoder. *IEEE Trans. Inform. Theory*, **22**:1–10, 1976.

3 T. Berger. *The Information Theory Approach to Communications*. Springer-Verlag, New York, 1977.

4 S.Y. Tung. *Multiterminal source coding*. PhD thesis, Cornell University, 1978.

5 S.S. Pradhan and K. Ramchandran. Distributed source coding using syndromes (discus): design and construction. In *Proceedings of the Data Compression Conference*, pages 158–167, 1999.

6 Y. Yang, V. Stankovic, Z. Xiong, and W. Zhao. On multiterminal source code design. *IEEE Trans. Inform. Theory*, **54**(5):2278–2302, 2008.

7 C. Berrou and A. Glavieux. Near optimum error correcting coding and decoding: turbo-codes. *IEEE Trans. Commun.*, **44**(10):1261, 1996.

8 D. MacKay. Good error-correcting codes based on very sparse matrices. *IEEE Trans. Inform. Theory*, **45**:399–431, 1999.

9 H.S. Witsenhausen and A.D. Wyner. Interframe coder for video signals, March 1980. US Patent 4,191,970.

10 R. Puri and K. Ramchandran. Prism: A new robust video coding architecture based on distributed compression principles. In *Proceedings of the Allerton Conference on Communication, Control, and Computing*, Allerton, IL, 2002.

11 A. Aaron, R. Zhang, and B. Girod. Wyner–Ziv coding of motion video. In *Proceedings of the Asilomar Conference on Signals, Systems and Computers*, volume 1, pages 240–244. IEEE, 2002.

12 T. Cover and J. Thomas. *Elements of Information Theory*. Wiley, New York, second edition, 2006.

13 J.K. Wolf. Data reduction for multiple correlated sources. In *Proceedings of the 5th Colloquium on Microwave Communication*, pages 287–295, 1973.

Distributed Source Coding: Theory and Practice, First Edition. Shuang Wang, Yong Fang, and Samuel Cheng.
© 2017 John Wiley & Sons Ltd. Published 2017 by John Wiley & Sons Ltd.
Companion Website: www.wiley.com/go/cheng/dsctheoryandpractice

14 A. Wyner. On source coding with side information at the decoder. *IEEE Trans. Inform. Theory*, **21**(3):294–300, 1975.

15 A.D. Wyner and J. Ziv. The rate-distortion function for source coding with side information at the decoder. *IEEE Trans. Inform. Theory*, **22**(1):1–10, 1976.

16 S. Cheng and Z. Xiong. Successive refinement for the Wyner–Ziv problem and layered code design. *IEEE Trans. Signal Processing*, **53**(8 Part 2):3269–3281, 2005.

17 N. Deligiannis, A. Sechelea, A. Munteanu, and S. Cheng. The no-rate-loss property of Wyner–Ziv coding in the z-channel correlation case. *IEEE Commun. Lett.*, **18**(10):1675–1678, 2014.

18 R. Zamir. The rate loss in the Wyner–Ziv problem. *IEEE Trans. Inform. Theory*, **42**(6):2073–2084, 1996.

19 Y. Oohama. Rate-distortion theory for Gaussian multiterminal source coding systems with several side informations at the decoder. *IEEE Trans. Inform. Theory*, **51**(7):2577–2593, 2005.

20 V. Prabhakaran, D. Tse, and K. Ramachandran. Rate region of the quadratic Gaussian CEO problem. In *International Symposium on Information Theory*, page 119. IEEE, 2004.

21 A.E. Gamal and Y.H. Kim. *Network Information Theory*. Cambridge University Press, 2012.

22 A. Wyner. Recent results in the Shannon theory. *IEEE Trans. Inform. Theory*, **20**:2–10, 1974.

23 N. Deligiannis, J. Barbarien, M. Jacobs, A. Munteanu, A. Skodras, and P. Schelkens. *Side-information-dependent correlation channel estimation in hash-based distributed video coding*, IEEE Transactions on Image Processing, vol. **21**, no. 4, pp. 1934–1949, 2012.

24 B. Rimoldi and R. Urbanke. Asynchronous Slepian–Wolf coding via source-splitting. In *ISIT '97*, Ulm, 1997.

25 T.P. Coleman, A.H. Lee, M. Medard, and M. Effros. On some new approaches to practical Slepian–Wolf compression inspired by channel coding. In *Proceedings of the Data Compression Conference*, pages 282–291, Snowbird, UT, 2004.

26 S.S. Pradhan and K. Ramchandran. Generalized coset codes for distributed binning. *IEEE Trans. Inform. Theory*, **51**(10):3457–3474, 2005.

27 V. Stankovic, A.D. Liveris, Z. Xiong, and C.N. Georghiades. On code design for the Slepian–Wolf problem and lossless multiterminal networks. *IEEE Trans. Inform. Theory*, **52**(4):1495–1507, 2006.

28 S. Cheng, S. Wang, and L. Cui. Adaptive Slepian–Wolf decoding using particle filtering based belief propagation. In *47th Annual Allerton Conference on Communication, Control, and Computing*, Urbana-Champaign, IL, 2009.

29 S. Cheng, S. Wang, and L. Cui. Adaptive nonasymmetric Slepian–Wolf decoding using particle filtering based belief propagation. In *IEEE ICASSP*, Dallas, TX, 2010.

30 A.W. Eckford and W. Yu. Rateless Slepian–Wolf codes. In *Proceedings of the Asilomar Conference on Signals, Systems and Computers*, pages 1757–1761, 2005.

31 J. Jiang, D. He, and A. Jagmohan. Rateless Slepian–Wolf coding based on rate adaptive low-density-parity-check codes. In *2007 IEEE International Symposium on Information Theory*, pages 1316–1320. IEEE, 2007.

32 R. Ma and S. Cheng. Hamming coding for multiple sources. In *2010 IEEE International Symposium on Information Theory*, pages 171–175. IEEE, 2010.

33 R. Ma and S. Cheng. The universality of generalized hamming code for multiple sources. *IEEE Trans. Commun.*, 59(10):2641–2647, 2011.

34 R. Ma and S. Cheng. Zero-error Slepian–Wolf coding of confined-correlated sources with deviation symmetry. *IEEE Trans. Inform. Theory*, 59(12):8195–8209, 2013.

35 M. Grangetto, E. Magli, and G. Olmo. Distributed arithmetic coding. *IEEE Commun. Lett.*, 11(11):883–885, 2007.

36 M. Grangetto, E. Magli, and G. Olmo. Distributed arithmetic coding for the Slepian–Wolf problem. *IEEE Trans. Signal Processing*, 57(6):2245–2257, 2009.

37 X. Artigas, S. Malinowski, C. Guillemot, and L. Torres. Overlapped quasi-arithmetic codes for distributed video coding. In *International Conference on Image Processing*, volume II, pages 9–12, 2007.

38 S. Malinowski, X. Artigas, C. Guillemot, and L. Torres. Distributed coding using punctured quasi-arithmetic codes for memory and memoryless sources. *IEEE Trans. Signal Processing*, 57(10):4154–4158, 2009.

39 J.J. Rissanen. Generalized Kraft inequality and arithmetic coding. *IBM J. Research and Development*, 20(3):198–203, 1976.

40 I. Witten, R. Neal, and J. Cleary. Arithmetic coding for data compression. *Commun. ACM*, 30(6):520–540, 1987.

41 Y. Fang. Distribution of distributed arithmetic codewords for equiprobable binary sources. *IEEE Signal Process. Lett.*, 16(12):1079–1082, 2009.

42 Y. Fang. DAC spectrum of binary sources with equally-likely symbols. *IEEE Trans. Commun.*, 61(4):1584–1594, 2013.

43 http://eqworld.ipmnet.ru/en/solutions/fe/fe1111.pdf.

44 http://en.wikipedia.org/wiki/Sinc_function.

45 D. Slepian and J. Wolf. Noiseless coding of correlated information sources. *IEEE Trans. Inform. Theory*, 19(4):471–480, 1973.

46 D. Varodayan, A. Aaron, and B. Girod. Rate-adaptive codes for distributed source coding. *Signal Process.*, 86(11):3123–3130, 2006.

47 http://www.stanford.edu/~divad/software.html.

48 M. Grangetto, E. Magli, and G. Olmo. Symmetric distributed arithmetic coding of correlated sources. In *Proceedings of the IEEE MMSP*, pages 111–114, 2007.

49 M. Grangetto, E. Magli, R. Tron, and G. Olmo. Rate-compatible distributed arithmetic coding. *IEEE Commun. Lett.*, **12**(8):575–577, 2008.

50 M. Grangetto, E. Magli, and G. Olmo. Decoder-driven adaptive distributed arithmetic coding. In *International Conference on Image Processing*, pages 1128–1131, 2008.

51 M. Grangetto, E. Magli, and G. Olmo. Distributed joint source-channel arithmetic coding. In *International Conference on Image Processing*, pages 3717–3720, 2010.

52 J. Conway and N.J.A. Sloane. *Sphere Packings, Lattices and Groups.* Springer, 1999.

53 Y. Yang, S. Cheng, Z. Xiong, and W. Zhao. Wyner–Ziv coding based on TCQ and LDPC codes. *IEEE Trans. Commun.*, **57**(2):376–387, 2009.

54 S.S. Pradhan, J. Chou, and K. Ramchandran. Duality between source coding and channel coding and its extension to the side information case. *IEEE Trans. Inform. Theory*, **49**:1181–1203, 2003.

55 D. Slepian and J. Wolf. Noiseless coding of correlated information sources. *IEEE Trans. Inform. Theory*, **19**:471–480, 1973.

56 J. Ascenso, C. Brites, and F. Pereira. Improving frame interpolation with spatial motion smoothing for pixel domain distributed video coding. In *5th EURASIP Conference on Speech and Image Processing, Multimedia Communications and Services*, pages 605–608, 2005.

57 A. Aaron, E. Setton, and B. Girod. Towards practical Wyner–Ziv coding of video. In *International Conference on Image Processing*, volume 3, pages III–869. IEEE, 2003.

58 C. Brites and F. Pereira. Correlation noise modeling for efficient pixel and transform domain Wyner–Ziv video coding. *IEEE Trans. Circuits Syst. Video Technol.*, **18**(9):1177–1190, 2008.

59 M. Dalai, R. Leonardi, and F. Pereira. Improving turbo codec integration in pixel-domain distributed video coding. In *2006 ICASSP Proceedings*, volume 2, pages II–II. IEEE.

60 M. Tagliasacchi, A. Trapanese, S. Tubaro, J. Ascenso, C. Brites, and F. Pereira. Exploiting spatial redundancy in pixel domain Wyner–Ziv video coding. In *2006 IEEE International Conference on Image Processing*, pages 253–256. IEEE, 2006.

61 A. Aaron, S. Rane, E. Setton, B. Girod, et al. Transform-domain Wyner–Ziv codec for video. In *Proceedings of SPIE Visual Communications and Image Processing*, volume 5308, pages 520–528, 2004.

62 L. Alparone, M. Barni, F. Bartolini, and V. Cappellini. Adaptively weighted vector-median filters for motion-fields smoothing. In *1996 IEEE International Conference on Acoustics, Speech, and Signal Processing*, volume 4, pages 2267–2270. IEEE, 1996.

63 A. Aaron, S. Rane, and B. Girod. Wyner–Ziv video coding with hash-based motion compensation at the receiver. In *International Conference on Image Processing*, volume 5, pages 3097–3100. IEEE, 2004.

64 R. Puri, A. Majumdar, and K. Ramchandran. Prism: A video coding paradigm with motion estimation at the decoder. *IEEE Trans. Image Processing*, 16(10):2436–2448, 2007.

65 M. Tagliasacchi, A. Trapanese, S. Tubaro, J. Ascenso, C. Brites, and F. Pereira. Intra mode decision based on spatio-temporal cues in pixel domain Wyner–Ziv video coding. In *2006 IEEE International Conference on Acoustics, Speech and Signal Processing*, volume 2, pages II–II. IEEE, 2006.

66 C. Brites, J. Ascenso, and F. Pereira. Improving transform domain Wyner–Ziv video coding performance. In *2006 IEEE International Conference on Acoustics, Speech and Signal Processing*, volume 2, pages II–II. IEEE, 2006.

67 B. Girod, A.M. Aaron, S. Rane, and D. Rebollo-Monedero. Distributed video coding. *Proceedings of the IEEE*, 93(1):71–83, 2005.

68 A. Aaron, S.D. Rane, E. Setton, and B. Girod. Transform-domain Wyner–Ziv codec for video. *Proceedings of SPIE*, 5308:520–528, 2004.

69 A. Aaron, S. Rane, and B. Girod. Wyner–Ziv video coding with hash-based motion compensation at the receiver. In *International Conference on Image Processing*, volume 5, pages 3097–3100, 2005.

70 P.F. Meyer, R.P. Westerlaken, R.K. Gunnewiek, and R.L. Lagendijk. Distributed source coding of video with non-stationary side-information. *Proceedings of SPIE*, 5960:857–866, 2005.

71 X. Fan, O.C. Au, and N.M. Cheung. Adaptive correlation estimation for general Wyner–Ziv video coding. In *International Conference on Image Processing*, pages 1409–1412. IEEE, 2009.

72 X. Huang and S. Forchhammer. Improved virtual channel noise model for transform domain Wyner–Ziv video coding. In *2009 IEEE International Conference on Acoustics, Speech and Signal Processing*, pages 921–924. IEEE, 2009.

73 A. Doucet and N. De Freitas. *Sequential Monte Carlo Methods in Practice*. Springer Verlag, 2001.

74 M. Bolic, P.M. Djuric, and S. Hong. Resampling algorithms for particle filters: A computational complexity perspective. *EURASIP Journal on Applied Signal Processing*, 15(5):2267–2277, 2004.

75 S. Chib and E. Greenberg. Understanding the Metropolis-Hastings algorithm. *American Statistician*, 49(4):327–335, 1995.

76 M.S. Arulampalam, S. Maskell, N. Gordon, and T. Clapp. A tutorial on particle filters for online nonlinear/non-Gaussian Bayesian tracking. *IEEE Trans. Signal Processing*, **50**(2):174–188, 2002.

77 X. Artigas, J. Ascenso, M. Dalai, S. Klomp, D. Kubasov, and M. Ouaret. The DISCOVER codec: architecture, techniques and evaluation. In *Picture Coding Symposium*. Citeseer, 2007.

78 G. Bjontegard. Calculation of average PSNR differences between rd-curves. *Doc. VCEG-M33 ITU-T Q6/16*, April 2001.

79 Z. Xiong, A. Liveris, and S. Cheng. Distributed source coding for sensor networks. *IEEE Signal Process. Magazine*, **21**:80–94, Sep. 2004.

80 T.P. Minka. Expectation propagation for approximate Bayesian inference. In *Uncertainty in Artificial Intelligence*, volume 17, pages 362–369, 2001.

81 Joint Video Team. Advanced video coding for generic audiovisual services. *ITU-T Rec. H*, 264:14496–10, 2003.

82 http://www.nasa.gov/mission_pages/stereo/mission/index.html.

83 http://24sevenpost.com/paranormal/ufo-spotted-visiting-sun-nasa-image.

84 G. Yua, T. Vladimirova, and M. Sweeting. Image compression systems on board satellites. *Elsevier Acta Astronautica*, **64**:988–1005, 2009.

85 ISO/IEC 15444-1. Information technology—JPEG2000 image coding system, part 1: Core coding system, 2000.

86 S.H. Seo and M.R. Azimi-Sadjadi. A 2-D filtering scheme for stereo image compression using sequential orthogonal subspace updating. *IEEE Trans. Circuits Syst. Video Technol.*, **11**(1):52–66, 2002.

87 A. Wyner and J. Ziv. The rate-distortion function for source coding with side information at the decoder. *IEEE Trans. Inform. Theory*, **22**:1–10, 1976.

88 Y. Wang, S. Rane, P. Boufounos, and A. Vetro. Distributed compression of zerotrees of wavelet coefficients. In *International Conference on Image Processing*, 2011.

89 A. Abrardo, M. Barni, E. Magli, and F. Nencini. Error-resilient and low-complexity onboard lossless compression of hyperspectral images by means of distributed source coding. *IEEE Trans. Geoscience and Remote Sensing*, **48**(4):1892–1904, 2010.

90 R. Nagura. Multi-stereo imaging system using a new data compression method. In *Proceedings of the IGARSS-2006 IEEE International Geoscience and Remote Sensing Symposium*, 2006.

91 A. Kiely, A. Ansar, A. Castano, M. Klimesh, and J. Maki. Lossy image compression and stereo ranging quality from Mars rovers. *IPN Progress Report*, pages 42–168, 2007.

92 Image compression algorithm altered to improve stereo ranging. Tech. Briefs. Available at http://www.techbriefs.com/component/content/article/2673, 2007.

93 L. Stankovic, V. Stankovic, S. Wang, and S. Cheng. Distributed video coding with particle filtering for correlation tracking. *Proceedings of EUSIPCO, Aalborg, Denmark*, 2010.

94 G. Petrazzuoli, M. Cagnazzo, and B. Pesquet-Popescu. Fast and efficient side information generation in distributed video coding by using dense motion representations. In *2010 18th European Signal Processing Conference*, pages 2156–2160. IEEE, 2010.

95 D. Varodayan, A. Mavlankar, M. Flierl, and B. Girod. Distributed grayscale stereo image coding with unsupervised learning of disparity. In *IEEE Data Compression Conference*, pages 143–152, 2007.

96 D. Varodayan, Y.C. Lin, A. Mavlankar, M. Flierl, and B. Girod. Wyner–Ziv coding of stereo images with unsupervised learning of disparity. In *Proceedings of the Picture Coding Symposium, Lisbon, Portugal*. Citeseer, 2007.

97 D. Chen, D. Varodayan, M. Flierl, and B. Girod. Distributed stereo image coding with improved disparity and noise estimation. In *IEEE ICASSP*, pages 1137–1140, 2008.

98 A. Zia, J.P. Reilly, and S. Shirani. Distributed parameter estimation with side information: A factor graph approach. In *IEEE ISIT*, pages 2556–2560, 2007.

99 V. Toto-Zarasoa, A. Roumy, and C. Guillemot. Maximum Likelihood BSC parameter estimation for the Slepian–Wolf problem. *IEEE Commun. Lett.*, 15(2):232–234, 2011.

100 N. Gehrig and P.L. Dragotti. Distributed compression of multi-view images using a geometrical coding approach. In *International Conference on Image Processing*, 2007.

101 X. Zhu, A. Aaron, and B. Girod. Distributed compression for large camera arrays. In *Proceedings of SSP-2003*, 2003.

102 A. Jagmohan, A. Sehgal, and N. Ahuja. Compression of light-field rendered images using coset codes. In *Proceedings of the Asilomar Conference on Signals, Systems and Computers*, 2003.

103 G. Toffetti, M. Tagliasacchi, M. Marcon, A. Sarti, S. Tubaro, and K. Ramchandran. Image compression in a multi-camera system based on a distributed source coding approach. In *Proceedings of Eusipco-2005*, 2005.

104 L. Stankovic, V. Stankovic, and S. Cheng. Distributed compression: Overview of current and emerging multimedia applications. In *International Conference on Image Processing*. IEEE, 2011.

105 C. Guillemot, F. Pereira, L. Torres, T. Ebrahimi, R. Leonardi, and J. Ostermann. Distributed monoview and multiview video coding: basics, problems and recent advances. *IEEE Signal Process. Magazine*, 24(5):67–76, 2007.

106 R. Zamir, S. Shamai, and U. Erez. Nested linear/lattice codes for structured multiterminal binning. *IEEE Trans. Inform. Theory*, **48**(6):1250–1276, 2002.

107 S. Wang, L. Cui, and S. Cheng. Adaptive Wyner–Ziv decoding using particle-based belief propagation. In *2010 IEEE Global Telecommunications Conference GLOBECOM*. IEEE, 2010.

108 G. Chambers, M.A. Rockey, R. Marshall, T. Russell, M. Weiner, and N. Kerkenbush. Achieving meaningful use going forward with the ehr and integrated service-lead initiatives. Technical report, DTIC Document, 2011.

109 O. Bouhaddou, J. Bennett, T. Cromwell, G. Nixon, J. Teal, M. Davis, R. Smith, L. Fischetti, D. Parker, Z. Gillen, et al. The Department of Veterans Affairs, Department of Defense, and Kaiser Permanente Nationwide Health Information Network Exchange in San Diego: Patient Selection, Consent, and Identity Matching. In *AMIA Annual Symposium Proceedings*, volume 2011, page 135. American Medical Informatics Association, 2011.

110 M. Hellman. An extension of the Shannon theory approach to cryptography. *IEEE Trans. Inform. Theory*, **23**(3):289–294, 1977.

111 D. Schonberg, C. Yeo, S.C. Draper, and K. Ramchandran. On compression of encrypted video. In *Proceedings of the Data Compression Conference*, pages 173–182. IEEE, 2007.

112 M. Johnson, P. Ishwar, V. Prabhakaran, D. Schonberg, and K. Ramchandran. On compressing encrypted data. *IEEE Trans. Signal Processing*, **52**(10):2992–3006, 2004.

113 D. Schonberg, S. Draper, and K. Ramchandran. On compression of encrypted images. In *2006 IEEE International Conference on Image Processing*, pages 269–272. IEEE, 2006.

114 D. Schonberg, S.C. Draper, and K. Ramchandran. On blind compression of encrypted correlated data approaching the source entropy rate. In *Proceedings of the 43rd Annual Allerton Conference on Communication, Control, and Computing*, 2005.

115 D. Schonberg, S.C. Draper, C. Yeo, and K. Ramchandran. Toward compression of encrypted images and video sequences. *IEEE Trans. Information Forensics and Security*, **3**(4):749–762, 2008.

116 C.E. Shannon. Communication in the presence of noise. *Proceedings of the IRE*, **37**(1):10–21, 1949.

117 S. Wang, L. Cui, L. Stankovic, V. Stankovic, and S. Cheng. Adaptive correlation estimation with particle filtering for distributed video coding. *IEEE Trans. Circuits Syst. Video Technol.*, **22**(5):649–658, 2012.

118 F. Kschischang, B. Frey, and H. Loeliger. Factor graphs and the sum-product algorithm. *IEEE Trans. Inform. Theory*, **47**:498–519, Feb. 2001.

119 S. Wang, X. Jiang, L. Ohno-Machado, L. Cui, S. Cheng, and Hongkai X. Privacy-preserving biometric system for secure fingerprint authentication. In *2012 IEEE Second International Conference on Healthcare Informatics, Imaging and Systems Biology (HISB)*, page 128, Sept. 2012.

120 Bob Barr. Biometrics pushed for medical records uses, April 2010.

121 D. Shapiro and S. Shapiro. Biometrics in pharma: Politics and privacy. In *Canadian Association for American Studies Conference*, 2010.

122 C.K. Leung, C.Y.L. Cheung, R.N. Weinreb, K. Qiu, S. Liu, H. Li, G. Xu, N. Fan, C.P. Pang, K.K. Tse, et al., Evaluation of retinal nerve fiber layer progression in glaucoma: a study on optical coherence tomography guided progression analysis. *Investigative Ophthalmol. Visual Sci.*, 51(1):217–222, 2010.

123 P. Ramani, P.R. Abhilash, H.J. Sherlin, N. Anuja, P. Priya, T. Chandrasekar, G. Sentamilselvi, and V.R. Janaki. Conventional dermatoglyphics -revived concept : A review. *Int. J. Pharma and Bio Sci.*, 2(3):B446–B458, 2011.

124 H. Chen. *Medical Genetics Handbook*. W.H. Green, St. Louis, MO.

125 B. Schneier. Inside risks: The uses and abuses of biometrics. *Communications of the ACM*, 42:136, 1999a.

126 T.C. Clancy, N. Kiyavash, and D.J. Lin. Secure smartcard based fingerprint authentication. In *Proceedings of the 2003 ACM SIGMM Workshop on Biometrics Methods and Applications*, pages 45–52. ACM, 2003.

127 S. Yang and I.M. Verbauwhede. Secure fuzzy vault based fingerprint verification system. In *Proceedings of the Asilomar Conference on Signals, Systems and Computers*, volume 1, pages 577–581. IEEE, 2004.

128 S.C. Draper, A. Khisti, E. Martinian, A. Vetro, and J. Yedidia. Secure storage of fingerprint biometrics using Slepian–Wolf codes. In *Information Theory and Applications Workshop in San Diego, CA*. UCSD, 2007.

129 S.C. Draper, A. Khisti, E. Martinian, A. Vetro, and J.S. Yedidia. Using distributed source coding to secure fingerprint biometrics. In *IEEE International Conference on Acoustics, Speech and Signal Processing*, volume 2, pages II–129. IEEE, 2007.

130 Y. Sutcu, S. Rane, J.S. Yedidia, S.C. Draper, and A. Vetro. Feature extraction for a Slepian–Wolf biometric system using LDPC codes. In *IEEE International Symposium on Information Theory 2008*, pages 2297–2301. IEEE, 2008.

131 A. Nagar, S. Rane, and A. Vetro. Alignment and bit extraction for secure fingerprint biometrics. In *SPIE Conference on Electronic Imaging*, pages 75410N–75410N. International Society for Optics and Photonics, 2010.

132 N.K. Ratha, J.H. Connell, and R.M. Bolle. Enhancing security and privacy in biometrics-based authentication systems. *IBM Systems Journal*, 40(3):614–634, 2001.

133 N.K. Ratha, J.H. Connell, and R.M. Bolle. A biometrics-based secure authentication system. In *Proceedings of the IEEE Workshop on Automatic Identification Advanced Technologies*, pages 70–73, 1999.

134 N.K. Ratha, J.H. Connell, and R.M. Bolle. Biometrics break-ins and band-aids. *Pattern Recognition Letters*, **24**(13):2105–2113, 2003.

135 C. Soutar. *Biometric system security*. White Paper. Bioscrypt, http://www.bioscrypt.com, 2002.

136 B. Schneier. *Applied Cryptography*, second edition. Wiley, 1996.

137 N.K. Ratha, S. Chikkerur, J.H. Connell, and R.M. Bolle. Generating cancelable fingerprint templates. *IEEE Trans. Pattern Analysis and Machine Intelligence*, **29**(4):561–572, 2007.

138 U. Uludag and A.K. Jain. Fuzzy fingerprint vault. In *Proceedings of the Workshop on Biometrics: Challenges Arising from Theory to Practice*, pages 13–16, 2004.

139 S. Yang and I. Verbauwhede. Automatic secure fingerprint verification system based on fuzzy vault scheme. In *IEEE International Conference on Acoustics, Speech, and Signal Processing 2005 (ICASSP'05)*, volume 5, pages v–609. IEEE, 2005.

140 Z. Xiong, A.D. Liveris, and S. Cheng. Distributed source coding for sensor networks. *IEEE Signal Process. Magazine*, **21**(5):80–94, 2004.

141 N. Gehrig and P.L. Dragotti. Symmetric and asymmetric Slepian–Wolf codes with systematic and nonsystematic linear codes. *IEEE Commun. Lett.*, **9**(1):61–63, 2005.

142 S. Cheng. Non-asymmetric Slepian–Wolf decoding using belief propagation. In *2009 43rd Annual Conference on Information Sciences and Systems*. IEEE, Baltimore, MD, 2009.

143 R.A. Rueppel. *Analysis and Design of Stream Ciphers*. Springer-Verlag, New York, 1986.

144 D. Schonberg, K. Ramchandran, and S.S. Pradhan. Distributed code constructions for the entire Slepian–Wolf rate region for arbitrarily correlated sources. In *Proceedings of the Data Compression Conference*, pages 292–301, 2004.

145 H. Jin, A. Khandekar, and R. McEliece. Irregular repeat-accumulate codes. In *Proceedings of the 2nd International Symposium on Turbo Codes and Related Topics*, pages 1–8, 2000.

146 D. Maltoni, D. Maio, A.K. Jain, and S. Prabhakar. *Handbook of Fingerprint Recognition*. Springer-Verlag, New York, 2009.

147 C.E. Shannon. A mathematical theory of communication. *Bell System Technical Journal*, **27**:379–423 and 623–656, 1948.

148 S. Kullback and R.A. Leibler. On information and sufficiency. *Annals of Mathematical Statistics*, **22**(1):79–86, 1951.

149 D.W. Scott and W.F. Szewczyk. From kernels to mixtures. *Technometrics*, **43**(3):323–335, 2001.

150 N. Deligiannis, F. Verbist, J. Slowack, R. v. d. Walle, P. Schelkens, and A. Munteanu, Progressively refined wyner-ziv video coding for visual sensors. *ACM Transactions on Sensor Networks (TOSN)*, vol. **10**, no. 2, p. 21, 2014.

151 F. Verbist, N. Deligiannis, S. M. Satti, P. Schelkens, and A. Munteanu. Encoder-driven rate control and mode decision for distributed video coding, *EURASIP Journal on Advances in Signal Processing*, vol. **2013**, no. 1, pp. 1–25, 2013.

Index

Distributed Source Coding: Theory and Practice, First Edition. Shuang Wang, Yong Fang, and Samuel Cheng.
© 2017 John Wiley & Sons Ltd. Published 2017 by John Wiley & Sons Ltd.
Companion Website: www.wiley.com/go/cheng/dsctheoryandpractice